Bridget S. Green · Bruce D. Mapstone ·
Gary Carlos · Gavin A. Begg
Editors

Tropical Fish Otoliths: Information for Assessment, Management and Ecology

Springer

Editors

Bridget S. Green
University of Tasmania
Tasmanian Aquaculture &
Fisheries Inst.
Crustacean Research Group
Hobart TAS 7001
Australia
Bridget.Green@utas.edu.au

Gary Carlos
University of Tasmania
Tasmanian Aquaculture &
Fisheries Inst.
Crustacean Research Group
Hobart TAS 7001
Australia

Bruce D. Mapstone
Centre for Australian Weather &
Climate Research, A Research
Partnership between CSIRO and the
Australian Bureau of Meteorology
GPO Box 1538 Hobart TAS
Australia

Gavin A. Begg
Bureau of Rural Sciences
Dept. Agriculture, Fisheries &
Forestry
Canberra ACT 2601
Australia

ISSN 1571-3075
ISBN 978-94-007-3678-8 e-ISBN 978-1-4020-5775-5
DOI 10.1007/978-1-4020-5775-5
Springer Dordrecht Heidelberg London New York

Cover photo: Section from a sagittal otolith of a 17 year old Pomacentrus moluccensis, a tropical damselfish growing to a maximum size of 9cm TL. Photo courtesy of Tony Fowler, SARDI South Australia.

Printed on acid-free paper

Springer is part of Springer Science+Business Media (www.springer.com)

PREFACE

When we commenced discussions about the need for this book there was a general lack of information on ageing tropical fishes and the broader use of information derived from calcified structures in tropical organisms. This deficiency was largely propagated by an historic misconception that tropical organisms could not be aged reliably from calcareous structures because of the assumed lack of seasonality in tropical environments. Since those initial discussions, significant research has occurred over the past decade on ageing and analysis of the composition and morphology of calcified structures, particularly otoliths, from tropical organisms. The aim of this book is to provide a compilation of this research on key concepts and practical techniques for deriving biological information from otoliths and other calcified structures in tropical aquatic organisms. Importantly, we do not attempt to replace the significant and expansive work that has occurred on temperate species which we direct readers to research (e.g., Secor et al. 1991, Panfili et al. 2002, Campana 2004), but to complement these volumes with a single, comprehensive reference on current practices for deriving and analysing information from calcified structures of tropical organisms.

The book is structured into three main areas, focused on age estimation in teleosts, age estimation in invertebrates and cartilaginous fishes, and growth and structure and chemical composition of otoliths and other calcified structures. Each chapter, which has been written as a stand alone reference, provides an overview of the relevant technologies and procedures including their uses and derived information. Integrated throughout each chapter are technical boxes that provide recipe-style detailed information on processing and analysis of calcified structures. The intent of the book structure is to provide information that is pertinent to both technical users and fisheries managers that require a broad overview of how otolith derived information can be applied and interpreted. We also trust that the book will provide a valuable teaching and training manual for those researching tropical aquatic organisms.

Contributions for the book were sought from an international field of scientific expertise. We thank them for their patience and commitment and encourage readers to follow up with individual authors on topics of interest.

Bridget S. Green
Bruce D. Mapstone
Gary Carlos
Gavin A. Begg
(Editors)

CONTENTS

CONTRIBUTORS

JOHN L. ACKERMAN
Bureau of Rural Sciences, Fisheries, Canberra, ACT, Australia

MIKE CAPPO
Australian Institute of Marine Science, Townsville, QLD, Australia

J. HOWARD CHOAT
James Cook University, Townsville, QLD, Australia

ANTHONY J. FOWLER
South Australian Research and Development Institute, Henley Beach, SA, Australia

JACOB P. KRITZER
Environmental Defense, New York, NY, USA

MARK G. MEEKAN
Australian Institute of Marine Science, Casuarina, NT, Australia

NATALIE MOLTSCHANIWSKYJ
University of Tasmania, Launceston, TAS, Australia

BEATRIZ MORALES-NIN
CSIC/UIB-IMEDEA, Esporles, Illes Balears, Spain

JACQUES PANFILI
Université Montpellier, Montpellier, France

SU SPONAUGLE
University of Miami, Miami, FL, USA

STEPHEN E. SWEARER
The University of Melbourne, Melbourne, VIC, Australia

SIMON R. THORROLD
Woods Hole Oceanographic Institution, Woods Hole, MA, USA

JAVIER TOMÁS
Université Montpellier, Montpellier, France
CSIC/UIB-IMEDEA, Esporles,Illes Balears, Spain

LAURENT VIGLIOLA
Institut de Recherche pour le Développement, Plouzané, France

1. INTRODUCTION TO OTOLITHS AND FISHERIES IN THE TROPICS

BRIDGET S. GREEN[1], BRUCE D. MAPSTONE[2*], GARY CARLOS[1], AND GAVIN A. BEGG[3]

[1]*Tasmanian Aquaculture and Fisheries Institute, University of Tasmania, TAS 7053, Australia*

[2]*Centre for Australian Weather & Climate Research, A Research Partnership between CSIRO and the Australian Bureau of Meteorology GPO Box 1538 Hobart TAS, Australia*

[3]*Bureau of Rural Sciences, Department of Agriculture, Fisheries and Forestry, Canberra, ACT 2601, Australia*

[*] *Current Address: Centre for Australian Climate & Weather Research, CSIRO, GPO Box 1538 Hobart, Tasmania, Australia*

1 Why otoliths are important to fisheries management

Total worldwide catch from wild harvest fisheries increased from 30 million to 70 million tonnes from 1955 to 1970 (Pitcher & Hart 1982) and more recently was estimated to reach 95 million tonnes in 2004 (FAO 2007). Fishing has had major impacts on target species and ecosystems and harvested populations have been declining globally at least since the late 1980s and probably from much earlier (Pauly 1998, Pauly et al. 2002). Fisheries and natural resource managers are faced with many challenges as fish stocks decline, overcapitalised fishing fleets from many countries roam the world's oceans in search of catches from diminishing stocks (Pitcher & Hart 1982) and illegal, unregulated and unreported (IUU) fishing undermines sustainability of fisheries worldwide. Technical advancements have increased fishing efficiency and sustained catch-rates on temperate fish stocks globally for many decades but many technological advances have reached tropical fisheries only in recent decades.

Most tropical fisheries are adjacent to developing nations where human populations are expected to increase by 50% in the next 25 years (Bryant 2002) and subsistence fishing is an important, in some cases almost exclusive, source of animal protein for local communities. It has been suggested that some tropical fish stocks (particularly those associated with coral reefs) are more vulnerable to exploitation than temperate fish stocks because of the multi-species, multi-gear and multi-sectoral nature of the fisheries that exploit them and the life history traits of the fish (see Russ 1991 for discussion). The diversity in life-history characteristics of tropical fish species coupled with increasing exploitation, mass-transport and international marketing add to the complexity of tropical fisheries management, which has adopted scientifically based national management practices only over recent decades. Emphasis on better understanding the impacts of fishing and associated harvest strategies is growing as management agencies and local governments invest more in promotion of sustainable fishing practices. Research is being expanded to parameterise target species' life history characteristics, identify stock structure, and understand migration patterns in order to protect harvested stocks and the key habitats on which they depend. Basic demographic

B.S. Green et al. (eds.), *Tropical Fish Otoliths: Information for Assessment, Management and Ecology*, Reviews: Methods and Technologies in Fish Biology and Fisheries 11,
DOI 10.1007/978-1-4020-5775-5_1,

data, particularly when applied to age-based stock assessments, have been central to many fisheries management decisions in temperate regions and increasingly are being sought for tropical species. Many, perhaps most, commonly targeted tropical species are often long-lived, relatively small-sized and reach maximum size at relatively young ages. These features limit the viability of traditional length-based stock assessment methods, making ageing fish essential for prudent management.

One of the biggest challenges faced by fisheries managers is to source information that describes the past and current status of fisheries and the fish stocks on which they depend and apply this information to ensure that future harvests are sustainable. The collection, preparation and interpretation of structural features of otoliths can provide for tropical fisheries much of the biological information that historically has been incorporated into fisheries management decision-making elsewhere. Biological information derived from species- or stock-level population studies increasingly is considered integral for fisheries assessment and management, to supplement analyses of basic catch and effort data and parameterise the complex models used in harvest strategy evaluations. Calcified structures in fish and many other aquatic animals provide valuable data from fisheries samples because they provide a means of estimating fish age, a key variable for almost all formal or informal fishery assessment procedures (Hilborn & Walters 1992). Box 1 provides a case study where the biological interpretation of otoliths has directly informed management decisions in a temperate species, both positively and negatively. This particular example was selected as it illustrates how a range of otolith processing techniques was used to parameterise stock assessments. Such diversity of applications, however, has yet to be applied generally to management of tropical fisheries.

Important information for ecological research and fisheries management is obtained from removal and processing of various calcified structures of many marine and freshwater species, including otoliths, statoliths, spines, fin-rays, vertebrae and other bones. A large amount of physiological and historical information is recorded within the chemical and physical makeup of such calcareous structures, including environmental information about where the animals have lived, daily, seasonal or annual characteristics of growth, migration patterns, home range, spatial distribution, stock structure and, of course, age. The wealth of biological, biochemical or genetic information that can be retrieved from such structures has made them central to many modern fishery assessment and management procedures (Hilborn & Walters 1992).

Otoliths are by far the most commonly sampled and analysed calcareous structures from fish. Accordingly, in this book we focus on the collection, processing and analysis of otoliths from tropical fish, though in Chapter 5 we also provide an overview of the use of other calcareous structures from other animals for similar purposes. Techniques and theory for processing otoliths in tropical marine fish have developed relatively recently due to an historic misconception that tropical organisms could not be aged reliably from calcareous structures (Pitcher & Hart 1982, Longhurst & Pauly 1987). The seasonality which produces the consistent, time-dependent banding in calcified structures necessary for derivation of age and age-based life-history features is weak in tropical oceans and cyclic interruptions to growth have been assumed to be less likely than in temperate environments (Pitcher & Hart 1982). Thus, current techniques and principles for processing calcified structures in tropical organisms have evolved rapidly and are under constant modification. Techniques involved differ markedly among laboratories and, more particularly, among applications to different species and life history stages. There is currently no single, comprehensive reference on current practice for describing the most appropriate techniques to retrieve fish biology or fisheries data from otoliths of tropical species. This book aims to fill in that need, complementing previous texts and collected papers on generic applications and processing otoliths from temperate species (e.g., Secor et al. 1991, Stevenson & Campana 1992, Secor et al. 1995, Fossum et al. 2000, Panfili et al. 2002, Campana 2004, Begg et al. 2005).

Box 1. Orange roughy *Hoplostethus atlanticus*: otoliths make or break sustainability

A new fishery for orange roughy opened in south eastern Australia in 1987. An initial stock assessment was done using acoustic surveys for biomass estimates and initial age estimates provided by examining the surfaces of whole, un-sectioned otoliths, with ageing 'validated' by comparison to length-frequency models. These data suggested a maximum age of 42 years and maturation at 20 years old (Mace et al. 1990). Investment into the fishery was large, resulting in a massive increase in fishing effort. It was demonstrated subsequently, through improved accuracy in otolith interpretation, that the initial age estimates on which harvest rates were based grossly underestimated the longevity of the fish. Later validation, which combined radiometric ageing and counts of proven annuli on sectioned otoliths, showed that orange roughy in fact lived to 125 years or more and matured at 25 years old (Smith et al. 1995). These differences profoundly affected estimates of stock productivity and, therefore, optimum harvest rates.

Orange roughy was also one of the first species where trace-element analysis of the otoliths was employed to examine stock structure. Edmonds et al. (1991) used otolith microchemistry to determine that there was minimal movement between distant stocks and recommended that these stocks be managed as discrete units. (See Chapter 8 for discussion on the uses of microchemistry in otoliths). A decade later, this otolith-described stock structure was validated using genetic microsatellite analysis and other techniques (Oke et al. 2002), illustrating that analyses of otolith compositions were able to provide fine-scale resolution of the stock structure before the genetic techniques were precise enough to do so.

The initial growth and investment in the orange roughy fishery had a large impact on the stocks. Well-validated otolith derived data was eventually used as a basis for stock assessment and management decisions but by 2004 biomass was estimated to be only 2–20% of the original pre-fished biomass. Severe restrictions have applied to the fishery since to allow the depleted populations to re-build.

2 The structure of this book

This book provides a compilation of key research ideas and practical methods for deriving biological information from otoliths in tropical marine fishes and from other calcified structures in other tropical aquatic animals. We hope to assist researchers and managers studying marine organisms in the tropics in training and technical transfer, providing informed choices of the best otolith processing protocols to get fisheries and biological information from specimens for application across a broad range of circumstances. We collate the expertise of researchers from around the world to provide a single source of technical and background knowledge from some of the leading researchers working with tropical otoliths in recent decades.

This chapter is intended to provide a general background about what sorts of information otoliths can provide fisheries biologists and managers, and some of the issues to be considered when sampling fish populations to get otoliths for analysis. Each of the subsequent chapters describes in detail how otoliths can be analysed to provide the different sorts of information we outline in the first chapter. Each chapter provides an overview of the state of relevant technologies and procedures combined with technical boxes that provide recipe-style detailed information on techniques, methods and equipment for those involved in processing otoliths.

The most frequent use of otolths is for estimating age of fishes, based on the assumed or verified periodicity of banding patterns observed in them. Choat et al. in Chapter 2 discuss the importance of validating that otoliths do act as reliable chronometers but also consider whether validation of the periodicity of increments remains essential for all species, given the considerable evidence that almost all fish so far considered do indeed deposit increments annually. Periodically deposited increments can be divided broadly into those deposited annually and those deposited daily. Fowler in Chapter 3 discusses annual increments, how they are used in determining population structure of a stock and why they are of particular interest in tropical fisheries management. Fowler gives a brief history of the application of otoliths to ageing adult fish, some of the problems that have occurred and how they might be avoided in future. In Chapter 4, Sponaugle explores daily otolith increments in the early life stages of tropical fish and describes the details of life-history that can be interpreted in fish less than one year old. Events such as birth, metamorphosis or settlement, and rates such as growth rate or development rate can be deduced from daily increments in otoliths.

Tropical fisheries target a range of organisms apart from fish, including (not exclusively) gastropods (e.g., trochus), bivalves (e.g., scallops, clams), beche-de-mer, crustaceans (e.g., crabs, crayfish and prawns) and cephalopods (squids and octopuses). The discussion of basic ageing is rounded off by Moltschaniwskyj and Cappo in Chapter 5, where they explore ageing in such organisms using alternatives to sectioned otoliths. They describe techniques used for processing and interpreting periodicity in structures in these organisms, as well as non-otolith structures (e.g. scales and spines) in fish.

Many environmental disturbances or changes alter the metabolism or growth of a fish and, since otolith structures are formed in real time as fish grow, these changes are generally recorded as a change in the widths or other features of increments in otoliths, mostly visible after the otolith has been ground and polished. Vigliola and Meekan describe in Chapter 6 how growth of individuals can be estimated from measurement of widths of increments in otoliths. They describe how fisheries biologists can use individual growth patterns recorded in the increments to retrospectively calculate growth trajectories of individuals and infer growth patterns for cohorts or populations when sample sizes are small. Vigliola and Meekan review the range of published back-calculation models available including the strengths and weaknesses of each method and provide an overview of recent field and experimental evidence of the accuracy of the models. Panfili et al. describe in Chapter 7 how the examination of the finer details of otolith microstructure, such as optical density and crystal alignment, can be used to identify the times of specific events in fishes' lives, including hatching, larval period, settlement, and reproduction. Thorrold and Swearer then review otolith microchemistry in Chapter 8, describing how elements from the water-mass in which fish live are incorporated into the crystalline matrix of the otolith. They describe how details of the daily, seasonal and annual record of the chemistry of the waters fish inhabited can be inferred through examining the microchemistry within otoliths using plasma source mass spectrometry. Finally, in Chapter 9, we consider the major challenges yet to be tackled in future research on the otoliths of tropical fishes and their application to the challenges of managing growing tropical fisheries.

3 Otolith formation

3.1 BACKGROUND

Fisheries managers rely on estimates of age and growth to determine the efficient use and sustainability of living marine resources. Quantitative studies of aquatic systems are challenging in comparison to terrestrial systems because fisheries species characteristically spend their lifetime obscured from direct observation. The logistical difficulties associated with working in aquatic habitats and the problems of continuous sampling throughout species' life histories, which recent studies of tropical species have shown often are measured in decades, add to the challenges. Researchers are generally unable to track individuals in a population throughout the individuals' lifespans, relying instead mostly on scientific data extricated from dead specimens often sampled patchily across both time and space.

Early attempts to employ simple traditional proxies such as length for age estimation were unsuccessful in tropical fisheries because length often bears little relation to age through much of the lifespans of many tropical species. There are certain logical requirements for providing useful age-related data from biological specimens and these considerations are discussed in Chapter 3. Most importantly, investigation must be limited to biological structures whose formation consistently reflects the passage of time and which contain physical or chemical features that can be resolved reliably to have formed with a specific periodicity. These structures should continue to be formed throughout the life span of the individual.

Many hard structures, mainly bones and scales in fish, have been investigated for ageing purposes in fisheries research. The disadvantage of using most integral biologically calcified structures (e.g., vertebrate skeletal bones and enamel, coral skeletons and shells of molluscs) for ageing purposes is that they have substantial living connections and so have the potential to be modified during growth after initial formation. Any such reworking will disturb the fidelity of historical records. Alternatively, structures such as scales are often damaged or lost and replaced completely, severing the continuity of any temporal record they contain.

The otoliths of teleost fish are peculiar among vertebrate hard structures because they are generally considered immune from modification once formed. Otoliths are a set of three pairs of calcified structures encapsulated within otic vesicles, a series of cavities on each side of the base of the skull, and comprise part of the inner ear sensory organs.

One of each pair of otoliths is formed in each side of the head and collectively they assist in balance, orientation (each of the three otoliths on each side of the skull is aligned in one of the three spatial dimensions) and auditory reception (Popper & Platt 1993). Significantly, otoliths are not formed by cellular activity but are precipitated in endolymph, an organic fluid retained within the otic vesicle. Otoliths grow throughout the individual's life but, being isolated from other tissues, they do not undergo the effects of replacement or cellular resorption that can occur in other bony parts. The three pairs of otoliths vary considerably in size and are named the sagittae (singular sagitta), lapilli (singular lapillus) and asterisci (singular asteriscus), from largest to smallest. The sagittae and lapilli generally form earlier in development than the asterisci, which in some species do not form until well after hatching. The sagittae are most often used for ageing and other analyses because they are the largest, earliest formed and easiest to extract of the three pairs.

3.2 OTOLITH STRUCTURE

Otoliths are comprised mostly of aragonite, a polymorph of calcium carbonate. All the precursors of otolith mineralisation are contained in the endolymph fluid, whose proteins mediate the size, shape and orientation of crystal formation (Payan et al. 1999). The growth of an otolith throughout the life of a fish is the result of continuous but irregular precipitation of aragonite. Biological mediation by proteins associated with the endolymph result in incremental accretion of the otolith by production of alternate concentric bands of mineral-deficient (less dense) and mineral-rich (more dense) zones as a solid structure is developed. Further discussion of these processes is provided by Panfili et al. in Chapter 7. Each pair of light and dense zones, together referred to as a growth band or growth increment, is generally formed over a consistent period (e.g., one increment per year) and so counting the increments gives an estimate of age once the period of formation is known. This incremental nature of otolith growth is the key to the use of otoliths for age estimation.

The biological origins of increment formation reflect environmental and developmental influences experienced by the growing fish. These influences may be as fundamental as daily and seasonal environmental rhythms or developmentally significant ontogenetic events such as first feeding and reproduction. A biological structure will accurately reflect the passage of time, however, only if internal physiological events linked to individual developmental processes and environmental circumstances do not disturb the regular routine of formation. It is therefore essential to demonstrate the time fidelity of otolith formation and internal features if they are to be used for ageing. The major emphasis on such verification to date has been on empirical experiments to validate accuracy of ageing techniques, rather than on more fundamental investigation of the processes of otolith formation.

The mechanism of otolith formation bears heavily on ageing techniques used at all scales of observation. Simple examinations of otolith mass often reveal well-defined linear relationships with specimen age and investigation of dimensional properties has generally shown that linear extension along the major (longitudinal) axis correlates predictably and linearly with age. Either of these results alone would suggest constant growth of the structure in either mass or extension. The two results together, however, seem counter-intuitive, as isometric growth should see these two metrics co-vary in a cubic relationship and so have different relationships with age of the otolith. Simultaneous analysis of growth along all major axes, however, shows that otolith growth is not isometric. Accretion is usually biased in one dimension and differential growth along different axes typically results in a laterally compressed, elliptical and distally concave otolith with consistent accretion mostly confined to one surface.

The peculiarity of this type of growth leads to some logistic benefits in the use of either otolith length or mass for age determination. Researchers have exploited the simple relationship between otolith weight and age to estimate age of fish from otolith weight alone. Ferreira and Russ (1994) suggested a mixed approach to ageing fish in a population, combining the faster measurement of otolith weight for the bulk of samples with slower, more technical methods such as age estimation by enumerating the internal growth increments for a representative sample to calibrate the age-weight relationship. Otolith weight has been the most widely used gross property of otoliths to age tropical species (e.g., Worthington et al. 1995a,b, Choat & Axe 1996, Luckhurst et al. 2000, Newman et al. 2000a,b, Pilling et al. 2003) across a range of spatial and temporal scales (Lou et al. 2005, 2007). The average weight of an otolith increases continuously with age and the derived relationship from a representative sample provides a rapid and economic proxy for estimating the age of fish which may be considered a better age predictor than other otolith measurements (Worthington et al. 1995b, Newman 2002). This is particularly useful for difficult-to-age fish, where growth increments in otoliths may be difficult or ambiguous to count routinely and where fish length is a poor proxy for age, as is typical for tropical fish species (Lou et al. 2005).

In practice, however, there are limitations that can restrict the apparent utility of these methods. For example, obtaining whole otolith weights may seem a simple procedure but its success depends on obtaining completely undamaged otoliths and otoliths are so delicate in many specimens that removal and handling without chipping or breaking is difficult and time consuming. Further, otoliths are often very small and so weighing them requires precise, and thus delicate and expensive, measuring equipment to obtain sufficient resolution for a meaningful weight-age relationship. Nevertheless, using otolith weight as a (calibrated) proxy for age may be extremely useful in resource limited situations such as in many tropical countries, providing a useful cost- and time-effective proxy for generating age-based metrics for species where otoliths are relatively large, robust and easily extracted and handled without being damaged.

Prevailing methods generally age individuals from the discernable internal micro-structural features, such as growth increments, created during the incremental deposition of otoliths. For example, the alternating mineral-poor and mineral-rich zones in each growth increment appear as differentiated relatively light and dark bands under microscopic examination with visible light, allowing relatively easy discrimination and counting of the increments. Increments are conventionally interpreted as either daily, with light and dark bands corresponding to daytime or night time formation, or annual, with light and dark bands corresponding to formation over different semi-annual seasons. Correct ageing, however, requires prior investigation and validation of the periodicity of increment formation. It should be emphasised that the appearance of a band as either light or dark is entirely dependent on the lighting source (Figure 1).

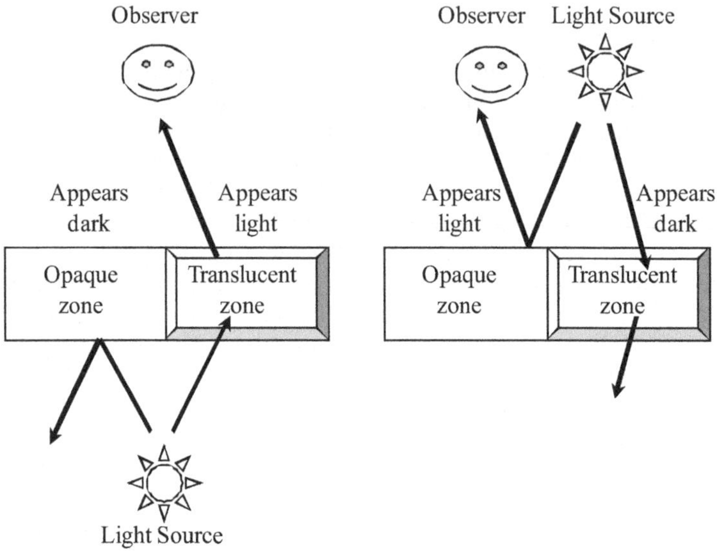

Figure 1. Illumination from different light sources can affect observations, and thus labelling, of the features of otoliths because the zones of an otolith with different mineral content will have different optical properties. The schematic zones in the diagram are labelled according to how they scatter visible light: 'opaque' zones exclude much light from passing through a thin section, while 'translucent' zones allow a greater fraction of visible light to pass through the sample. An observer viewing an otolith under transmitted light (*left*) will label sectors with these optical properties as dark and light respectively because of the fate of light (*black arrows*) hitting the zones from the side opposite to the observer. In contrast, if the otolith is illuminated from the same side as the observer (*right*), the specimen would be viewed via reflected light and the appearance of each region would be reversed.

Practical experiments to validate the periodicity of formation of otolith growth increments, and hence their use for ageing, have centred on following animals for known times, either as tagged individuals at liberty or maintained in aquaria. Raising (or tagging) marine species from eggs over long periods generally is not practical (but see Box 2) and techniques which artificially 'tag' otoliths internally at known times have been developed so that the features of the otolith formed between the time of tagging and death can be examined. The most common internal tagging method involves injection of the antibiotic oxytetracycline or tetracycline, which is incorporated into the growing otolith within a short time after injection. The incorporated oxytetracycline fluoresces under ultraviolet light and so allows unambiguous identification of the point in otolith growth corresponding to the time of injection and, hence, identification of all growth increments formed between then and when the fish was killed and the otoliths extracted for examination. Examination of these post-marking increments facilitates validation of increment periodicity. Some issues related to the perceived need for routine validation of ageing methods are explored by Choat et al. in Chapter 2.

Ontogenetic changes and environmental fluctuations are generally considered the disturbances experienced throughout a fish's life most likely to interrupt the function of otoliths as biological chronometers. Thorough empirical validation, therefore, requires investigations covering the entire lifespan of a species, in addition to a wide range of environmental variation likely to be experienced by individuals during life. Fish otoliths generally have shown remarkable chronological robustness in the face of variable internal and external environments, although major ontogenetic events such as hatching and first feeding also leave discernable marks in the otolith record (Fowler & Short 1998, Sponaugle, Chapter 4, this volume).

The optical discontinuities that present as the characteristic daily or annual banding in otoliths are internal features but even a whole otolith will commonly appear banded when viewed under low magnification with suitable illumination. There might be no need for further treatment of the otolith and a direct count of bands visible from the whole otolith might be used to infer the age of the specimen provided such externally visible banding corresponds to the expected periodicity (e.g., annual formation). Further specimen preparation is needed, however, if the desired age resolution is beyond that visible from a whole otolith (for example daily ageing is required or banding deep within the structure is obscured by the otolith's bulk), or all growth increments are not visible externally. Cutting and polishing thin transverse sections of a sagitta is usually enough to expose all increments to any level of resolution optically discernable under high powered light microscopy. Details of the process of preparation of otolith specimens for sectioning and microscopic analysis for different purposes can be found in Chapters 3 (Fowler), 4 (Sponaugle) and 6 (Vigliola & Meekan), whilst sectioning and preparation procedures for non-otolith structures from other marine animals are described by Moltschaniwskyj and Cappo in Chapter 5.

Optical examinations of prepared sections of fish otoliths provide excellent records of the passage of an individual through time. It is to this level of analysis that the majority of otolith work in marine research laboratories around the world is targeted, with an estimated 1 million otoliths processed per year (Campana & Thorrold 2001).

3.3 BEYOND THE BAND: READING MORE THAN AGE FROM THE OTOLITH STRUCTURE

More recent quantitative analyses of the finer details of otolith shape, surface features, internal microstructure and chemical composition have revealed patterns which can be characteristically linked to specific and even intra-specific differences in populations or environmental histories of individuals. This work has led to the use of otoliths beyond the simple provision of age estimates and into areas of stock discrimination, ontogenetic migration and exploration of environmental influences on life history and ecological relationships among species.

3.3.1 Otolith morphology

Small differences in otolith shape have been used to determine that fish have grown for much of their lives in different geographic regions from other fish of the same species, and that different groups of fish may be effectively isolated from each other into relatively distinct 'stocks' that may need to be managed separately (Smith 1992,

DeVries et al. 2002, Bergenius et al. 2006). Otolith shape also can be used to infer ecological relationships of harvested species through collection of otoliths from stomach contents and subsequent identification of the species from which they came (Baker & Sheaves 2005). Otolith shape is relatively species-specific and so can be used to determine predator-prey relationships of harvested species from stomach contents and so go some way to describing the ecology of the species and potential ecosystem impacts of fishing. Environmental effects, particularly mediated through growth rate (Smith 1992, Campana & Casselman 1993) and feeding history (Gagliano & McCormick 2004), are postulated to be the main drivers of differences in otolith shape by influencing the development of otolith crystalline microstructure.

Otolith morphology can be characterised by linear measurements (e.g., length, width), internal otolith measurements (e.g., nucleus length, increment width), two-dimensional size measurements (e.g., area, perimeter) and outline shape analysis. Fourier analysis has been used recently to model otolith shape, where a series of cosine and sine curves are produced from the coordinates of a traced outline (Smith 1992, Campana & Casselman 1993, Begg & Brown 2000, Bergenius et al. 2006). Importantly, potential confounding sources of variation among otolith samples, such as size, age, sex and year-class differences, need to be accounted for prior to attributing differences in otolith morphology to environmental effects or as signalling different stocks. Otolith shape analysis is a useful technique in that a large number of samples can be examined quite rapidly, particularly using modern image processing procedures. Shape analysis techniques that use image processing procedures are relatively inexpensive, enable images to be readily reanalysed and audit trails to be established, and offer improvements in speed and ease of use over more traditional linear morphometric techniques (Cadrin & Friedland 1999).

3.3.2 Otolith microstructure

Methods of analysing otolith structures as a whole treat otoliths only as chronometers – that is, a device recording only an absolute age and resulting in a single value for each specimen sampled. This approach requires multiple specimens to provide longitudinal information on a population. Extensive observations on internal microstructures have allowed an extra dimension to be added to this basic analytical approach. It is a simple extrapolation from validation experiments which delineate the formation of certain internal microstructures in time to relate particular periods to specific life-history events or identify specific regions within an otolith related to known life-history events. For example, knowing the time of imparting an artificial mark or, more usefully, the known time of capture of any specimen allows us to back-track through the otolith to internal features that would have formed during specific events (e.g., at the age of maturation or a period of known environmental disturbance). Identifying distinctive features of the otoliths corresponding with known events might allow those features to be used as proxies for such events in otoliths from other individuals. Such work opens up an array of possibilities for investigating many aspects of a fish's otherwise concealed life history. For example, growth patterns in early years can be back-calculated by measuring the widths of relatively well demarcated otolith increments deposited early in the fish's life, as discussed in more detail by Vigliola and Meekan in Chapter 6.

3.3.3 Otolith microchemistry

The introduction of microchemical composition analysis has added another dimension to otolith studies. The general carbonate structure of otoliths is composed of aragonite but, as with any crystalline material, trace elements can be incorporated into the aragonite lattice during formation. Analysis of the elemental or isotopic composition of an otolith often can be correlated to levels of such elements in habitats occupied by individual fish. Measuring a variety of trace elements simultaneously will enhance the ability to discriminate specific elemental signatures relating to various water bodies. Analysis of the chemical composition of an otolith by discrete sampling of microchemical trace elements across a section can reveal seasonal or inter-annual patterns in otolith chemistry that signal developmental migrations through different waters. The uses of otolith microchemistry are explored further by Thorrold and Swearer in Chapter 8. Further developments of this archival information can also offer improvements in age validation of wild caught fish, a striking example being the use of radioactive isotopes from historical nuclear weapons testing incorporated onto otoliths as markers of known past events from which subsequent increment formation can be calibrated (Kalish 1993, further discussed by Moltschaniwskyj & Cappo in Chapter 5).

4 Collection of otoliths – sampling considerations

Estimates of age and other information from otoliths are required for several purposes. The objectives for which age data are needed are central to the design of field sampling strategies for collecting the fish from which otoliths are extracted and analysed. For example, estimating particular demographic transition points such as longevity, age at maturity, age at sex-change or age at recruitment to a fishery may require different sampling strategies from estimating longitudinal or population demographic parameters such as mortality or growth. Similarly, sampling to derive age data that are representative of a harvested stock over a fishery ground may be approached differently from sampling to compare age parameters among stocks or sub-populations in a meta-population. Here we discuss some general factors that should be considered when designing field sampling strategies to collect fish and otoliths for different objectives.

We do not seek to provide here a catalogue or review of the gears by which fish are collected. There are many collection gears available (e.g., nets, lines, traps, etc.) and which are most appropriate in a particular situation is probably best known by the researchers working in that context. Nor do we provide a technical reference for the formal statistical design of optimum or efficient sampling programs, since formal sampling design procedures are covered in numerous other sources (e.g., Green 1979, Bernstein & Zalinski 1983, Hurlbert 1984, Schweigert et al. 1985, Andrew & Mapstone 1987, Fowler 1987, Quinn II & Deriso 1999). Rather, we discuss here some of the main conceptual and procedural considerations for deploying the chosen collection gear in space and time to produce samples that are most likely to provide a robust and rigorous basis for answering the questions of interest. This discussion is deliberately conceptual and intended to identify key issues to be considered by researchers intending to sample fish for ageing and other purposes, managers either commissioning or reviewing research to inform management decisions, or funders of research needing to consider whether the proposed research appears rigorous.

Sound sampling design is a particularly important consideration because errors made at the stage of specimen collection usually cannot be corrected by re-analysis or re-treatment of the otoliths or data. Sampling design mistakes can only be rectified by additional field sampling, usually at considerable cost. If a field sampling design is rigorous and robust, however, and errors occur in the interpretation of otoliths or analyses of derived data, those errors often can be corrected by re-interpreting the otoliths or re-analysing the data, without requiring further field sampling.

4.1 BIAS, PRECISION AND POWER

There are several important considerations in the design of any field sampling program. The most important is whether the sampling design is adequate to provide data that will address the primary questions of interest with certainty and without ambiguity – that is, whether the sampling design is 'fit for purpose'. We discuss some of the issues in this category below for some of the common types of questions addressed with age data.

Three important issues that can be addressed, at least in part, when designing a sampling program are bias, precision and statistical power. Bias refers to systematic or consistent departure of parameter estimates from the true value of the parameter (Cochran & Cox 1957, Cochran 1963, Andrew & Mapstone 1987). Biases in estimates can result in inferential errors, coming to the wrong conclusions about key questions, or under- or over-estimating the ecological status of a population or a stock's capacity for sustained harvest. Biases that can arise during processing otoliths once collected are discussed in later chapters, so here we consider ways in which sampling strategies can be designed to reduce risks of bias attributable to specimen collection. Bias generated by procedural errors in collection or inappropriate sampling design is perhaps the most difficult problem to correct without completely re-doing the sampling.

Precision refers to the repeatability of estimates or the expected consistency among repeated estimates of a parameter, whether such estimates are biased or unbiased (Cochran & Cox 1957, Cochran 1963, Andrew & Mapstone 1987). Precision (or its complement, imprecision) is a key measure of uncertainty in estimates and underpins uncertainties in inferences from analyses. Imprecision can arise from several sources, ranging from characteristics of the sampling design through to the analysis or interpretation of data. Imprecision related to field sampling can be most easily addressed by ensuring that sufficient numbers of specimens are collected during field sampling (Kritzer et al. 2001).

Statistical power is a characteristic of hypothesis testing and is the probability that we will correctly infer that an alternative (to the Null hypothesis) is true – that is, correctly detect patterns in space or time or in response to impacts (such as fishing) when they actually exist (Andrew & Mapstone 1987, Peterman 1990, McAllister & Peterman 1992, Mapstone 1995). The formal analysis of power is dependent on the specific hypothesis being tested and the form of analysis applied to the data in hand and so, for age data, will vary depending on whether age data are being used to estimate mortality, demographic transitions (e.g., maturity), harvest capacity, or some other metric (e.g., growth metrics). Statistical power, therefore, is mainly associated with the analyses applied to data derived from field sampling or extracted from collected specimens. Here also, however, an inadequate sampling design will severely constrain the options for powerful analyses of derived data. There has been considerable attention in the literature to statistical power since the late 1980s but very little of that attention has been directed at analyses of age data or derived metrics and it is beyond the scope of this overview to develop such analyses.

Below we discuss some general principles of designing field sampling that are likely to minimise potential bias, improve precision and provide good prospects of resolving real patterns in sampled populations (i.e., support analyses that have good statistical power). We discuss the characteristics of field sampling designs that we expect will contribute to powerful analyses in general and note those key deficiencies in sampling that would diminish the statistical power of analyses of resultant data.

4.2 SAMPLING FOR DIFFERENT TYPES OF INFORMATION

Most age-based metrics are derived to either represent or characterise a population, estimate characteristics of the harvested part of a population, identify particular point-in-time properties of a population (e.g., age at maturity), use in combination with other data to infer population characteristics (e.g., age-length keys) or make comparisons among potentially different populations.

4.2.1 Characterising a single population

Perhaps the most frequent objective for acquiring age data is to characterise the growth and mortality characteristics of individuals in an exploited population in order to derive sustainable harvest strategies through stock assessments or management strategy evaluations. Age data usually are applied in these cases to describe the dynamics of a population over the extent of a fishery ground or domain under a given jurisdiction or management regime. It is important, therefore, when sampling individuals from which to extract otoliths that samples are gained over most or all of the fishing domain and in a way that represents the structure of the population being harvested. Variations in population dynamics within the fishery domain usually would not be considered explicitly, with average growth and mortality profiles applied to the whole ground.

Characterising a population implies that the sample of specimens from which age data are derived effectively has the same properties as the wild population. Hence, we would aim to collect a sample that has the same proportion of individuals in each size or age or growth category as exists in the wild population from which the sample was taken. Such a sample would be considered a representative sample, because it represented accurately all the features of the wild population from which it came. Sampling that selects individuals with given characteristics from the wild population in proportion to their abundance will result in a representative sample that should adequately characterise the wild population. Sampling specimens at random would achieve such a representative sample, as would sampling with a gear that had the same prospect of catching any individual, irrespective of size, age, sex, etc. and irrespective of the presence of other individuals. The latter, completely 'unselective' gear is a holy grail of sampling that is rarely available. Some approaches to compensating for gear selectivity biases are discussed later. Representative samples are the most generally useful and robust samples and, if sufficient specimens are collected, will often suffice for each of the other purposes described below.

Hurlbert (1984) and Andrew and Mapstone (1987) discussed in some detail the need for adequate distribution of sampling effort over the region for which estimates of a parameter were being derived and some of the common mistakes made in sampling that might result in biased estimates. Whilst it usually is impossible to sample from every subset of a domain (e.g., from every coral reef within a reef fishery), it is important to disperse sampling across the range of the domain to provide a reasonable chance of capturing the range of circumstances that might influence age estimates or derived parameters (e.g., mortality, growth) within the domain. In theory, allocation of sampling effort across the domain should be randomised, stratified according to known or expected environmental strata, or allocated in a regular pattern (e.g., by transects or according to a regular grid). In practice, however, such sampling is often difficult and costly to achieve and substitutes for such designed allocation of sampling effort are used instead, such as sampling from the catches from fishing fleets or markets or opportunistic sampling during other research. Opportunistic sampling from catches or collections for other research is unavoidable in many circumstances and may be the only viable option for sampling where research and monitoring resources are scarce or research capacity is limited. It might also be the case in many tropical fisheries that catch sampling is the only feasible way to gather information from the range of fish populations being harvested over patchy, heterogeneous habitats (e.g., coral reef archipelagos). The information so provided will be particularly important where other information is scarce or absent and is by far preferable to no information. It is important to recognise, however that these are compromise solutions, and wherever feasible, at least some designed sampling should be done to verify that the compromise sampling is delivering relatively unbiased and credible results.

4.2.2 Sampling catches to represent a fishery harvest

Collecting samples from fishery catches can provide a convenient way of assessing the characteristics of a population being harvested. It is important to realise, however, that fishing of any sort is itself a selective process that results in a biased 'sample' of the wild population, even where the sample accurately represents the harvestable part of the stock. For example, fishers often will use gears that preferentially select (e.g., by size) the fish that have the highest market value or fish in specific areas that are convenient and accessible to them. Even samples taken from catches in an unbiased way will 'inherit' certain biases from the way in which the catch was collected by the fishery. Thus, some steps must be taken to minimise the risk that the specimens collected from the catch are subject to additional biases associated with sampling of that catch. For example, taking only specimens of low market value (because they are of least value to the fisher) incurs the risk that the resultant age metrics are biased because the low value fish may have particular characteristics (e.g., small or large size) not shared by higher valued fish. Similarly, collecting specimens from only a small sub-set of fishers in a single port or from a single large catch might add bias to the data if, for example, the sampled fishers took catch from only certain areas of the fishing grounds and specimens in those areas differed from elsewhere.

One approach to avoiding such biases is to ensure that specimens are selected randomly from the catch of each of many fishers believed to fish in different places or from a random selection of all fishers working on the fishing grounds. This approach will minimise the risk of sampling bias and improve the prospects that the derived age-based metrics will capture adequately the characteristics of the harvested portion of the population over the range it is being harvested. Selecting a sequence of numbers from a table of random numbers and buying or sampling the specimens corresponding to those numbers as a catch is counted is one easy way to randomly sample from a catch. Taking every *n*th specimen should also result in an approximately random sample with respect to the population being harvested, provided the catch has not been previously sorted by size or location of capture. Where the catch is sorted (e.g., by size or area where caught) before being sampled, then sampling should be tailored to benefit from this prior sorting. For example, if the catch is sorted by size, it will be advantageous for representative sampling to collect the same proportion of individuals from each of the sorted size classes. Where catch is landed according to the area it was caught, then the total number of specimens to be sampled can be split among the areas to ensure good cover of the fishing grounds in the collected sample.

4.2.3 Sampling for longevity or ontogeny

Identifying particular point-in-time features or transition points for a population often involves specifically collecting individuals with particular characteristics (e.g., in a particular size-range within which maturity is believed to be reached) or collecting specimens with roughly equal frequency across a range of categories (e.g., across a range of sizes classes), irrespective of the frequency with which individuals occur in those categories in nature. This is targeted sampling. Such samples are generally easier to obtain than representative samples because there is less concern about sampling biases but these samples would be unlikely to be representative of the structure or characteristics of the wild population and would almost certainly provide a biased characterisation of the wild population if used as though they were representative samples. Hence, it is important to be clear about the objectives and limitations of such targeted sampling.

Targeted sampling to resolve specific questions about the age at which certain life-history events occur is in many ways simpler than representative sampling. For example, if a primary purpose of sampling is to estimate longevity, then in most cases selecting the largest individuals available also will be most likely to select the oldest individuals. it may be acceptable to age relatively few 'large' specimens to estimate maximum age or longevity where a species is believed to have approximately indeterminate growth. A caveat here for tropical fish, however, is that many species have 'table-top' growth by which individuals reach their approximate maximum size at a relatively early age (as young as 15% of expected longevity, see Choat et al., Chapter 2, & Fowler, Chapter 3, of this volume). It will likely be necessary to sample and age many 'large' individuals to get a good estimate of maximum age or longevity for such species (Kritzer et al. 2001).

Targeted sampling to estimate the age of transitional (rather than terminal) life-history events is more difficult because identifying the 'target' (e.g., size range) for sampling depends on having at least some basic prior information about the biology of the species. One moderately efficient strategy in these cases is to sample relatively few individuals over a wide range of characteristics (e.g., size) initially, analyse the data (e.g., age, reproductive status) from those samples and then focus subsequent intensive sampling on the approximate sizes at which key ontogenetic events of interest (e.g., approximate size at maturity or sex change) seem to occur. It is usually preferable to derive such metrics from a rigorous representative sample of a fished population but the cheaper and logistically simpler targeted strategy might be acceptable and more feasible in some situations. For example, targeted sampling might be sufficient where resources are limited and the work is to be used to establish relatively coarse and simple regulations, such as setting a minimum legal size limit or gear regulations to allow individuals to reproduce in at least one or two years before being available for harvest.

4.2.4 Unstructured sampling

Unstructured or ad hoc samples are those collected with little specific attention to either being rigorously representative of the population from which they come, other than in some general sense, or to targeting particular groups of individuals. Such samples are often collected as initial surveys of populations to gain a snapshot of the general properties of a population (e.g., age and size range) without seeking to describe its characteristics in detail. Such data also can be used to construct an age-length key from which to convert future length data to age (or vice versa) because such a key does not convey anything about the relative abundance of different length or age classes in the population. It is important to note, however, that age-length keys will be biased if sampling disproportionately favours individuals of particular sizes within an age class. For example, if fast-growing individuals are more likely to be sampled than slower growing individuals, perhaps because fish become vulnerable to the sampling gear used at a particular size, then the age-length key will be biased for those age classes in which some individuals have not reached the vulnerable size. Unstructured or ad hoc sampling has been fairly common in fisheries research, especially where research resources are scarce, and have contributed significantly to fisheries management in many situations, based largely on the premise that ‘some information is better than nothing’. It is important to recognise the significant constraints, uncertainties and potential biases such sampling impose on data interpretation, however, and strive wherever possible to structure sampling appropriately to address specific questions of interest.

4.2.5 Sampling for comparisons

Many tropical demersal fish are relatively sedentary after the larval stage of life. Further, many of the target species for tropical fisheries are associated with coral reefs or coastal fringing reefs, which often have a very patchy distribution. These characteristics – sedentary fish associated with patchy habitats – mean that many tropical fish exist as meta-populations – collections of very many local sub-populations, often distributed over considerable areas. It is generally considered that there is likely to be relatively high rates of genetic exchange among the local sub-populations via dispersal of planktonic larvae, but this assumption is increasingly being challenged as we gain better understanding of larval behaviours (Jones et al. 1999, Swearer et al. 1999, James et al. 2002).

Such species often are being harvested over a wide area (i.e., from many sub-populations) within which local sub-populations might have quite different demographic characteristics and so different capacities for harvest. It may be necessary, therefore, to consider spatially structured management strategies or setting overall strategies that will be effective in spite of variations in dynamics of local populations. It will be important in such situations to sample across the broad range of sub-populations and compare the growth, mortality and other demographic features among them in order to devise management strategies appropriate to the amount of variation or homogeneity in population dynamics of the target species. It usually will not be either possible or sensible to sample and compare all sub-populations with all others, and some spatial stratification into groups of sub-populations will be required (e.g., grouping by different island groups, at different latitudes, or in different provinces or regions). Sampling for age data in these circumstances requires collecting either representative samples from all strata or samples that have the same biases (e.g., because of sampling gear or depth) in all strata so that comparisons among strata are not confounded by differences in sampling biases. It usually will also simplify analyses if samples of approximately equal size (numbers of specimens) are collected from each of the strata, and it is particularly important, if possible, to avoid having very uneven sample sizes among strata.

5 Application of otolith information in fisheries management

Information from otoliths combined with data such as fish length, weight and reproductive condition can be used to derive a range of parameters that are useful for managing harvest of a fish stock. The most basic information an otolith provides is the number of growth increments formed before the fish was sampled, providing an estimate of the fish's age at the time of capture (provided the periodicity of increment formation is known). This information becomes most useful in a management framework with the addition of extra details of the fish's biology and with the application of some basic mathematics to describe the structure and dynamics of the population considered to comprise the harvested stock.

The combination of fish length and age data can be used to describe a general trajectory of life-time growth of average individuals in the population and calculate growth rates. Integrating sex and reproductive condition with growth data allows for development of an age-fecundity relationship and identification of sex-specific growth. Survivorship, and therefore mortality rates, can be calculated by combining age estimates from otoliths with counts of the number of fish per age class in a sample, as can the number of fish entering into an age-class each year (age-specific recruitment). These analyses provide fisheries analysts and managers with a range of information that describes the exploitable fish stock. Such information collected from a previously unexploited fish stock also provides an estimate of potential productivity from which sustainable yield can be calculated. The number of eggs provided from each cohort can be derived from the age-specific fecundity relationship and total egg production calculated and used to predict recruitment for the next year from a stock-recruitment relationship (Beverton & Holt 1957, Ricker 1975). This age-fecundity relationship can provide an indication of the relative importance of reproductive fish of different ages to the yield of the next cohort.

Relating age to the number of fish caught estimated from catch and effort monitoring provides catch-at-age data and comparison of these data among years of fishing will allow identification of shifts in the demography of the exploited fish stock, potentially highlighting growth overfishing. These models can be made more dynamic by incorporating variability, such as age-specific gear-selectivity, vulnerability to fishing and age-specific weight gain and variable natural mortality. More detailed discussion of these methods can be found in Hilborn and Walters (1992), Quinn II and Deriso (1999) and Haddon (2001) and we recommend these or similar texts for exposition of stock assessment methods and modelling details. We refer to the above derivations simply to illustrate the nature of the information available when data from otoliths are incorporated formally or informally into assessments of the status and harvest of fish populations. Below we describe some specific uses of otoliths in gathering information useful for fisheries management.

Interpretation of otolith microchemistry in relation to known environmental chemical signals has been used to infer: connectivity (or lack of connectivity) between stocks (e.g., tropical shad, *Tenualosa toli*, Milton et al. 1997, tropical serranids, Bergenius et al. 2005b); separation between juvenile nursery grounds and adult grounds (e.g., yellowtail snapper, *Ocyurus chrysuru,* Verweij et al. 2008); differences between alternative nursery grounds (e.g. bluefin tuna, *Thunnus thynnus*, Rooker et al. 2003); and evidence of relatively closed self-seeding populations (e.g., damselfish, *Pomacentrus amboinensis,* Jones et al. 1999, Swearer et al. 1999, Box 2). Thorrold and Swearer in Chapter 8 provide a more comprehensive description of such applications.

Box 2. Self-seeding fish populations

A basic assumption over the history of marine fisheries science has been that dispersive larvae are relatively passive particles carried away from natal waters by prevailing currents. Most recruitment to a particular location has been assumed to be from sources elsewhere. Two influential papers in which otolith microchemistry was applied to larval ecology have challenged this theory by demonstrating the presence of self-seeding populations in which significant fractions of larval fish settling into the populations were spawned from those same populations. Swearer et al. (1999) demonstrated that trace element analysis of otoliths from newly settled fish could identify recruits that had been retained within the lee of an island in St Croix, US Virgin Islands, and therefore formed a self-recruiting population. Jones et al. (1999) approach was simpler. They tagged over 10 million eggs of a coral reef species with oxytetracycline, which leaves a fluorescent mark in growing calcareous structures, and examined the otoliths of the corresponding generation of settling fishes when they recruited to the reef where the eggs were marked. A small proportion of the new recruits examined exhibited oxytetracycline marks and so this study directly demonstrated self-recruitment in a tropical coral reef fish population. Both papers, based on analyses of otoliths of marine fish larvae, have provided novel data on the early life ecology of reef fish that may prove important to tropical fisheries management. The historic view on larval dispersal has been challenged, and modified, in light of these studies, as has our perception of dispersal and recruitment rates among sub-populations within species meta-populations.

Back-calculation techniques have been used for tropical species to determine the date of settlement of fish from the plankton to benthic habitats by counting daily increments from the otolith margin back to an identifiable settlement mark. This approach, when applied to a large number of fish, can provide information on temporal variation in numbers of fish recruiting to a population (Doherty 1991). Reconstructing settlement patterns and recruitment pulses can be used to estimate persistence of cohorts in populations and infer the relative importance of recruitment or post-recruitment processes to population sizes (Victor 1983). Counting daily increments inside the settlement mark to the nucleus of an otolith also has been used to estimate the duration of the larval phase preceding settlement (Brothers et al. 1983, Victor 1991). Analyses of these early life growth increments have been correlated to environmental conditions and perturbations that occurred during early life-history stages (Wilson & Meekan 2001, Meekan et al. 2003, Bergenius et al. 2005a) and so provide insights to parameters that might be used to predict recruitment. Some of these applications are discussed in more detail by Sponaugle in Chapter 4 and Vigliola and Meekan in Chapter 6

Thus, otoliths can be used as simple chronographs but also as structures that record events in real time, archiving records of life events such as growth, location of formation, environmental events and stock associations, among others. Properly collected, processed and analysed, therefore, otoliths provide invaluable information for scientists and managers to improve understanding and management of tropical fisheries.

Acknowledgements

We thank the authors and reviewers who have contributed or reviewed the chapters that follow for their hard work, persistence and patience. Funding for the editors was provided, at various times, by the Australian Fisheries Research and Development Corporation, the Great Barrier Marine Park Authority, James Cook University, and the Australian Cooperative Research Centres (CRC) Programme through the CRC for the Great Barrier Reef World Heritage Area and the Antarctic Climate & Ecosystems CRC.

References

Andrew NL, Mapstone BD (1987) Sampling and the description of spatial pattern in marine ecology. Oceanog Mar Biol Ann Rev 25:39–90

Baker R, Sheaves M (2005) Redefining the piscivore assemblage of shallow estuarine nursery habitats. Mar Ecol Prog Ser 291:197–213

Begg GA, Brown RW (2000) Stock identification of haddock *Melanogrammus aeglefinus* on Georges Bank based on otolith shape analysis. Trans Am Fish Soc 129:935–945

Begg GA, Campana SE, Fowler AJ, Suthers IM (2005) Otolith research and application: current directions in innovation and implementation. Mar Freshwater Res 56:477–483

Bergenius MAJ, Begg GA, Mapstone BD (2006) The use of otolith morphology to indicate the stock structure of common coral trout (*Plectropomus leopardus*) on the Great Barrier Reef, Australia. Fish Bull 104:498–511

Bergenius MAJ, Mapstone BD, Begg GA, Murchie CD (2005a) The use of otolith chemistry to determine stock structure of three epinepheline serranid coral reef fishes on the Great Barrier Reef, Australia. Fish Res 72:253–270

Bergenius MAJ, McCormick MI, Meekan MG, Robertson DR (2005b) Environmental influences on larval duration, growth and magnitude of settlement of a coral reef fish. Mar Biol 147:291–300

Bernstein BB, Zalinski J (1983) An optimum sampling design and power tests for environmental biologists. J Environ Manage 16:35–43

Beverton RJH, Holt SJ (1957) On the dynamics of exploited fish populations. U.K. Ministry of Agriculture and Fisheries, London

Brothers EB, Williams DMcB, Sale PF (1983) Length of larval life in twelve families of fishes at One Tree Lagoon, Great Barrier Reef, Australia. Mar Biol 76:319–324

Bryant PJ (2002) Biodiversity and conservation: A hypertext book. School of Biological Sciences, University of California, Irvine, Irvine, CA 92697, USA

Cadrin SX, Friedland KD (1999) The utility of image processing techniques for morphometric analysis and stock identification. Fish Res 43:129–139

Campana SE (2004) Photographic atlas of fish otoliths of the Northwest Atlantic Ocean. NRC Research Press, Ottawa, Ontario

Campana SE, Casselman JM (1993) Stock discrimination using otolith shape analysis. Can J Fish Aquat Sci 50:1062–1083

Campana SE, Thorrold SR (2001) Otoliths, increments and elements: keys to a comprehensive understanding of fish populations? Can J Fish Aquat Sci 58:30–38

Choat JH, Axe LM (1996) Growth and longevity in acanthurid fishes; an analysis of otolith increments. Mar Ecol Prog Ser 134:15–26

Cochran WG (1963) Sampling techniques, 2nd Edition. Wiley, New York

Cochran WG, Cox GM (1957) Experimental designs, 2nd Edition. Wiley, New York

DeVries DA, Grimes CB, Prager MH (2002) Using otolith shape analysis to distinguish eastern Gulf of Mexico and Atlantic Ocean stocks of king mackerel. Fish Res 57:51–62

Doherty PJ (1991) Spatial and temporal patterns in recruitment. In: Sale PF (Ed) Ecology of fishes on coral reefs. Academic Press, San Diego

Edmonds JS, Caputi N, Morita M (1991) Stock discrimination by trace – element analysis of otoliths of orange roughy (Hoplostethus atlanticus), a deep-water marine teleost. Aust J Mar Freshwater Res 42:383–389

FAO (2007) The state of world fisheries and aquaculture 2006, FAO Fisheries and Aquaculture Department, Food and agriculture organization of the United Nations

Ferreira BP, Russ GR (1994) Age validation and estimation of growth rate of the coral trout, *Plectropomus leopardus*, (Lacepede 1802) from Lizard Island, Northern Great Barrier Reef. Fish Bull 92:46–57

Fossum P, Kalish J, Moksness E (2000). Special issue: 2nd International Symposium on Fish Otolith Research and Application, Bergen, Norway, 20–25 June 1998. Fish Res 46:1–2

Fowler AJ (1987) The development of sampling strategies for population studies of coral reef fishes: a case study. Coral Reefs 6:49–58

Fowler AJ, Short DA (1998) Validation of age determination from otoliths of the King George whiting *Sillaginodes punctata* (Perciformes). Mar Biol 130:577–587

Gagliano M, McCormick MI (2004) Feeding history influences otolith shape in tropical fish. Mar Ecol Prog Ser 278:291–296

Green R (1979) Sampling design and statistical methods for environmental scientists. Wiley, New York

Haddon M (2001) Modelling and quantitative methods in fisheries. Chapman and Hall, Boca Raton, Florida

Hilborn R, Walters CJ (1992) Quantitative fisheries stock assessment: choice, dynamics and uncertainty. Chapman & Hall, New York

Hurlbert SH (1984) Pseudoreplication and the design of ecological field experiments. Ecol Monog 54:187–211

James MK, Armsworth PR, Mason LB, Bode L (2002) The structure of reef fish metapopulations: modelling larval dispersal and retention patterns. Proc R Soc Lond 269:2079–2086

Jones GP, Milicich MJ, Emslie MJ, Lunow C (1999) Self-recruitment in a coral reef fish population. Nature 402:802–804

Kalish JM (1993) Pre- and post-bomb radiocarbon in fish otoliths. Earth Planet Sci Lett 114:549–554

Kritzer JP, Davies CR, Mapstone BD (2001) Characterizing fish populations: effects of sample size and population structure on the precision of demographic parameter estimates. Can J Fish Aquat Sci 58:1557–1568

Longhurst A, Pauly D (1987) Ecology of tropical oceans. Academic Press, San Diego, CA, USA

Lou DC, Mapstone BD, Russ GR, Begg GA, Davies CR (2007) Using otolith weight–age relationships to predict age based metrics of coral reef fish populations across different temporal scales. Fish Res 83:216–227

Lou DC, Mapstone BD, Russ GR, Davies CR, Begg GA (2005) Using otolith weight–age relationships to predict age based metrics of coral reef fish populations at different spatial scales. Fish Res 71:279–294

Luckhurst BE, Dean JM, Reichert M (2000) Age, growth and reproduction of the lane snapper *Lutjanus synagris* (Pisces: Lutjanidae) at Bermuda. Mar Ecol Prog Ser 203:255–261

Mace PM, Fenaughty JM, Coburn RP, Doonan IJ (1990) Growth and productivity of orange roughy (*Hoplostethus atlanticus*) on North Chatham Rise. NZ J Mar Freshwater Res 24:105–119

Mapstone BD (1995) Scalable decision rules in environmental impact assessment: effect size, Type I, and Type II errors. Ecol Apps 5:401–410

McAllister MK, Peterman RM (1992) Experimental design in the management of fisheries: a review. N Am J Fish Manage 12:1–18

Meekan MC, Carleton JH, McKinnon, AD, Flynn K, Furnas M (2003) What determines the growth of tropical reef fish larvae in the plankton: food or temperature? Mar Ecol Prog Ser 256:193–204

Milton DA, Chenery SR, Farmer MJ, Blaber SJM (1997) Indentifying the spawning estuaries of the tropical shad, terubok *Tenualosa toli*, using otolith microchemistry. Mar Ecol Prog Ser 153:283–291

Newman SJ (2002) Growth rate, age determination, natural mortality and production potential of the scarlet seaperch, *Lutjanus malabaricus* Schneider 1801, off the Pilbara coast of northwestern Australia. Fish Res 58:215–225

Newman SJ, Cappo M, Williams D (2000a) Age, growth and mortality rates and corresponding yield estimates using otoliths of the tropical red snappers, *Lutjanus erythropterus*, *L. Malabaricus* and *L. Sebae*, from the central Great Barrier Reef, Australia. Fish Res 48:1–14

Newman SJ, Cappo M, Williams D (2000b) Age, growth and mortality of the stripey, *Lutjanus carponotatus* (Richardson) and the brown–stripe snapper, *L. vitta* (Quoy and Gaimard) from the central Great Barrier Reef, Australia. Fish Res 48:263–275

Oke CS, Crozier RH, Ward RD (2002) Stock structure of Australian populations of orange roughy analysed using microsatellites. Report No. FRDC 1997/118, James Cook University, Townsville, Australia

Panfili J, Pontual H de, Troadec H, Wright PJ (2002) Manual of fish sclerochronology. Ifremer-IRD coedition, Brest, France, 464pp

Pauly D (1998) Fishing down marine food webs. Science 279:860–863

Pauly D, Christensen V, Guenette S, Pitcher T, Sumaila UR, Walters CJ, Watson R, Zeller D (2002) Towards sustainability in world fisheries. Nature 418:689–695

Payan P, Edeyer A, de Pontual H, Borelli G, Boeuf G, Mayer-Gostan N (1999) Chemical composition of saccular endolymph and otolith in fish inner ear: lack of spatial uniformity. Am J Physiol Regul Integr Comp Physiol 277:R123–R131

Peterman RM (1990) Statistical power analysis can improve fisheries research and management. Can J Fish Aq Sci 47:2–15

Pilling GM, Grandcourt EM, Kirkwood GP (2003) The utility of otolith weight as a predictor of age in the emperor *Lethrinus mahsena* and other tropical fish species. Fish Res 60:493–506

Pitcher TJ, Hart PJB (1982) Fisheries ecology. Croon Helm Ltd, London

Popper A, Platt C (1993) Inner ear and lateral line. In: Evans D (Ed) The physiology of fishes. CRC Press, Boca Raton, FL, pp. 99–136

Quinn II TJ, Deriso RB (1999) Quantitative fish dynamics. Oxford University Press, Oxford

Ricker WE (1975) Computation and interpretation of biological statistics of fish populations. Fish Res Bd Can 191:1–382

Rooker JR, Secor DH, Zdanowicz VS, DeMetrio G, Relini LO (2003) Identification of Atlantic bluefin tuna (*Thunnus thynnus*) stocks from putative nurseries using otolith chemistry. Fish Oceanogr 12:75–84

Russ G (1991) Coral reef fisheries: effects and yields. In: Sale PF (Ed) The ecology of fishes on coral reefs. Academic Press, New York, pp 601–636

Schweigert JF, Haegele CW, Stocker N (1985) Optimal sampling design for herring spawn surveys in the Strait of Georgia, B.C. Can J Fish Aquat Sci 42:1806–1814

Secor DH, Dean JM, Campana SE (1995) Recent Developments in Fish Otolith Research. University of South Carolina Press, Columbia SC

Secor DH, Dean JM, Laban EH (1991) Manual for otolith removal and preparation for microstructural examination. Belle W. Baruch Institute for Marine Biology and Coastal Research, University of South Carolina, Columbia, SC

Smith DC, Fenton GE, Robertson SG, Short SA (1995) Age determination and growth of orange roughy (*Hoplostehus atlanticus*): a comparison of annulus counts with radiometric ageing. Can J Fish Aquat Sci 52:391–401

Smith MK (1992) Regional differences in otolith morphology of the deep slope red snapper *Etelis carbunculus*. Can J Fish Aquat Sci 49:795–804

Stevenson DK, Campana SE (1992) Otolith microstructure examination and analysis. Can Spec Publ Fish Aquat Sci 117:126pp

Swearer S, Caselle J, Lea D, Warner R (1999) Larval retention and recruitment in an island population of a coral–reef fish. Nature 402:799–802

Verweij MC, Nagelkerken I, Hans I, Ruseler SM, Mason PRD (2008) Seagrass nurseries contribute to coral reef fish populations. Limnol Oceanogr 53:1540–1547

Victor B (1983) Recruitment and population dynamics of a coral reef fish. Science 219:419–420

Victor BC (1991) Settlement strategies and biogeography of reef fishes. In: Sale PF (Ed) The ecology of fishes on coral reefs. Academic Press, New York, pp 231–260

Wilson DT, Meekan MG (2001) Environmental influences on patterns of larval replenishment in coral reef fishes. Mar Ecol Prog Ser 222:197–207

Worthington DG, Fowler AJ, Doherty PJ (1995a) Variation in the relationship between otolith weight and age: implications for the estimation of two tropical damselfish (*Pomacentrus moluccesis* and *P. wardi*). Can J Fish Aquat Sci 52:233–242

Worthington DG, Fowler AJ, Doherty PJ (1995b) Determining the most efficient method of age determination for estimating the age structure of a fish population. Can J Fish Aquat Sci 52:2320–2326

2. AGEING IN CORAL REEF FISHES: DO WE NEED TO VALIDATE THE PERIODICITY OF INCREMENT FORMATION FOR EVERY SPECIES OF FISH FOR WHICH WE COLLECT AGE-BASED DEMOGRAPHIC DATA?

J.H. CHOAT[1*], J.P. KRITZER[1,2], AND J.L. ACKERMAN[1,3]

[1] *James Cook University, School of Marine Biology, Townsville, QLD 4811, Australia*

[2] *Environmental Defense, 257 Park Avenue South, New York, NY 10010, USA*

[3] *Bureau of Rural Sciences, Fisheries & Marine Science Program, GPO Box 858, Canberra, ACT 2601, Australia*

**Corresponding Author: J.H.Choat john.choat@jcu.edu.au*

1 Preamble

The purpose of this chapter is to consider the question "Is it necessary to validate the periodicity of increment formation in every species of fish for which we seek age-based demographic data"? The focus is on coral reef fishes. Four issues require consideration. Firstly, validation programs are expensive in terms of resources and time. This is especially important for coral reef fishes as resources available to tropical fisheries are often very limited. Secondly, many modern techniques used to validate the accuracy of age estimates require field and laboratory infrastructure that may not be available to fisheries laboratories serving coral reefs. Thirdly, the great majority of validation studies have confirmed the annual periodicity of increment formation. Fourthly, opportunities to study undisturbed populations of reef fishes from which reference age data can be derived are limited due to over-fishing and habitat alteration. We argue for a more strategic approach to age-based studies in coral reef fishes.

We firstly acknowledge the substantial effort and expertise devoted to ensuring appropriate standards of accuracy in the ageing of fishes. As a result of this effort, there are now numerous examples where the increments observed in sagittal otoliths have been demonstrated to form annually. At some stage we might assume that such increments are annual in nature without the requirement that the assumption of annual formation be validated in every species for which age data are published. We argue that the time to accept this assumption has now arrived. This does not imply that we should cease validation studies. We would make a distinction between studies of individual species of widespread commercial importance, especially if they are the targets of new fisheries, and those of species assemblages for which the focus is ecological and evolutionary processes. Demanding validation of age estimates for all species before publication in the latter type of study will distort the allocation of scarce resources and result in substantial delays in dissemination of age-based data.

B.S. Green et al. (eds.), *Tropical Fish Otoliths: Information for Assessment, Management and Ecology*,
Reviews: Methods and Technologies in Fish Biology and Fisheries 11,
DOI 10.1007/978-1-4020-5775-5_2,

2 The historical setting

Teleost fishes are the most readily aged of all vertebrates. This reflects the unique properties of their otoliths, including continuous patterns of growth and the absence of metabolic reworking of otolith material once deposited (Campana & Thorrold 2001). Age-based information is of crucial importance to the understanding of life history features of fishes (Beverton & Holt 1957), especially as they exhibit highly plastic patterns of growth (Gust et al. 2002, Swain et al. 2003). Research over the last two decades has demonstrated that thin sections of sagittal otoliths provide the most consistent and reliable records of age (Secor et al. 1995, Campana 2001, Begg et al. 2005), especially for fish with ages exceeding 20 years. The increasing availability of age-based information now makes it possible to develop comprehensive studies of life-histories within this most diverse and abundant group of all vertebrates.

Age-based data have been rapidly accumulating over the past two decades. The process has been highly uneven with respect to the lineages of fish that have been examined, however, as well as the environmental settings where the studies have been carried out. Teleost fishes, like most other groups of living organisms, show strong latitudinal gradients in species richness. Our ignorance is greatest for the diverse assemblages characteristic of low latitude shallow water environments.

History strongly influences the course of research enterprises, and the study of fish population dynamics is no exception. It is important to trace the course of fisheries science over the latter half of the last century in order to understand why we are ignorant about the demography of large groups of fishes. Pauly (1998a) identified the publication of the seminal study of Beverton and Holt (1957) as the genesis of modern fisheries science and noted the critical importance of age-based demographic information in this development. The dynamics of those fishes that supported extensive multi-national fisheries in northern temperate and boreal waters (mainly gadids, clupeoids and pleuronectids) became a research priority (Beverton 1992). The success of Beverton and Holt (1957) as a blue-print for fisheries science was due to two factors. Firstly, the existence of a substantial archive of fisheries data, scientific infrastructure and expertise developed in association with the industrial-scale fisheries of northern temperate and boreal waters. Secondly "coldwater fish could straightforwardly be aged by reading annuli on otoliths" (Pauly 1998a). In the same study, Pauly provided a comprehensive account of the difficulties faced by tropical fisheries biologists, including ageing studies, and advocated length-based approaches as an alternative. Pauley presented a convincing case for the difficulties associated with tropical fisheries biology, partly based on the argument that identification of growth increments in calcareous structures was easier in temperate than in tropical species (Munro 1983, Gjosaeter et al. 1984, Fowler 1995), a fact usually associated with the stronger seasonal cycling characteristic of higher latitudes. Subsequent research demonstrated, however, that tropical fish could be aged through the reading of annuli in otoliths.

A problem in fisheries biology has been the underestimation of age in many commercially important species (Campana 2001), especially those inhabiting deeper water (Berkeley et al. 2004, Cailliet et al. 2001, Munk 2001). Campana (2001) also emphasized the methodological issues where artefacts of otolith preparation frequently led to underestimation of ages and thus resulted in artificially high estimates of growth and mortality rates. Campana also summarized the protocols by which the frequency of growth increments and absolute ages in fishes were validated. The manifest impacts of fishing on long-lived species demanded greater accuracy in the assignment of ages to individuals and focused attention on the need to identify artefacts in otolith preparation, with particular emphasis on those that resulted in underestimation of ages.

Advances in the methods used in validating age estimates have resulted in greater accuracy in the interpretation of age, including for long-lived, deep water species. The most significant result of these advances has been provided by the opportunity to use combinations of methods (e.g., bomb radiocarbon, marginal increment analysis, chemical marking of otoliths) to confirm age estimations in a wide variety of species. Coupled with this has been the improvement of the preparation and optical resolution of otolith sections and the capacity to store and transmit high quality images of otolith preparations. The validation of different methods (Campana 2001) and the associated technical advances have resulted in increasing confidence that thin sections of otoliths are providing the basis for accurate age interpretation. These advances occurred in an environment dominated by extensive commercial fisheries supported by well developed sampling and laboratory infrastructure.

The most influential paper in the context of validation studies was that of Beamish and McFarlane (1983) who emphasized the importance of validation. They argued for a comprehensive protocol, including validation of all age classes for each species studied, and warned that extrapolation of the results from one population even to other populations of the same species was dangerous. The messages in this widely cited study (263 citations by July 2003) were reinforced by the comprehensive review of Campana (2001) who also prioritized different validation protocols.

The challenges of validation in most cases were successfully met during the two decades spanning the publication of these studies (Secor et al. 1995, Campana 1999, 2001). In almost every case, the results have confirmed the hypothesis that structures observed in sectioned sagittae represented annual increments, allowing estimates of age from direct counts. The support for annual periodicity of increments is very strong, even though there may be a bias due to the non-reporting of negative results. The most impressive evidence comes from two sources. Firstly, there have been multiple studies of widely distributed species in which controversies have been resolved via the application of independent methods (Baker & Wilson 2001, Kalish 2001, Cass-Calay & Bahnick 2002, Fischer et al. 2005). Secondly, with increases in the sample sizes of fish aged from sagittal otoliths, congruent demographic patterns are emerging among phylogenetically-related groups of species, including those of deep water (Cailliet et al. 2001, Munk 2001) and coral reef environments (Choat & Robertson 2002).

Access to large vessels, modern laboratory infrastructure and innovative research approaches based on the analysis of bomb radiocarbon and the use of radiochemical dating, (Kalish 1993, Campana 1999) have been crucial. The new methods demanded a high level of technical skill and expensive instrumentation, but they allowed fisheries scientists to validate a wide range of species including those from deep water environments (Cailliet et al. 2001). These studies resolved some controversial issues with respect to fish life spans and confirmed that the deep sea was dominated by populations of very long-lived fishes with slow growth and extended generation times.

A more strategic approach to the study of ageing in tropical teleost populations is now required. At present, ageing and demographic studies proceed on a case by case basis in which detailed investigations are carried out when the exploitation of a particular species becomes an issue. The logic and future directions for this type of research have been debated in only in a few instances. We pose two questions: (1) Do the benefits of validating the age of every species studied outweigh the need for a more comprehensive demographic picture based on the assumption that formation of growth increments in otoliths are annual?; and (2) How many more species must be validated before annual periodicity in increment formation is accepted as a credible assumption? The purpose of this chapter is to present the case that attention to single-species detail should be traded off against studies aimed at providing broader perspectives on prevailing patterns of demography and age-structure among large groups of species, based on the assumption of annual periodicity in otolith increment formation.

There are two further issues that should be considered in the context of reef fish ageing. Firstly, molecular tools have allowed us to develop an evolutionary perspective on reef fish life histories and demography. The opportunities for evaluating ecological and evolutionary hypotheses in this diverse assemblage of vertebrates are exciting. Secondly, over-fishing and habitat destruction are increasingly impacting tropical fish assemblages (Jackson et al. 2001, Pauly et al. 1998). Both the opportunities for comprehensive ecological and evolutionary research and the urgency generated by increasing over-fishing suggest that delays in analyzing age-based dynamics of tropical fishes will be problematic. If, as suggested by Bell (2001), chronologies must be validated *before* "ecological and evolutionary studies *become* possible" then it is unlikely that comprehensive demographic studies will ever see the light of day. Time is short and resources limited, especially for tropical species. Understanding of reef fish demography lags behind that of temperate species. There is a lot of time to make up.

2.1 WHAT DID BEAMISH AND MCFARLANE (1983) ACTUALLY SAY AND HOW DID IT IMPACT ON SUBSEQUENT STUDIES?

The core issues of Beamish and McFarlane (1983) were illustrated by two examples of the consequences of non-validation. One was a freshwater fish (*Catostomus*) and the second a deepwater marine fish (*Sebastes*). In both instances, initial ageing was based on the examination of growth increments in scales. Subsequent investigation of sectioned fin rays in *Catostomus* and sectioned sagittal otoliths in *Sebastes* showed that in each case the initial estimates of growth and mortality rates were significantly greater than the true rates. The assumption of rapid growth and high natural mortality rates had serious consequences when applied to long-lived species, especially for *Sebastes*.

Two primary messages emerged. Firstly, *underestimation* of true ages was the most serious problem facing fisheries biologists and, secondly, scales were an inappropriate structure for age estimation. The concerns of Beamish and McFarlane (1983) and Campana (2001) have been met over the past two decades, however, in that ageing studies recognize the reality of extended life spans in fishes and focus on sagittal otoliths to assess age. Long-lived fishes from a range of environments are being identified and reported with increasing frequency (Cailliet et al. 2001, Reznick et al. 2002). The literature has confirmed repeatedly that sectioned sagittal otoliths accurately record older ages in many species of teleosts (Table 1). Modern image-capture and analysis techniques have improved vastly the capacity to identify and measure increments in sectioned sagittae. Figure 1 illustrates the capacity to obtain consistent estimates of increment widths near the otolith margin in long-lived reef fish species. Otolith increments can be clearly identified and analyzed in tropical species, providing reliable estimates of growth processes. The assumption that increments identified in sectioned sagittae of long lived fish are annual is a realistic working hypothesis.

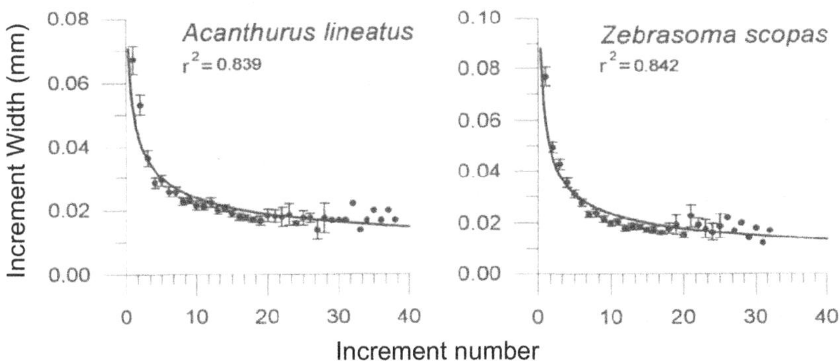

Figure 1. The relationship between increment width (mm) and increment number in two long-lived species of acanthurid fishes (Choat & Axe 1996). N=5 individuals measured for each species.

There remains the possibility, however, that the record of annual growth increments will not accurately reflect the true age of the fish. Why would this occur, and if so, how frequently? Firstly, rings may be deposited annually but may sometimes be difficult to detect. Secondly, the rings may be always detectable but deposited more frequently or less frequently than each year. For example, while the pattern of increment formation may be annual, there might be some years in which deposition does not occur. Individual increment formation may be modified or suppressed either in individuals or in populations, possibly in response to climatic forcing (Meekan et al. 1999). Moreover, there is no doubt that in some species growth increments are difficult to identify (Figure 2) and measure, and that there is a gradient in increment clarity from low to high latitudes (Figure 3). These alternatives cannot be distinguished (Francis 1995), but unless they occur consistently they are unlikely to have major effect on the estimation of life spans, especially in long-lived species.

Figure 2. Transverse sections of sagittal otoliths of two species of serranid fishes (**A**) *Epinephelus polyphokadian* and (**B**) *Cephalopolis argus* sampled from the same latitude, 18°S on the Great Barrier Reef. *E. polyphokadian* consistently had clear annual growth increments whereas increments in *C. argus* were invariably difficult to read.

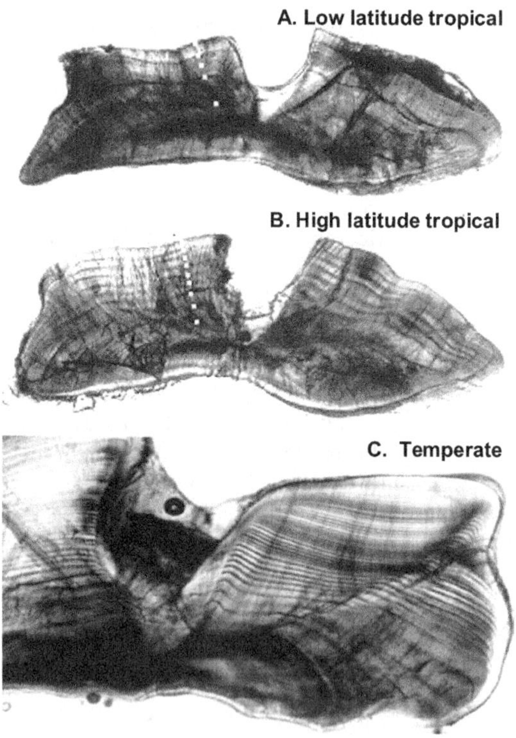

Figure 3. Gradient of clarity in the display of growth increments in sectioned sagittae from low to high latitudes. **A**. Sectioned sagitta from the tropical parrot fish *Sparisoma viride* from latitude 9°N and **B**, from 23°N. **C** is the sectioned sagitta of a temperate water reef fish *Girella tricuspidata* from 36°S.

A much greater problem arises in situations where increments are clearly displayed but do not reflect an annual signal. These will not simply truncate or extend age distributions, they will give an erroneous picture of population dynamics. Examples are the European hake *Merluccius merluccius* (Morales-Nin et al. 1998), *Pagrus pagrus* growing in culture (Machias et al. 1998) and the reported non-annual periodicity of increment formation in lutjanids (Milton et al. 1995) (see below).

Table 1 provides examples of three important groups of exploited reef fishes in which recent validations of the annual periodicity of increment formation have been completed. These are serranids (groupers), lutjanids (snappers) and sparids (porgies). The first two occur mainly in coral reef waters, while the third group also extends into temperate reef and estuarine environments. The majority of validations were accomplished using marginal increment analysis (MIA), a reflection of the logistic and infrastructure issues in tropical fisheries. The table identifies widely distributed species within each family that have been subject to at least two independent validation protocols. These are *Epinephelus flavolimbatus, Lutjanus campechanus, L. erythrograma, L. malabaricus, L. sebae, L. griseus* and *Pagrus auratus*. In each case, the validation of annual periodicity of increments visualized in sectioned sagittae was confirmed independently through bomb radiocarbon analysis.

Lutjanus campechanus, important in the commercial and recreational fisheries of the south-eastern USA, was aged from sectioned sagittae and validated by MIA in three independent studies (Wilson & Nieland 2001, Patterson et al. 2001, White & Palmer 2004). The accuracy of annuli revealed in sectioned sagittae was confirmed by bomb radiocarbon analysis (Baker & Wilson 2001) and radiometric analyses (Baker et al. 2001). A similar confirmatory procedure was carried out for *E. flavolimbatus* with sagittal increments validated (MIA) by Manickchand-Heileman and Phillip (2000) later confirmed by bomb radiocarbon analysis (Cass-Calay & Bahnick 2002).

What are the challenges to studies that clearly confirm the one ring – one year hypothesis? One of the most explicit is provided by Milton et al. (1995). They found differences in counts of the increments observed in sectioned compared with whole otoliths of the tropical lutjanids *Lutjanus erythropterus, L. malabaricus,* and *L. sebae* from unexploited populations in the Gulf of Carpentaria, Australia. Increments in sectioned sagittae were 1.6–2.4 times the number found in whole otoliths. Pb-210/Ra-226 radioactive disequilibria of both whole and cored otoliths were measured to obtain independent estimates of age. The whole-otolith counts agreed better with the radiometric age in samples whose sectioned and whole-otolith ages differed by more than 4 years. The conclusions of this paper were particularly important as the species were subject to commercial exploitation elsewhere and independent studies based on sectioned sagittae indicated that each species (in contrast to the radiometric assessments) was long lived with a relatively slow growth rate.

Table 1. Species of the families Serranidae, Lutjanidae and Sparidae from tropical and warm temperate environments in which periodicity of increment formation has been validated recently. Shading indicates species in which annual periodicity has been confirmed independently by bomb radiocarbon analysis. Ages were derived from sectioned sagittae unless otherwise indicated with * following the method, in which case whole sagittae were used. *Legend*: MIA – Marginal increment analysis; OTC – Oxytetracycline injection; δ^{14} C – Bomb Radiocarbon.

Serranidae

Species	Method	References
Epinephelus adscensionis	MIA	Potts and Manooch (1995)
Epinephelus cruentatus	MIA	Potts and Manooch (1999)
Epinephelus flavolimbatus	MIA	Manickchand-Heileman and Phillip (2000)
Epinephelus flavolimbatus	δ^{14} C	Cass-Calay and Bahnick (2002)
Epinephelus fulvus	MIA	Potts and Manooch (1999)
Epinephelus fuscoguttatus	OTC MIA	Pears et al. (2005)
Epinephelus guttatus	OTC	Sadovy et al. (1992)
Epinephelus guttatus	MIA	Potts and Manooch (1995)
Epinephelus itajara	MIA	Bullock et al. (1992)
Epinephelus merra	OTC	Pothin et al. (2004)
Epinephelus niveatus	MIA	Wyanski et al. (2000)
Epinephelus octofasciatus	δ^{14} C	Kalish (2001)
Epinephelus striatus	MIA	Bush et al. (1996)
Cephalopholis boenak	MIA	Chan and Sadovy (2002)
Cephalopholis cyanostigma	OTC	Mosse (2001)
Plectropomus maculatus	OTC	Ferreira and Russ (1992)
Plectropomus leopardus	OTC	Ferreira and Russ (1994)
Myctoperca microlepis	MIA*	McErlean (1963)
Myctoperca microlepis	MIA	Hood and Schlieder (1992)
Myctoperca bonaci	MIA	Crabtree and Bullock (1998)
Myctoperca interstitialis	MIA	Manickchand-Heileman and Phillip (2000)
Polyprion oxygeneios	OTC	Francis et al. (1999)
Polyprion americanus	MIA	Peres and Haimovici (2004)
Serranus cabrilla	MIA	Tserpes and Tsimenides (2001)
Centropristis striata	MIA	Hood et al. (1994)

Table 1. (Continued)

Lutjanidae

Species	Method	Reference
Lutjanus adetti	OTC	Newman et al. (1996)
Lutjanus analis	MIA	Mason and Manooch (1985)
Lutjanus analis	MIA	Burton (2002)
Lutjnaus argentimaculatus	OTC	Cappo et al. (2000)
Lutjnaus argentimaculatus	OTC	Russell et al. (2003)
Lutjanus bohar	OTC	Marriott and Mapstone (2006)
Lutjanus campechanus	MIA	Patterson et al. (2001)
Lutjanus campechanus	MIA	Wilson and Nieland (2001)
Lutjanus campechanus	δ^{14} C	Baker and Wilson (2001)
Lutjanus campechanus	Radiometric	Baker et al. (2001)
Lutjanus campechanus	MIA	White and Palmer (2004)
Lutjanus erythropterus	Radiometric*	Milton et al. (1995)
Lutjanus erythropterus	OTC	Cappo et al. (2000)
Lutjanus erythropterus	δ^{14} C	Kalish (2001)
Lutjanus fulviflamma	MIA	Kamukuru et al. (2005)
Lutjanus griseus	MIA	Burton (2001)
Lutjanus griseus	δ^{14} C	Fischer et al. (2005)
Lutjanus johnii	δ^{14} C	Kalish (2001)
Lutjanus kasmira	Daily Rings	Morales-nin and Ralston (1990)
Lutjanus malabaricus	Radiometric*	Milton et al. (1995)
Lutjanus malabaricus	OTC	Cappo et al. (2000)
Lutjanus malabaricus	δ^{14} C	Kalish (2001)
Lutjanus malabaricus	MIA	Newman (2002)
Lutjanus peru	MIA	Rocha-Olivares (1998)
Lutjanus quinquilineatus	OTC	Newman et al. (1996)
Lutjanus sebae	Radiometric*	Milton et al. (1995)
Lutjanus sebae	OTC	Cappo et al. (2000)
Lutjanus sebae	C14	Kalish (2001)
Lutjanus sebae	MIA	Newman and Dunk (2002)
Lutjanus synagris	MIA	Manickchand-Dass (1987)
Lutjanus synagris	MIA	Luckhurst et al. (2000)
Aprion virescens	MIA	Pilling et al. (2000)
Pristipomoides multidens	MIA	Newman and Dunk (2003)
Rhomboplites aurorubens	MIA	Hood and Johnson (1999)
Ocyurus chrysurus	MIA	Manooch and Drennon (1987)

Table 1. (Continued)

Sparidae

Species	Method	Reference
Acanthopagrus berda	MIA	James et al. (2003)
Acanthopagrus bifasciatus	MIA	Grandcourt et al. (2004)
Acanthopagrus butcheri	MIA	Sarre and Potter (2000)
Archosargus probatocephalus	MIA	Beckman et al. (1991)
Archosargus probatocephalus	MIA	Dutka-Gianelli and Murie (2001)
Argyrops spinifer	MIA	Grandcourt et al. (2004)
Argyrozona argyrozona	OTC	Brouwer and Griffiths (2004)
Dentex dentex	MIA	Machias et al. (2002)
Diplodus vulgaris	MIA*	Goncalves et al. (2003)
Diplodus vulgaris	MIA*	Pajuelo and Lorenzo (2003)
Diplodus sargus	MIA	Pajuelo and Lorenzo (2002a)
Diplodus annularis	MIA*	Pajuelo and Lorenzo (2002b)
Lithognathus aureti	M/Recap	Holtzhausen and Kirchner (2001)
Lithognathus mormyrus	MIA	Lorenzo et al. (2002)
Lithognathus mormyrus	MIA	Pajuelo et al. (2002)
Pagrus auratus	δ^{14} C	Kalish (2001)
Pagrus auratus	OTC	Ferrell et al. (1992)
Pagrus auratus	OTC	Francis et al. (1992)
Pagrus pagrus	Culture	Machias et al. (1998)
Pagrus pagrus	MIA	Hood and Johnson (2000)
Polysteganus undulosus	MIA	Chale-Matsau et al. (2001)
Rhabdosargus sarba	MIA/OTC	Radebe et al. (2002)
Sarpa salpa	MIA	van der Walt and Beckley (1997)
Sarpa salpa	MIA	Villamil et al. (2002)
Sparodon durbanensis	MIA	Buxton and Clarke (1991)

A substantial amount of additional work on these species using both marginal increment analysis and oxytetracycline (OTC) marking (Newman et al. 2000, Cappo et al. 2000, Newman 2002, Newman & Dunk 2002) on both the east and west coasts of Australia confirmed that sectioned sagittae provided estimates of greater longevities and slower growth rates than those obtained from radiometric analyses and whole otoliths. Confirmation of the estimates of age structure and growth rates derived from MIA and OTC marking was provided by Kalish (2001), who validated the accuracy of these age estimates by analysis of bomb radiocarbon for all three species. This example demonstrates the benefits of using a combination of methods to resolve problems where artefacts of otolith analysis have resulted in discrepancies (in this case underestimation) of ages (Campana 2001). The value of bomb radiocarbon analysis has been confirmed clearly by Kalish (2001), who reported validation of the accuracy of annual increments in sagittae in 23 species from a wide range of environments including the deep sea, open ocean and temperate and tropical reefs. In addition, Cailliet et al. (2001) demonstrated that radiometric ageing confirmed the age estimates of four species of the genus *Sebastes* derived from sagittal growth increments.

The only problematic examples in the Kalish (2001) study concerned species from the open ocean and deeper waters. These included species in which the environment of juveniles was variable with respect to δ^{14} C (*Pristipomoides multidens),* otolith structure was difficult to interpret (*Hyperoglyphe antarctica*) and species in which non-otolith structures were used in ageing (*Xiphius gladius*). Although problems may arise in deep water groups (e.g., trachicthids, oreosomatids) with respect to the pattern of penetration of δ^{14} C by depth and with interpretation of otolith morphology, bomb radiocarbon analysis has confirmed longevity in many deep sea species.

2.2 WHY HAS THE AGE-BASED ANALYSIS OF TROPICAL FISH POPULATIONS MOVED RELATIVELY SLOWLY?

The life history features of stocks in shallow water tropical regions are poorly known compared to the temperate and deep water fishery stocks. This situation is being rectified through increasing studies of larger, commercially important species.

Demographic analysis of coral reef fishes has had a mixed history. Whilst Beamish and McFarlane's (1983) paper has received numerous citations, the FAO report on ageing tropical fish by Gjosaeter et al. (1984) largely has been forgotten. Gjosaeter and co-workers, however, made a pertinent point. They stressed that whilst the importance of considering the different validation methods should not be underrated, worthwhile studies can still be performed without rigorous validation. They claimed that even "rather rough indications [of age] may be sufficient". In fact, the record was more informative than suggested by Gjosaeter et al. (1984). McErlean (1963) showed clearly that tropical serranids could be aged through analysis of sagittal otoliths. Subsequent studies on the dynamics of tropical fish (e.g., Munro 1983) did not expand on these findings, mainly because otolith structure in tropical fishes is often difficult to interpret compared with that seen in high latitude fishes.

Age-based demographic studies of tropical fishes lag behind their temperate counterparts. This reflects to some extent the view that it is not possible to age fishes from low latitudes reliably except through daily increments (e.g.: Polunin & Roberts 1996). Examples to date, however, demonstrate that fish at low latitudes retain a reliable record of age in sagittal otoliths (Choat et al. 2003, Robertson et al. 2005). In addition, the view that tropical fishes have shorter life spans and higher growth and mortality rates than their temperate equivalents (Pauly 1994, 1998b) seems to have diverted attention from age-based to size-based models of reef fish demography. Surprisingly, the view that there are few examples of age based studies on tropical marine fishes and that tropical species generally have fast growth rates is still being promulgated (Henderson 2005). It is now clear, however, that the perciform assemblages that constitute reef fish fauna harbour a great deal of ecological diversity, including a wide range of life spans, growth rates and size structures – including long-lived and slow-growing species (Choat & Robertson 2002).

2.3 DO REEF FISHES PRESENT SPECIAL CHALLENGES TO AGE-BASED STUDIES AND VALIDATION?

The answer is yes, both in terms of their biological features and the nature of tropical fisheries biology. Local and regional diversity in reef fishes is greater than that encountered at higher latitudes (Helfmann et al. 1997). Reef fish assemblages are highly diverse at local scales and characterized by complexes of closely related species, though many species are rare at local scales (Figure 4). Consequently, most reef fisheries are usually multi-specific (Polunin et al. 1996). There is increasing evidence of demographic variation at a variety of scales over the geographic range of a species, reflecting the influence of habitats, environmental gradients and evolutionary history (Kritzer 2002, Williams et al. 2003, Robertson et al. 2005). Many tropical species are longer lived than anticipated, attaining ages in excess of 30 years or more (Choat & Robertson 2002), although the majority of reef fishes alive today were recruited post-1970, after the cessation of nuclear atmospheric testing, so diminishing the utility of bomb radiocarbon methods for validation. Flat-topped growth curves, in which size and age are decoupled for the majority of the life span, characterize many lineages so that size-based estimates of population processes are not informative (Robertson et al. 2005, Marriott et al. 2007).

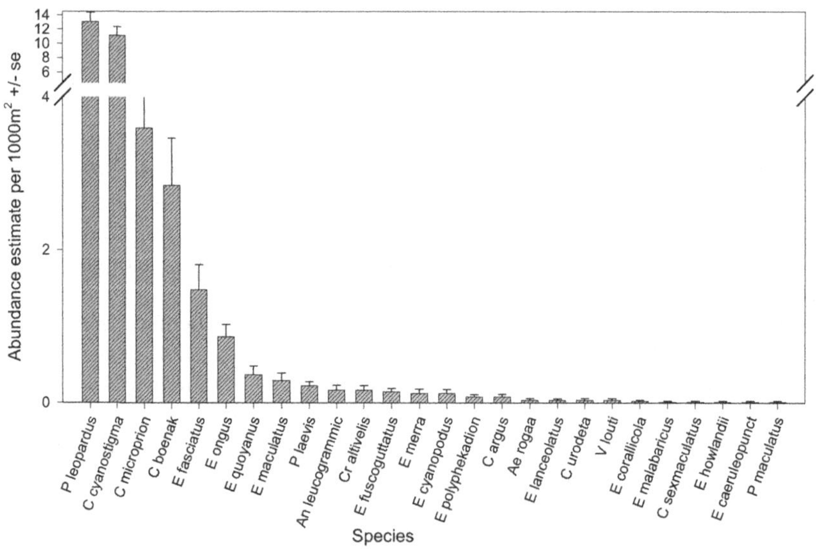

Figure 4. Abundance estimates of serranid fishes from northern midshelf reefs of the Great Barrier Reef Marine Park. With the exception of *Plectropomus leopardus* (highly characteristic of the GBR), larger commercially important species are rare with abundances of less than 1 individual per 1,000 m^2 (Pears 2006).

The early life history of reef fishes provides a number of challenges for tropical fisheries biologists. Knowledge of gyres and other oceanographic features has provided a framework for understanding the pattern and magnitude of recruitment variation in commercially important species in temperate environments (Cushing 1975, Iles & Sinclair 1982, Sinclair 1988). The environment for pre-settlement reef fishes, however, is complex (Cowen 2002), with variable and relatively unpredictable movement of water masses. Different groups of fishes exhibit very different capacities for active movement and directional swimming during the "larval" phase (Stobutzki & Bellwood 1997, Jones et al. 1999). The result is that patterns of recruit variation are difficult to predict in coral reef systems.

Coral reef fishes have been exploited since prehistoric times (Wing & Wing 2001), but historically their fisheries have been subsistence or artisanal. The highly dispersed nature of coral reefs means that fisheries research in many localities has substantial logistical costs and is often expeditionary in nature. The recent development of industrial level coral reef fisheries (largely associated with the live reef fish trade) has been competitive, highly exploitative, and lacking a concomitant development of research infrastructure (Sadovy & Vincent 2002). The distribution of coral reef fisheries across broad, politically complex geographical regions means that they lack a shared information infrastructure. FISHBASE (Froese & Pauly 2000) is an exception.

3 Validation protocols and coral reef fishes

The most comprehensive listing of validation protocols is in Campana (2001) where 16 methods are listed ordered by scientific value. Not all of these are appropriate for reef fishes. We deal with the most realistic protocols below, and provide comments on the methodological issues that arise in each case.

Release of known age fish into the wild. The most effective method involves the release of hatchery reared chemically mass-marked fish into the natural environment. There have been no instances in which reef fish have been successfully established on reefs through the release of cultured juveniles of known age, however, despite the success of re-seeding sessile invertebrates on coral reefs. A key problem is the difficulty associated with closing the life-cycle via culture experiments with groups such as serranids, lutjanids, haemulids and labrids. The high mortality rates experienced by newly recruited reef fishes and the possibility that cultured fish released back into the environment would suffer enhanced mortalities (Carr et al. 2004, Fairchild & Howell 2004, Masuda et al. 2003) suggest that this will not be effective for reef fishes. It is no surprise that success has been primarily with fresh water species (Campana 2001).

Bomb radiocarbon. This method is unlikely to be effective for future work on reef fishes because the hatch dates of most reef fish alive today do not extend back to the 1960s, when nuclear tests provided signature isotopes that could be incorporated into otoliths. Moreover, the costs of the method will be beyond research budgets of many tropical institutions and most workers dealing with tropical fish species. Archived otoliths might be used to obtain specimens with the correct temporal window for bomb radiocarbon dating, though very little archival material from the appropriate periods exists, because of the prevailing views during the 1970s and 1980s that otoliths would not provide an acceptable basis for ageing tropical fish. The method has been most valuable in confirming age estimates in long-lived commercially important species.

Mark-recapture of chemically-tagged wild fish. This has proved to be the most rigorous and cost-effective method for validating increment periodicity in reef fishes. Most examples are Australian (Ferreira & Russ 1992, 1994, Lou 1992, Choat & Axe 1996, Choat et al. 1996, Newman et al. 1996, Cappo et al. 2000, Hernaman et al. 2000) or from tropical Atlantic waters (Bullock et al. 1992, Sadovy et al. 1992, Crabtree et al. 1995, 2002, Crabtree & Bullock 1998, Luckhurst et al. 2000, Choat et al. 2003, Robertson et al. 2005). Drawbacks lie in the initial capture of fish in a condition that makes tagging worthwhile and the difficulty of securing adequate recaptures. The local rarity of many species (Figure 4) and the structural complexity of the reef environment means that initial tagging rates generally will be low. Coral reef fish cannot usually be caught alive in large numbers and must be handled with great care to avoid eye infections and skin lesions. Any injury to the fish associated with capture (by line fishing and traps especially) or tagging usually results in high initial mortalities after release through predation. Logistic difficulties associated with capture means that tagging rates are low (compared with temperate fishes), which influences recapture rates, although site-fidelity of many coral reef fishes may make recapture easier. It is desirable to leave tagged fish as long as possible in the field, but recapture rates even for long-lived species are usually very low, reflecting both mortality and tag loss. Moreover, tagging locations of reef fish are often widely dispersed and in remote locations. In many instances the cost-effectiveness of recaptures may be a significant problem. The primary method for recapture to date has been selective spearing following visual identification of tagged individuals. Difficulties with initial capture and subsequent recapture compound rapidly with increase in size of the fish. Despite these problems, there is an increasing number of successful validations of annual increment periodicity for coral reef fishes (Choat & Axe 1996, Cappo et al. 2000, Robertson et al. 2005, Marriott & Mapstone 2006). A further problem is not so much logistics, but the fact that researchers with access to reef environments seem disinclined to embark on tagging and recapture programs for the purpose of validation.

Radiochemical dating. This method may be used to effectively distinguish between divergent age-estimates, at a high cost per otolith. This is not a major issue in coral reef fish studies, however, given that perciform otoliths are relatively easily read.

Progression of length modes and length frequency analysis. This is not useful for coral reef fishes as many species have size and age decoupled and table-topped growth curves, with no evidence of length modes over the majority of the life cycle.

Capture of wild fish with natural date-specific markers. This is an underused resource. Such markers arise primarily through influences of temperature anomalies on fish growth patterns (Meekan et al. 1999). Strong possibilities exist for use with long-lived fish showing growth responses to known temperature anomalies (Nakano et al. 2004, Black et al. 2005).

Marginal increment analysis. This is the most commonly used protocol but there are potential problems associated with indistinct marginal conditions, especially for long-lived fishes in which growth increments may be compressed near the otolith margins. An additional problem is that sampling of specific age classes may be difficult in a number of reef fish taxa in which the relationship between size and age is so obscure that it is not possible to pre-select age classes for otolith processing. Most importantly, the protocol requires monthly samples, which may be difficult to obtain for many species in reef environments distant from transport or research hubs. Similar issues have been raised with respect to MIA in deep sea fishes (Cailliet et al. 2001).

Captive rearing of chemically tagged fishes. This method has been "generally discounted" as a reliable means for validating annulus formation but, with exceptions such as Machias et al. (1998), little evidence to support its dismissal has been provided. The important issue is to determine whether increment periodicity is modified in captivity, even though growth in captivity is likely to vary from growth in the wild. This remains the most realistic possibility for a wide range of tropical species as culture technologies improve, especially for those of large size (Cappo et al. 2000).

All validation protocols involve some expense. The least expensive are those based on the analysis of size structures, e.g., progression of length modes and length frequency analysis, but these are compromised due to the uncertainty of the relationship between size and age, or their complete decoupling, in many reef species. Other protocols such as the mark-recapture of chemically tagged fish must bear the cost of the tagging field work, which may be considerable as it involves not only the initial sampling but episodes aimed at recapture of tagged individuals. Given the complex nature of reef environments and the local rarity of many species, this usually involves dedicated sampling carried out at the expense of other activities. Furthermore, lessons from those studies that have successfully used this method of validation indicate that it is necessary to keep the number of species targeted for tagging to a minimum. Captive maintenance of chemically tagged fish offers a cost effective alternative provided suitable large scale aquaria or field enclosures are available. Bomb radiocarbon usually requires 10–15 otoliths for analysis with a cost in the order of US$1,000 per otolith. Although given a low priority by Campana (2001), marginal increment analysis remains the method of choice in most coral reef fisheries enterprises, as the major requirement is simply samples of a number of age-classes collected on a monthly basis.

A true analysis of costs highlights two major problems that inhibit the widespread application of age validation in coral reef fishes. Firstly, while many tropical maritime nations support excellent fisheries groups, they frequently lack the infrastructure of modern vessels and sophisticated laboratories found in temperate and boreal maritime nations. Secondly, research budgets are often not sufficient to cover the costs of novel analytical procedures.

4 What are the problems if we don't validate?

Analyses of otoliths for individuals and species of reef fishes are now routine. Given the difficulties of validation studies in coral reef environments, we must consider problems that may arise if we estimate and publish age-based demographic information without validation of the periodicity of increment formation in every species studied. The costs of validation of reef fishes become prohibitive when whole species assemblages are considered, because of the high diversity in tropical reef fish assemblages. It is unlikely that this will be accomplished in the more speciose lineages such as serranids, pomacentrids, labrids and acanthurids. Will this invalidate comparative studies on demography and life histories in such lineages? We argue that it will not. The major concern (underestimation of ages due to the use of inappropriate structures) of Beamish and McFarlane (1983) and Campana (2001) largely have been dealt with. Contrary examples to the one increment per year hypothesis are very rare. Bomb radiocarbon analysis, in providing estimates of age independent of our visualization of otolith increments, has overwhelmingly confirmed that counts of sagittal increments provide accurate estimates of age in long-lived species. Even in those taxa where accuracy is still questioned, extended life spans have been confirmed (Kalish 2001).

The issue driving the need for validation is not inaccuracies at the level of 12–13 as opposed to 14–15 increments. Reading errors will invariably introduce this level of variation into our estimates, regardless of whether a validation program has been undertaken or not. Miscounts of a small number of increments or occasional or localized disruptions to the cycle of increment formation due to climatic or metabolic variation are unlikely to influence demographic conclusions for species living in excess of 15 years. The critical issues are: (i) is the level of underestimation likely to be sufficient to result in counts of approximately 15 as opposed to 40 increments? (ii) Are schedules of increment formation as visualized in thin sagittal sections non-annual in nature? The published record strongly suggests that neither of these circumstances is likely in shallow water tropical species. Validated age estimates are more critical where estimates of biomass rely on age-length keys for fisheries stock assessments, where the key is used to convert lengths into age classes (Jones 1992). This emphasizes the distinction between the use of age data for fisheries estimates and for investigation of ecological and evolutionary processes.

It remains difficult to get demographic work on fishes published without validation procedures for each species, despite the consistent results from numerous studies over the last two decades. The philosophy that validation of ages in all species must be accomplished before general treatments of life histories can be developed is well established in the reviewing community. This has a particularly negative effect on the analysis of demographic and life-history trends for coral reef fishes. For reasons given above, this requirement is not only prohibitively time-consuming, but beyond the budgets of most tropical fisheries workers and biologists. Given the fact that we have an accessible protocol for estimating life spans (sectioning sagittal otoliths) for which accuracy has been confirmed in the majority of confirmatory studies, the question should be turned around. What problems will accrue if publication of age-based demographic work must await the validation of periodicity of increment formation in every species investigated?

Problems might occur in three areas of investigation.

A. **Fisheries management**. Although a large number of successful validations have occurred (primarily serranids, lutjanids and sparids), the number of species validated or even aged is trivial compared with the number of taxa harvested in tropical multi-species fisheries.

B. **Evaluating the efficacy of Marine Protected Areas**. Demographic information is fundamental to understanding responses of different groups of reef fishes to protective measures (Sale et al. 2005).

C. **The evolution of life histories in reef fishes.** The diversity of perciform reef fishes, coupled with the increasing availability of phylogenetic analyses, provides a significant opportunity to analyze contrasting patterns of size structure, longevity, growth rates and reproductive tactics manifested within and among different clades of reef fishes.

5 The consequences of miscounts of increments or an irregular pattern of increment formation

As a preliminary exploration of what can happen when the number of increments shown in the otolith (or counted by the researcher) differs from the true age of the fish, we compared true and estimated values for the von Bertalanffy growth coefficient, K, and mean asymptotic length, L_∞, and the total mortality rate, Z, from replicate samples of four hypothetical fish populations. Each population was generated to exhibit a different combination of either a short or long lifespan and either a steep or gradual growth trajectory (Figure 5). The four populations capture much of the range of life history variation exhibited among coral reef fishes (Kritzer et al. 2001, Choat & Robertson 2002). Two general types of discrepancies between real and estimated age were considered – overestimation and underestimation – and three different degrees of error were considered for each (Appendix 1). Our simulations combined population- and individual-level error by using a common function for the degree of error for an entire population, but choosing the specific error for each specimen individually. See Appendix 1 for more detail on our simulation approach.

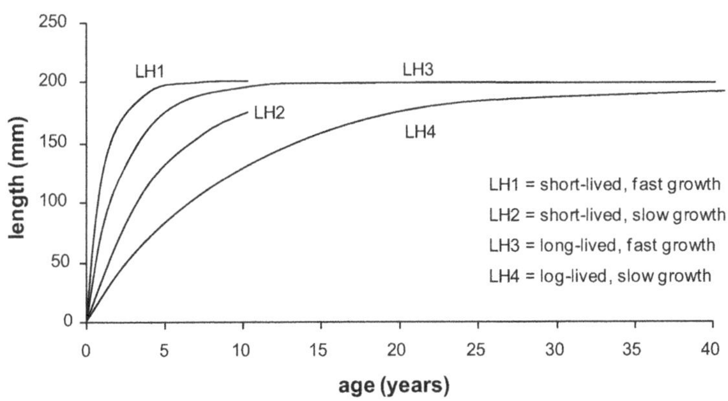

Figure 5. Four life history (LH) types considered in simulations to examine the effects of discrepancies in otolith increment periodicity on estimation of growth and mortality parameters. See Appendix 1 for descriptions of the 4 LH types.

The simulations showed that age estimation error had little effect on values of L_∞ for any of the life history types considered. Percent differences in L_∞ between the true and estimated values were all on the order of 10^{-5} or 10^{-6}, and are therefore not shown. This suggests that L_∞ estimates are far more dependent upon length characteristics and are largely independent of age values. The result supports Pauly's (1984) approach to obtaining a preliminary estimate of L_∞ as the average of the 10 largest fish in a sample without reference to age.

Age error had greater effects on estimates of K when the growth trajectory was less steep (i.e., slower growth; Figure 6A). This is to be expected. There is less scope for the ascending slope of the growth curve to vary when maximum body size is reached more quickly and fewer age classes will have an effect on the parameter value. In contrast, when growth continues for much of the life of the fish, there is considerable room for the curve to become more or less steep as size-age pairs move about within the plot by changing the age value. There is also evidence of an interaction between the growth trajectory and longevity (Figure 6A). A general, though not consistent, trend was that age error had greater effects on the longer-lived species, presumably because common proportional errors are much larger in an absolute sense when fish get older.

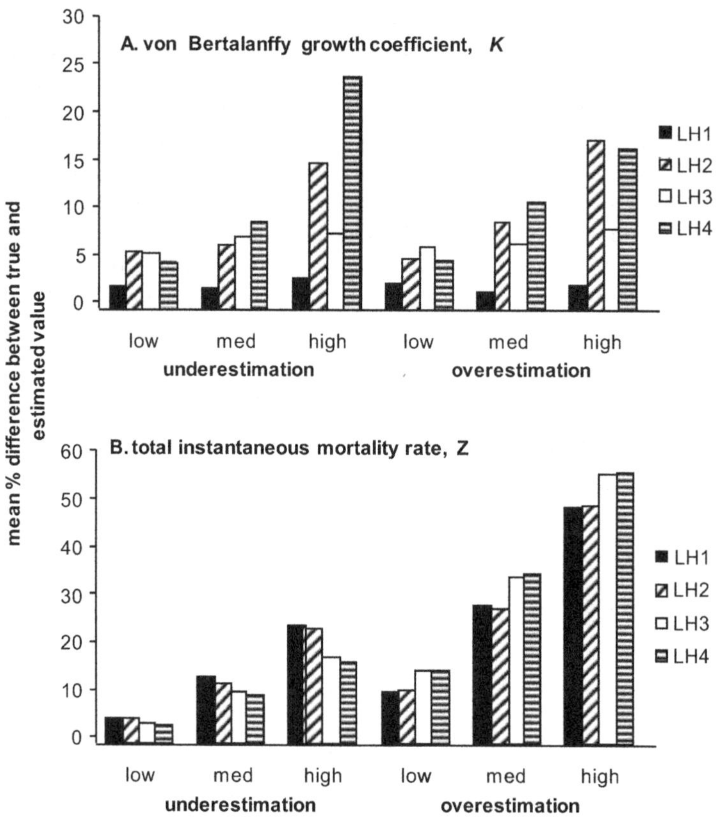

Figure 6. Mean difference between true and estimated values of two key demographic parameters, as percentage of the true value, due to different types and degrees of error in age estimation because of variation in increment periodicity (Appendix 1) for four life history (LH) types (Figure 5, Appendix 1).

Estimation of mortality using an age-based catch curve is independent of length, so results for the species with common longevity were similar (Figure 6B). The results showed that underestimation of age had greater effects on shorter-lived species, whereas overestimation had greater effects on longer-lived species. The pattern was not strong, however, and differences were similar for all species for a common pattern of age error. Overestimation of age seemed to have proportionally greater effects than underestimation. Potentially doubling the number of increments observed in an otolith (i.e., the high overestimation scenario) caused an approximately 50% difference in mortality estimates for all species but halving the number of increments observed only resulted in an approximately 20% difference (Figure 6B). This result is due more to the fact that doubling the age estimate results in addition of more age classes than are lost by halving an age estimate. The low overestimation and medium underestimation used similar percentage changes (+20% and –25%, respectively; Figure A1, Appendix 1) and caused similar differences in parameter estimates (Figure 6B).

The mean percent differences reported in Figure 6 are based on absolute values and therefore do not indicate the direction of parameter estimation error. Directionality was predictable, however, based upon the type of error introduced. Overestimation of age decreased values of both K and Z, while underestimation increased values of both.

These simulations can be expanded in a myriad of ways. Age errors can be size-specific, age-specific, or occur at specific times within a population (simulating environmental events that cause increment anomalies, and not affecting age classes born after the event). Marriott and Mapstone (2006) have recently explored the analogous question of what happens when different criteria for choosing an age estimate based upon multiple readings are applied. Their analysis likewise considered the effects of not getting the age of the fish right, although with a focus on observer error as opposed to biologically-induced discrepancies. Also, different demographic parameters (see Kritzer et al. 2001) or results of stock assessment models using the age and demographic information could be considered in future analyses.

Our initial results provide some important insights, despite the potential for extension of the study. Our scenarios were far from conservative even though we did not consider the cases where increment deposition was consistently different from annual (i.e., two increments per year or one increment per 2 years as the norm rather than the exception). We only considered errors in one direction in each scenario and did not allow some ages to be overestimated and others to be underestimated within a sample, which might have resulted in errors offsetting one another and therefore smaller relative differences from the true parameter values for the sample. Also, what we defined as "low", "medium" and "high" all entailed incidence of age discrepancies much greater than suggested by the consistency of the one increment – one year pattern reported in existing validation studies (Table 1), non-reporting of other patterns notwithstanding. For instance, our "medium" level of overestimation still allowed 25% of the population to show 38% more increments than the true age of the fish, likely closer to a "high" level of error. Despite this high level of error introduced, many estimates in our simulations were within approximately 10% of the true values (Figure 6). This is comparable to the degree of error that can be introduced simply by effects of sample size typically used in many studies of reef fish (Kritzer et al. 2001).

Irregular deposition cannot be distinguished from miscounting or failing to detect increments (Francis 1995), as noted earlier, but the two scenarios have very different implications for future improvement of age estimation of tropical fishes. Counting errors can be corrected by continued refinement of the techniques by which otoliths are prepared and analyzed (e.g., microscope power, image analysis tools) and vigilant observer training and calibration. Biological anomalies, on the other hand, call for greater understanding of the mechanisms that cause deviations from the one increment – one year pattern. Given that parameter estimation seems robust to likely degrees of error, we argue that the burden of proof needs to be shifted not toward defending annual periodicity, but rather toward demonstrating where, why, and how frequently it does not occur. This will be a far more stimulating line of ecological and physiological research than repeated pedestrian age validation. Furthermore, we will learn more from a handful of studies describing mechanisms by which annual periodicity does not occur than we will from numerous studies for numerous species re-confirming the general annual pattern.

6 What is the way forward?

The literature on the age-based demography of fishes is focused mainly on temperate, boreal and deep sea fishes. Validation studies in these groups have been aided by access to sophisticated infrastructure and to extensive data bases including otolith archives. By comparison, research into the demography of coral reef fish is in its infancy and lacks the comparative data base that has guided the analysis of temperate and boreal fish populations over the last 4–5 decades. A more structured approach to reef fish demography is required which includes the establishment of otolith archives. The capacity of otoliths to record the influence of past climatic events and variations in ocean chemistry demonstrates that archives would have a critical role in predicting responses of fish populations to environmental change.

Reef fish demography is at an exciting stage, with the variety of life histories in perciform assemblages becoming apparent. A more coherent approach to population biology of tropical fish should occur in three phases: (i) wide dissemination of otolith based analyses of growth and age structure in a broad range of species and populations of coral reef fishes, without the requirement to validate increment periodicity for every species; (ii) validation procedures that confirm the temporal meaning of the increments and understanding of the mechanisms by which the timing of increment formation can be altered, especially in commercially important species; and (iii) studies on selected species to provide quantitative estimates of accuracy (Francis 1995). This approach requires greater collaboration among tropical fish biologists.

The first of these is controversial due to the insistence that age information must be accompanied by validation of at least the temporal meaning of increments. Analysis of demography of tropical parrot fishes is a case in point. This group, comprising 80 species (Parenti & Randall 2000), is arguably one of the most ecologically important components of the reef fish fauna (Hughes 1994, Bellwood et al. 2003, Mumby et al. 2004) and is heavily over-fished at many localities (Jackson et al. 2001). It is unclear whether members of this group will respond rapidly to protective measures or whether it is one of those in which the recovery process will be prolonged (Russ & Alcala 2004). Demographic information is required to resolve this issue but, while age-based information may be readily obtained (Choat & Robertson 2002), to date annual increment formation has been validated for only six species. It is a moot point as to whether it is better to retain non-validated age data until we have confirmed the temporal meaning of increment formation or to publish the material acknowledging that the working hypothesis is that one ring is formed each year. We argue for the latter.

The case of the largest parrot fish *Bolbometopon muricatum* is instructive. This species is a critically important functional component of the reef fish fauna but is heavily over-fished over much of its range (Bellwood et al. 2003, Hamilton 2004). It is now considered to be threatened (Donaldson & Dulvy 2004) and issues such as growth and mortality rates and generation times are important in managing existing populations through harvest regulation and the application of marine protected areas. Otoliths of this species show well defined increments. A large body of data on the age-based demography of this species now exists (Choat & Robertson 2002, Hamilton 2004). This information is of potential value to coral reef conservation and management enterprises. In addition, there is renewed interest in the demography and life history patterns of labroid fishes following analyses that provide a fresh perspective on evolutionary relationships within this group (Clements et al. 2004, Westneat & Alfaro 2005). Attempts to validate periodicity of increment formation in this very large, mobile and ecologically important reef fish, however, have failed. Tagged and tetracyclined individuals in the Solomons processed after night-time capture were killed and eaten by sharks the next day following release. Individuals maintained in external enclosures starved to death (Hamilton 2004), an indication of the lack of knowledge of their nutritional ecology (Choat & Clements 1998).

The problems encountered in validating age and growth rates in parrot fishes reflect four general problems. (i) Individual species distributions extend over thousands of kilometers and a wide range of habitats. Population parameters are not informative if applied to whole species. (ii) High local diversity is associated with the local rarity of many species. Obtaining sufficient individuals at a locality is frequently a problem. (iii) Local fisheries are usually multi-specific, targeting complex assemblages of ecologically similar fish at a given site. (iv) Many species are large and highly mobile or rare and cryptic.

The fact that such a small proportion of tropical fish species has been successfully validated confirms the importance of these issues but there have been few attempts to develop alternative approaches. We advocate the development of wide-scale multi-species marking programs using either natural or anthropogenic dated marks in otoliths. One approach is to analyze natural date specific markers such as temperature anomalies that influence short term growth and leave signals in otolith increments. Meekan et al. (1999) present a recent example where Galapagos pomacentrids displayed checks in their otoliths that corresponded to the timing of the 1982–1983 El Niño and suggested a reduction in growth over this period. Similar growth reductions corresponding with the same El Niño were observed in the otoliths of species of *Sebastes* (Woodbury 1999) and in the otoliths of Pacific pomacentrids that corresponded to the 1997–1998 El Niño (Nakano et al. 2004). The most promising approach is suggested by Black et al. (2005) where cross validation methods developed by dendrochronologists have been successfully applied to populations of long-lived fishes. In addition evidence is becoming stronger that other metrics of otoliths, such as sagitta weight – age relationships, can be used reliably as a proxy for direct estimates of age (Lou et al. 2005).

The possibility of marking egg clutches with chemical makers that will incorporate a date specific pre-dispersal mark into embryonic fish has been confirmed, providing a rigorous estimate of age that will be retained in the adult fish. The study of Jones et al. (1999), designed to assess the dispersal of larval pomacentrid fishes, is an additional example of mass chemical marking that may provide cost-effective age validation. In addition, an exploratory technique involving marking reproductive females with elemental markers that would be incorporated into eggs and subsequently larvae as a date-specific marker is also now available (Thorrold et al. 2006). These techniques hold promise for large scale, multispecies validation studies. Innovative funding proposals are required as many agencies are reluctant to fund what they see as confirmatory activities, such as OTC marking of yet another set of species.

Campana (2001) made the important point that validation of increment periodicity in very young and very old age classes was a priority. We believe that well prepared and analyzed sagittal otoliths of perciformes will allow identification of the oldest age-classes. Most somatic growth occurs within the first 15% of the life span in many species of reef fishes (Choat & Robertson 2002). Identifying the first growth increment becomes an important issue under these circumstances. Errors in identification of the first three annual increments can lead to substantial changes in growth parameters, especially the VBGF parameter *K*. Ideally, regular sampling of recruits over an annual interval with analysis of daily increments will allow the time of the first annual growth increment to be confirmed.

A difficulty in the ageing of coral reef fishes is that many workers have unduly optimistic expectations as to the clarity of growth increments visualized in sectioned sagittae. Perciformes sampled from high latitudes invariably display clearer growth increments than those from low latitudes (Figure 3). We are frequently approached by researchers commencing demographic studies on tropical reef fishes claiming that they are unable to detect recognizable growth increments in sectioned sagittae. In each instance, however, we have been able to detect increment structures that correspond to those in related and validated species. A useful approach is to examine populations along latitudinal gradients, determining the structure of increments in high latitude populations and using these as a template to help establish the usually problematic increments laid down at young ages (Robertson et al. 2005). Various forms of validation studies may be attempted then but it is unlikely that validation will be achieved in all populations and age classes along latitudinal gradients.

The bottom line in such studies is that there are no short cuts. It is likely that hundreds of otolith sections from a variety of habitats must be examined for each species before low latitude populations can be aged with confidence. We can understand why initial attempts to age low latitude reef fishes were discouraging and led to attempts to use size-based approaches. Both basic and applied studies are making assumptions about growth, mortality rates and longevity in reef fishes, however, and there is an increasing need for more age-based data to be widely disseminated to encourage further studies in age-based demography, including validation.

Unfortunately, demographic studies on reef fishes are increasingly unpopular, driven by the perception that most reef fishes are endangered or threatened. Analysis of otoliths means that fish must be killed, most effectively by selective spearing that can accommodate variation attributable to identity, size classes, habitats and location. Given the expeditionary nature of coral reef research, this can result in numbers of dead dissected fish that result from intensive episodes of sampling over short time periods. Research-driven mortality is usually trivial in terms of the numbers removed relative to the numbers present and the prevailing natural and fishing mortality rates. Sample sizes needed for demographic studies of reef fish can be and have been strategically selected by quantitative analysis of precision, reducing impacts and increasing cost-effectiveness (Kritzer et al. 2001). Still, demographic studies frequently get a hostile reception not only from management and conservation agencies but sometimes within the scientific community during the peer review process. A better understanding of the benefits of age-based analyses with respect to management and conservation will flow from a more comprehensive data base on fish life histories and population biology.

7 Conclusions

The concerns driving arguments for a comprehensive approach to age-validation in fishes have arisen historically from situations in which under-estimation of age has resulted in over-fishing (Beamish & McFarlane 1983, Campana 2001). The widespread use of sectioned sagittal otoliths as an ageing tool has shown that many species do have extended life spans and low natural mortality rates. Harvesting of such species should proceed only with precautionary safeguards.

Validation of the temporal pattern of increment formation and age structures of many species has strongly confirmed that the primary increments visualized in sectioned sagittae are indeed annual. Bomb radiocarbon analyses (Campana 2001, Kalish 2001) have been crucial in this process.

Coral reef fishes pose particular problems for the validation process. Evidence exists that otoliths from some coral reef fishes may have annual increments that are difficult to detect or may not be deposited every year (Fowler 1995), but there is little evidence that increment formation reflects a metabolic or environmental cycle that is not annual. If this was found to be a common situation then clearly the utility of otoliths as an ageing tool would be severely compromised. This has not been the case to date.

We argue that it is unrealistic to attempt validation of every species for which demographic information is sought, given the logistic difficulties associated with age-validation in reef fishes, their biological characteristics, and the record of validation studies to date. We contend that unvalidated otolith age information is more valuable if made available to the scientific community with the caveat that the temporal pattern of increment formation is only assumed to be annual. Retaining data until validation is achieved would serve little purpose. Hopefully, publication would encourage well-funded groups concerned with reef fish management and conservation to embark on their own validation studies in order to provide a stronger basis for remedial management. We agree with Fowler (1995), however, that the biological processes that underlie the formation of macrostructures in otoliths are still poorly understood and require additional experimentally-based research.

A more strategic approach is warranted for reef fish demographic research, regardless of the controversies that different approaches to validation may generate. As validation studies are logistically expensive, decisions must be made with respect to the effort devoted to validation as opposed to more general age-based demographic studies. Validating species simply because they happen to be a research target is not a good guide for deploying funds and effort. We do not deny the importance of validation studies where the sustainability of commercially important fisheries is under consideration, but this involves only one component of the reef fish research agenda. Numerous studies focus on ecological and evolutionary processes for which demographic data are important. For these studies, many of which involve whole lineages and species assemblages of tropical fishes, the demands for comprehensive validation are unrealistic in terms of both deployment of resources and timely publication.

Appendix 1: Age error simulation method

Populations of four hypothetical species were established to compare effects of age errors on different life history (LH) types. Species differed in longevity and growth trajectory. Growth trajectories were varied by changing the growth coefficient, K, in the von Bertalanffy growth function (VBGF) while keeping the mean asymptotic length, L_{∞}, and x-intercept, t_0, constant. Total instantaneous mortality rate, Z, was calculated from the maximum age of each species using the equation of Hoenig (1983). Attributes of the four populations were as detailed in Table A1.

Table A1. Demographic attributes of the four hypothetical species modeled to assess the effects of age errors arising from violation of the assumption of annual increment formation in otoliths.

Species	Max. age (yr)	Mortality rate, Z (yr^{-1})	VBGF K (yr^{-1})	VBGF L_{∞} (mm)	VBGF t_0 (yr)
LH1	10	0.42	0.8	200	–0.2
LH2	10	0.42	0.2	200	–0.2
LH3	40	0.10	0.4	200	–0.2
LH4	40	0.10	0.1	200	–0.2

Each population had a stable age distribution, with the frequency in each age class from 0 up to the maximum age determined by Z. Demographic parameters were estimated from replicate samples drawn from each population, the size of which was infinite (i.e., sampling was done with replacement). Each sample was 200 specimens for LH1 and LH2, and 400 specimens for LH3 and LH4. Sample sizes were chosen to be small enough to minimize computing time but large enough to minimize sample size effects on parameter estimates (Kritzer et al. 2001) and focus on effects of age errors.

A length was assigned to each specimen in each sample. The length was the mean length for that age class in that population, determined by the underlying VBGF for each life history type, modified by a random normal variate from a distribution with a mean of 1 and a CV of 0.15. Length was assumed to be measured without error. The true age of each specimen was known. The estimated age was determined by selecting a random number, r, between 0 and 1 for each specimen, and entering it into the equation:

$$\text{estimated age} = \text{true age} + (\text{true age} \times r \times m),$$

where m is a slope parameter that scaled the degree of error in age estimation. Modified ages were rounded to the whole number, since age values are rarely assigned as fractions in actual studies. Values of m for different scenarios are provided in Table A2 and illustrated in Figure A1.

The extreme values for each type of error were selected to simulate the possibility of half and twice as many increments shown (or counted) as the true age of the fish. However, we did not apply those extreme degrees of error, or any degree of error, across the entire population, instead allowing for individual variation in age discrepancy under a population-wide function.

Table A2. Error scenarios modeled for each of the four populations and slope parameters for assignment of variation to age estimates in each error scenario.

Type of error	Degree of error	m
Underestimation	Low	–0.1
Underestimation	Medium	–0.25
Underestimation	High	–0.5
Overestimation	Low	0.2
Overestimation	Medium	0.5
Overestimation	High	1.0

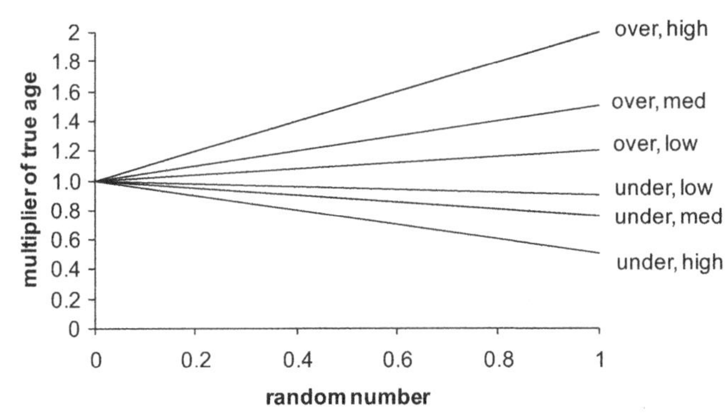

Figure A1. Functions used to modify true ages of fish to simulate deviations from annual deposition of otolith increments and examine effects on estimation of demographic parameters. Two general types of error were considered, overestimation and underestimation, with three degrees of error considered within each. Each function was used for a separate set of simulations.

For each sample, a VBGF was fitted using both the set of true ages and the set of estimated ages by non-linear least-squares regression of length on age. Total mortality rate was estimated by fitting an age-based catch curve to log-transformed age frequency data. Effects of age error were examined by calculating the difference between the value of each parameter for the set of true ages and the set of estimated ages, expressed as a percentage of the parameter value for the true ages. The percentage difference between the true ages and estimated ages for each sample was recorded as the absolute value and used to calculate the mean. The mean percentage difference was calculated for 100 replicate samples for each life history type under each type and degree of age discrepancy.

Acknowledgements

The following assisted us with provision of material and assistance with otolith preparation and interpretation: L. Axe, R. Pears, W. Robbins, R. Robertson, K. Clements, E. Laman Trip, K. Ranatunga, S. Mutz, and T. Molea. Financial and logistic support was provided by the Queensland Government/Smithsonian Institution Collaborative Research Program on Reef Fishes, Seychelles Fishing Authority, Lizard Island Research Station and the James Cook University internal funding scheme. The ideas in this chapter were developed in discussions with G. Russ, R. Robertson, B. Mapstone, R. Hamilton, G. Begg, R. Pears, and W. Robbins.

References

Baker MS, Wilson CA (2001) Use of bomb radiocarbon to validate otolith section ages of red snapper *Lutjanus campechanus* from the northern Gulf of Mexico. Limnol Oceanogr 46:1819–1824

Baker MS, Wilson CA, VanGent DL (2001) Testing assumptions of otolith radiometric aging with two long-lived fishes from the northern Gulf of Mexico. Can J Fish Aquat Sci 58:1244–1252

Beamish RJM, McFarlane GA (1983) The forgotten requirement for age validation in fisheries biology. Trans Amer Fish Soc 112:735–743

Beckman DW, Stanley AL, Render JH, Wilson CA (1991) Age and growth-rate estimation of sheepshead *Archosargus probatocephalus* in Louisiana waters using otoliths. Fish Bull 89:1–8

Begg GA, Campana SE, Fowler AJ, Suthers IM (2005) Otolith research and application: current directions in innovation and implementation. Mar Freshwater Res 56:477–483

Bell MA (2001) Fish do not lie about their age but they might lose count. TREE 16:599–600

Bellwood DR, Hoey AS, Choat JH (2003) Limited functional redundancy in high diversity systems: resilience and ecosystem function on coral reefs. Ecol Lett 6:281–285

Berkeley SA, Hixon MA, Larson RJ, Love MS (2004) Fisheries sustainability via protection of age structure and spatial distribution of fish populations. Fisheries 29:23–32

Beverton RJH (1992) Patterns of reproductive strategy parameters in some marine teleost fishes. J Fish Biol 41:137–160

Beverton RJH, Holt SJ (1957) On the dynamics of exploited fish populations. Fish Invest Min Agric Fish Food GB 19:1–533

Black AB, Boehlert GW, Yoklavich MM (2005) Using tree-ring crossdating techniques to validate annual growth increments in long-lived fishes. Can J Fish Aquat Sci 62:2277–2284

Brouwer SL, Griffiths MH (2004) Age and growth of *Argyrozona argyrozona* (Pisces: Sparidae) in a marine protected area: an evaluation of methods based on whole otoliths, sectioned otoliths and mark-recapture. Fish Res 67:1–12

Bullock LH, Murphy MD, Godcharles MF, Mitchell ME (1992) Age, growth, and reproduction of jewfish *Epinephelus itajara* in the Eastern Gulf of Mexico. Fish Bull 90:243–249

Burton ML (2001) Age, growth, and mortality of gray snapper, *Lutjanus griseus*, from the east coast of Florida. Fish Bull 99:254–265

Burton ML (2002) Age, growth and mortality of mutton snapper, *Lutjanus analis*, from the east coast of Florida, with a brief discussion of management implications. Fish Res 59:31–41

Bush PGE, Ebanks GC, Lane ED (1996) Validation of the ageing technique of the Nassau grouper (*Epinephelus striatus*) in the Caymen Islands. In: Arreguin-Sanchez F, Munroe JL, Balgos MC, Pauly D (Eds) Biology, fisheries and culture of tropical groupers and snappers. ICLARM, Manila, pp 150–158

Buxton CD, Clarke JR (1991) The biology of the white musselcracker *Sparodon durbanensis* (Pisces, Sparidae) on the Eastern Cape Coast, South-Africa. Afr J Mar Sci 10:285–296

Cailliet GM, Andrews AH, Burton EJ, Watters DL, Kline DE, Ferry-Graham LA (2001) Age determination and validation studies of marine fishes: do deep-dwellers live longer? Exp Gerontol 36:739–764

Campana SE (1999) Chemistry and composition of fish otoliths: pathways, mechanisms and applications. Mar Ecol Prog Ser 188:263–297

Campana SE (2001) Accuracy, precision and quality control in age determination, including a review of the use and abuse of age validation methods. J Fish Biol 59:197–242

Campana SE, Thorrold SR (2001) Otoliths, increments, and elements: keys to a comprehensive understanding of fish populations? Can J Fish Aquat Sci 58:30–38

Cappo M, Eden P, Newman SJ, Robertson S (2000) A new approach to validation of periodicity and timing of opaque zone formation in the otoliths of eleven species of *Lutjanus* from the central Great Barrier Reef. Fish Bull 98:474–488

Carr JW, Whoriskey F, O'Reilly P (2004) Efficacy of releasing captive reared broodstock into an imperilled wild Atlantic salmon population as a recovery strategy. J Fish Biol 65:38–54

Cass-Calay SL, Bahnick BM (2002) Status of the yellowedge grouper fishery in the Gulf of Mexico. Sustainable Fisheries Division Contribution No. SFD-02/03-172, Southeast Fisheries Science Center, Sustainable Fisheries Division, Miami

Chale-Matsau JR, Govender A, Beckley LE (2001) Age, growth and retrospective stock assessment of an economically extinct sparid fish, *Polysteganus undulosus*, from South Africa. Fish Res 51:87–92

Chan TTC, Sadovy Y (2002) Reproductive biology, age and growth in the chocolate hind, *Cephalopholis boenak* (Bloch, 1790) in Hong Kong. Mar Freshwater Res 53:791–803

Choat JH, Axe LM (1996) Growth and longevity in acanthurid fishes: an analysis of otolith increments. Mar Ecol Prog Ser134:15–26

Choat JH, Axe LM, Lou DC (1996) Growth and longevity in fishes of the family Scaridae. Mar Ecol Prog Ser 145:33–41

Choat JH, Clements KD (1998) Vertebrate herbivores in marine and terrestrial environments: a nutritional ecology perspective. Annu Rev Ecol Evol S 29:375–403

Choat JH, Robertson DR (2002) Age-based studies. In: Sale PF (Ed) Coral reef fishes: diversity and dynamics in a complex system. Academic Press, San Diego, pp 57–80

Choat JH, Robertson DR, Ackerman JL, Posada JM (2003) An age-based demographic analysis of the Caribbean stoplight parrotfish *Sparisoma viride*. Mar Ecol Prog Ser 246:265–277

Clements KD, Alfaro ME, Fessler JL, Westneat MW (2004) Relationships of the temperate Australasian labrid fish tribe Odacini (Perciformes; Teleostei). Mol Phylogenet Evol 32:575–587

Cowen RK (2002) Larval dispersal and retention and consequences for population productivity. In: Sale PF (Ed) Coral reef fishes: diversity and dynamics in a complex system. Academic Press, San Diego, pp 149–170

Crabtree RE, Bullock LH (1998) Age, growth, and reproduction of black grouper, *Mycteroperca bonaci*, in Florida waters. Fish Bull 96:735–753

Crabtree RE, Cyr EC, Dean JM (1995) Age and growth of tarpon, *Megalops atlanticus*, from South Florida waters. Fish Bull 93:619–628

Crabtree RE, Hood PB, Snodgrass D (2002) Age, growth, and reproduction of permit (*Trachinotus falcatus*) in Florida waters. Fish Bull 100:26–34

Cushing DH (1975) Marine ecology and fisheries. Cambridge University Press, London

Donaldson TJ, Dulvy NK (2004) Threatened fishes of the world: *Bolbometopon muricatum* (Valenciennes 1840) (Scaridae). Environ Biol Fish 70:373pp

Dutka-Gianelli J, Murie DJ (2001) Age and growth of sheepshead, *Archosargus probatocephalus* (Pisces: Sparidae), from the northwest coast of Florida. Bull Mar Sci 68:69–83

Fairchild EA, Howell WH (2004) Factors affecting the post release survival of cultured juvenile *Pseudopleuronectes americanus*. J Fish Biol 65:69–87

Ferreira BP, Russ GR (1992) Age, growth and mortality of the inshore coral trout *Plectropomus maculatus* (Pisces, Serranidae) from the central Great Barrier Reef, Australia. Mar Freshwater Res 43:1301–1312

Ferreira BP, Russ GR (1994) Age validation and estimation of growth-rate of the coral trout, *Plectropomus leopardus*, (Lacepede 1802) from Lizard Island, Northern Great Barrier Reef. Fish Bull 92:46–57

Ferrell DJ, Henry GW, Bell JD, Quartararo N (1992) Validation of annual marks in the otoliths of young snapper, *Pagrus auratus* (Sparidae). Mar Freshwater Res 43:1051–1055

Fischer AJ, Baker MS, Wilson CA, Nieland DL (2005) Age, growth, mortality, and radiometric age validation of gray snapper (*Lutjanus griseus*) from Louisiana. Fish Bull 103:307–319

Fowler AJ (1995) Annulus formation in otoliths of coral reef fish – a review. In: Secor DH, Dean JM, Campana SE (Eds) Recent developments in fish otolith research. University of South Carolina Press, Columbia, pp 45–64

Francis MP, Mulligan KP, Davies NM, Beentjes MP (1999) Age and growth estimates for New Zealand hapuku, *Polyprion oxygeneios*. Fish Bull 97:227–242

Francis R, Paul LJ, Mulligan KP (1992) Aging of adult snapper (*Pagrus auratus*) from otolith annual ring counts – validation by tagging and oxytetracycline injection. Mar Freshwater Res 43:1069–1089

Francis RICC (1995) The analysis of otolith data – mathematicians perspective (What, precisely is your model?). In: Secor DH, Dean JM, Campana SE (Eds) Recent developments in fish otolith research, Vol 19. University of South Carolina Press, Columbia, pp 81–96

Froese RP, Pauly D (2000) FishBase 2000: concepts, design and data sources. ICLARM, Los Baños, Laguna, Philipinnes, 344pp

Gjosaeter JD, Dayarante P, Bergstad OA, Gjosaeter H, Sousa MI, Beck IM (1984) Ageing tropical fish by growth rings in otoliths. Report No. FAO Fisheries Circular No.776, FAO, Rome

Goncalves JMS, Bentes L, Coelho R, Correia C, Lino PG, Monteiro CC, Ribeiro J, Erzini K (2003) Age and growth, maturity, mortality and yield-per-recruit for two banded bream (*Diplodus vulgaris* Geoffr.) from the south coast of Portugal. Fish Res 62:349–359

Grandcourt EM, Al Abdessalaam TZ, Francis F, Al Shamsi AT (2004) Biology and stock assessment of the Sparids, *Acanthopagrus bifasciatus* and *Argyrops spinifer* (Forsskal, 1775), in the Southern Arabian Gulf. Fish Res 69:7–20

Gust NC, Choat JH, Ackerman JR (2002) Demographic plasticity in tropical reef fishes. Mar Biol 140:1039–1051

Hamilton RJ (2004) The demographics of bumphead parrotfish (*Bolbometopon muricatum*) in lightly and heavily fishes regions in the western Solomon Islands. PhD Thesis, University of Otago

Helfmann GSC, Collette BB, Facey DE (1997) The diversity of fishes. Blackwell Science, Malden

Henderson PA (2005) The growth of tropical fishes. In: Val A, Val V, Randall D (Eds) The physiology of tropical fishes. Elsevier, pp 85–101, New York

Hernaman V, Munday PL, Schlappy ML (2000) Validation of otolith growth-increment periodicity in tropical gobies. Mar Biol 137:715–726

Hoenig JM (1983) Empirical use of longevity data to estimate mortality rates. Fish Bull 81:898–903

Holtzhausen JA, Kirchner CH (2001) Age and growth of two populations of West Coast steenbras *Lithognathus aureti* in Namibian waters, based on otolith readings and mark-recapture data. Afr J Mar Sci 23:169–179

Hood PB, Godcharles MF, Barco RS (1994) Age, growth, reproduction, and the feeding ecology of Black-Sea bass, *Centropristis striata* (Pisces, Serranidae), in the eastern Gulf of Mexico. Bull Mar Sci 54:24–37

Hood PB, Johnson AK (1999) Age, growth, mortality, and reproduction of vermilion snapper, *Rhomboplites aurorubens*, from the eastern Gulf of Mexico. Fish Bull 97:828–841

Hood PB, Johnson AK (2000) Age, growth, mortality, and reproduction of red porgy, *Pagrus pagrus*, from the eastern Gulf of Mexico. Fish Bull 98:723–735

Hood PB, Schlieder RA (1992) Age, growth, and reproduction of gag, *Mycteroperca microlepis* (Pisces, Serranidae), in the Eastern Gulf of Mexico. Bull Mar Sci 51:337–352

Hughes TP (1994) Catastrophes, phase-shifts, and large-scale degradation of a Caribbean coral reef. Science 265:1547–1551

Iles TD, Sinclair M (1982) Atlantic herring: stock discreteness and abundance. Science 215:627–633

Jackson JBC, Kirby MX, Berger WH, Bjorndal KA, Botsford LW, Bourque BJ, Bradbury RH, Cooke R, Erlandson J, Estes JA, Hughes TP, Kidwell S, Lange CB, Lenihan HS, Pandolfi JM, Peterson CH, Steneck RS, Tegner MJ, Warner RR (2001) Historical overfishing and the recent collapse of coastal ecosystems. Science 293:629–638

James NC, Mann BQ, Beckley LE, Govender A (2003) Age and growth of the estuarine-dependent sparid *Acanthopagrus berda* in northern KwaZulu-Natal, South Africa. Afr Zool 38:265–271

Jones CM (1992) Development and application of the otolith increment technique. In: Stevenson DK, Campana SE (Eds) Otolith microstructure examination and analysis. Can Spec Publ Fish Aquat Sci 117:1–11

Jones GP, Milicich MJ, Emslie MJ, Lunow C (1999) Self-recruitment in a coral reef fish population. Nature 402:802–804

Kalish J (2001) Use of the bomb radiocarbon chronometer to validate fish age. Report No. FRDC Project 93/109, Fish Res and development corporation, Canberra Australia

Kalish JM (1993) Pre- and post-bomb radiocarbon in fish otoliths. Earth Planet Sci Lett 114:549–554

Kamukuru AT, Hecht T, Mgaya YD (2005) Effects of exploitation on age, growth and mortality of the blackspot snapper, *Lutjanus fulviflamma*, at Mafia Island, Tanzania. Fish Manag Ecol 12:45–55

Kritzer JP (2002) Variation in the population biology of stripey bass *Lutjanus carponotatus* within and between two island groups on the Great Barrier Reef. Mar Ecol Prog Ser 243:191–207

Kritzer JP, Davies CR, Mapstone BD (2001) Characterizing fish populations: effects of sample size and population structure on the precision of demographic parameter estimates. Can J Fish Aquat Sci 58:1557–1568

Lorenzo JM, Pajuelo JG, Mendez-Villamil M, Coca J, Ramos AG (2002) Age, growth, reproduction and mortality of the striped seabream, *Lithognathus mormyrus* (Pisces, Sparidae), off the Canary Islands (Central-east Atlantic). J Appl Ichthyol 18:204–209

Lou DC (1992) Validation of annual growth bands in the otolith of tropical parrotfishes (*Scarus schlegeli* Bleeker). J Fish Biol 41:775–790

Lou DC, Mapstone BD, Russ GR, Davies CR, Begg GA (2005) Using otolith weight – age relationships to predict age-based metrics of coral reef fish populations at different spatial scales. Fish Res 71:279–294

Luckhurst BE, Dean JM, Reichert M (2000) Age, growth and reproduction of the lane snapper *Lutjanus synagris* (Pisces: Lutjanidae) at Bermuda. Mar Ecol Prog Ser 203:255–261

Machias A, Maraveyia E, Pavlidis M, Somarakis S, Divanach P (2002) Validation of annuli on scales and otoliths of common dentex (*Dentex dentex*). Fish Res 54:287–294

Machias A, Tsimenides N, Kokokiris L, Divanach P (1998) Ring formation on otoliths and scales of *Pagrus pagrus*: a comparative study. J Fish Biol 52:350–361

Manickchand-Dass S (1987) Reproduction, Age and Growth of the Lane Snapper, *Lutjanus synagris* (Linnaeus), in Trinidad, West Indies. Bull Mar Sci 40:22–28

Manickchand-Heileman SC, Phillip DAT (2000) Age and growth of the yellowedge grouper, *Epinephelus flavolimbatus*, and the yellowmouth grouper, *Mycteroperca interstitialis*, off Trinidad and Tobago. Fish Bull 98:290–298

Manooch CS, Drennon CL (1987) Age and growth of yellowtail snapper and queen triggerfish collected from the United States Virgin Islands and Puerto Rico. Fish Res 6:53–68

Marriott RJ, Mapstone BD (2006) Consequences of inappropriate criteria for accepting age estimates from otoliths, with a case study for a long-lived tropical fish. Can J Fish Aquat Sci 63:2259–2274

Marriott RJ, Mapstone BD, Begg GA (2007) Age-specific demographic parameters, and their implications for management of the red bass, *Lutjanus bohar* (Forsskal 1775): a large, long-lived reef fish. Fish Res 83:204–215

Mason DL, Manooch CS (1985) Age and growth of mutton snapper along the east coast of Florida. Fish Res 3:93–104

Masuda R, Keller K, Ziemann DA, Ogle J (2003) Association with underwater structures in hatchery-reared and wild red snapper *Lutjanus campechanus* juveniles. J World Aquacult Soc 34:140–146

McErlean AJ (1963) A study of the age and growth of the gag, *Mycteroperca microlepis* Goode and Bean (Pisces: Serranidae), on the west coast of Florida. Fla Board Conserv Mar Res Lab Tech Ser 41:1–29

Meekan MG, Wellington GM, Axe L (1999) El Niño Southern Oscillation events produce checks in the otoliths of coral reef fishes in the Galapagos Archipelago. Bull Mar Sci 64:383–390

Milton DA, Short SA, Oneill MF, Blaber SJM (1995) Aging of 3 species of tropical snapper (Lutjanidae) from the Gulf of Carpentaria, Australia, using radiometry and otolith ring counts. Fish Bull 93:103–115

Morales-Nin B, Ralston S (1990) Age and growth of *Lutjanus kasmira* (Forskal) in Hawaiian waters. J Fish Biol 36:191–203

Morales-Nin B, Torres GJ, Lombarte A, Recasens L (1998) Otolith growth and age estimation in the European hake. J Fish Biol 53:1155–1168

Mosse JW (2001) Population biology of *Cephalopholis cyanostigma* (Serranidae) of the Great Barrier Reef, Australia. PhD Thesis, James Cook University

Mumby PJ, Edwards AJ, Arias-Gonzalez JE, Lindeman KC, Blackwell PG, Gall A, Gorczynska MI, Harborne AR, Pescod CL, Renken H, Wabnitz CCC, Llewellyn G (2004) Mangroves enhance the biomass of coral reef fish communities in the Caribbean. Nature 427:533–536

Munk KM (2001) Maximum ages of groundfishes in waters off Alaska and British Columbia and considerations of age determination. Alaska Fish Res Bull 8:12–21

Munro JL (Ed) (1983) Caribbean coral reef fishery resources, Vol 7. ICLARM, Manila

Nakano K, Takemura A, Nakamura S, Nakano Y, Iwama GK (2004) Changes in the cellular and organismal stress responses of the subtropical fish, the Indo-Pacific sergeant, *Abudefduf vaigiensis*, due to the 1997–1998 El Niño Southern Oscillation. Environ Biol Fish 70:321–329

Newman SJ (2002) Growth rate, age determination, natural mortality and production potential of the scarlet seaperch, *Lutjanus malabaricus* Schneider 1801, off the Pilbara coast of north-western Australia. Fish Res 58:215–225

Newman SJ, Cappo M, Williams DM (2000) Age, growth, mortality rates and corresponding yield estimates using otoliths of the tropical red snappers, *Lutjanus erythropterus, L. malabaricus* and *L. sebae*, from the central Great Barrier Reef. Fish Res 48:1–14

Newman SJ, Dunk IJ (2002) Growth, age validation, mortality, and other population characteristics of the red emperor snapper, *Lutjanus sebae* (Cuvier, 1828), off the Kimberley coast of north-western Australia. Estuar Coast Shelf S 55:67–80

Newman SJ, Dunk IJ (2003) Age validation, growth, mortality, and additional population parameters of the goldband snapper *(Pristipomoides multidens)* off the Kimberley coast of northwestern Australia. Fish Bull 101:116–128

Newman SJ, Williams DM, Russ GR (1996) Age validation, growth and mortality rates of the tropical snappers (Pisces: Lutjanidae) *Lutjanus adetii* (Castelnau, 1873) and *L. quinquelineatus* (Bloch, 1790) from the central great barrier reef, Australia. Mar Freshwater Res 47:575–584

Pajuelo JG, Lorenzo JM (2002a) Growth and age estimation of *Diplodus sargus cadenati* (Sparidae) off the Canary Islands. Fish Res 59:93–100

Pajuelo JG, Lorenzo JM (2002b) Age and growth of the annular seabream, *Diplodus annularis* (Pisces: Sparidae), from the Canarian archipelago (central-east Atlantic). Ciencias Marinas 28:1–11

Pajuelo JG, Lorenzo JM (2003) The growth of the common two-banded seabream, *Diplodus vulgaris* (Teleostei, Sparidae), in Canarian waters, estimated by reading otoliths and by back-calculation. J Appl Ichthyol 19:79–83

Pajuelo JG, Lorenzo JM, Mendez M, Coca J, Ramos AG (2002) Determination of age and growth of the striped seabream *Lithognathus mormyrus* (Sparidae) in the Canarian archipelago by otolith readings and backcalculation. Scientia Marina 66:27–32

Parenti PR, Randall JE (2000) An annotated checklist of the species of the labroid fish families Labridae and Scaridae. Ichyth Bull J.L.B. Smith Institute of Ichthyology 68:1–97

Patterson WF, Cowan JH, Wilson CA, Shipp RL (2001) Age and growth of red snapper, *Lutjanus campechanus*, from an artificial reef area off Alabama in the northern Gulf of Mexico. Fish Bull 99:617–627

Pauly D (1984) Fish population dynamics in tropical waters: a manual for use with programmable calculators. International Center for Living Aquatic Resource Management, Manilla

Pauly D (1994) On the sex of fish and the gender of scientists. Chapman and Hall, London

Pauly D (1998a) Beyond our original horizons: the tropicalization of Beverton and Holt. Rev Fish Biol Fisheries 8:307–334

Pauly D (1998b) Tropical fishes: patterns and propensities. J Fish Biol 53:1–17

Pauly D, Christensen V, Dalsgaard J, Froese R, Torres F (1998) Fishing down marine food webs. Science 279:860–863

Pears RJ (2006) Comparative demography and assemblage structure of serranid fishes: implications for conservation and fisheries management. PhD Thesis, James Cook University

Pears RJ, Choat JH, Mapstone BD, Begg GA (2005) Demography of a large grouper, *Epinephelus fuscoguttatus*, from the Great Barrier Reef: implications for harvest limits and size regulations. Mar Ecol Prog Ser 307:259–272

Peres MB, Haimovici M (2004) Age and growth of southwestern Atlantic wreckfish *Polyprion americanus*. Fish Res 66:157–169

Pilling GM, Millner RS, Easey MW, Mees CC, Rathacharen S, Azemia R (2000) Validation of annual growth increments in the otoliths of the lethrinid *Lethrinus mahsena* and the lutjanid *Aprion virescens* from sites in the tropical Indian Ocean, with notes on the nature of growth increments in *Pristipomoides filamentosus*. Fish Bull 98:600–611

Polunin NVC, Roberts CM (Eds) (1996) Reef fisheries. Chapman & Hall, London

Polunin NVC, Roberts CM, Pauly D (1996) Developments in tropical reef fisheries science and management. In: Polunin NVC, Roberts CM (Eds). Reef fisheries. Chapman & Hall, London, pp 361–377

Pothin K, Letourneur Y, Lecomte-Finiger R (2004) Age, growth and mortality of the tropical grouper *Epinephelus merra* (Pisces, Serranidae) on Reunion Island, SW Indian Ocean. Vie et Milieu 54:193–202

Potts JC, Manooch CS (1995) Age and growth of red hind and rock hind collected from North Carolina through the dry Tortugas, Florida. B Mar Sci 56:784–794

Potts JC, Manooch CS (1999) Observations on the age and growth of graysby and coney from the southeastern United States. T Am Fish Soc 128:751–757

Radebe PV, Mann BQ, Beckley LE, Govender A (2002) Age and growth of *Rhabdosargus sarba* (Pisces: Sparidae), from KwaZulu-Natal, South Africa. Fish Res 58:193–201

Reznick D, Ghalambor C, Nunney L (2002) The evolution of senescence in fish. Mech Ageing Dev 123:773–789

Robertson DR, Ackerman JL, Choat JH, Posada J, Pitt J (2005) Ocean surgeonfish *Acanthurus bahianus*. I. The geography of demography. Mar Ecol Prog Ser 295:229–244

Rocha-Olivares A (1998) Age, growth, mortality, and population characteristics of the Pacific red snapper, *Lutjanus peru*, off the southeast coast of Baja California, Mexico. Fish Bull 96:562–574

Russ GR, Alcala AC (2004) Marine reserves: long-term protection is required for full recovery of predatory fish populations. Oecologia 138:622–627

Russell DJM, Fletcher AS, Ovenden JR, Street R (2003) Biology, management and genetic stock structure of mangrove jack, (*Lutjanus argentimaculatus*) in Australia. Report No. FRDC Project Number 1999/122, DPI Queensland

Sadovy Y, Figuerola M, Roman A (1992) Age and growth of red hind *Epinephelus guttatus* in Puerto Rico and St Thomas. Fish Bull 90:516–528

Sadovy Y, Vincent ACJ (2002) Ecological issues and the trades in live reef fishes. In: Sale PF (Ed) Coral reef fishes. Dynamics and diversity in a complex ecosystem. Academic Press, San Diego, pp 391–419

Sale PF, Cowen RK, Danilowicz BS, Jones GP, Kritzer JP, Lindeman KC, Planes S, Polunin NVC, Russ GR, Sadovy YJ, Steneck RS (2005) Critical science gaps impede use of no-take fishery reserves. TREE 20:74–80

Sarre GA, Potter IC (2000) Variation in age compositions and growth rates of *Acanthopagrus butcheri* (Sparidae) among estuaries: some possible contributing factors. Fish Bull 98:785–799

Secor DH, Dean JM, Campana SE (1995) Recent developments in fish otolith research. University of South Carolina Press, Columbia

Sinclair M (1988) Marine populations. University of Washington Press, Seattle

Stobutzki IC, Bellwood DR (1997) Sustained swimming abilities of the late pelagic stages of coral reef fishes. Mar Ecol Prog Ser 149:35–41

Swain DP, Sinclair AF, Castonguay M, Chouinard GA, Drinkwater KF, Fanning LP, Clark DS (2003) Density versus temperature-dependent growth of Atlantic cod (Gadus morhua) in the Gulf of St. Lawrence and on the Scotian Shelf. Fish Res 59:327–341

Thorrold SR, Jones GP, Planes S, Hare JA (2006) Transgenerational marking of embryonic otoliths in marine fishes using barium stable isotopes. Can J Fish Aquat Sci 63:1193–1197

Tserpes G, Tsimenides N (2001) Age, growth and mortality of *Serranus cabrilla* (Linnaeus, 1758) on the Cretan shelf. Fish Res 51:27–34

van der Walt BA, Beckley LE (1997) Age and growth of Sarpa salpa (Pisces: Sparidae) off the east coast of South Africa. Fish Res 31:241–248

Villamil MM, Lorenzo JM, Pajuelo JG, Ramos A, Coca J (2002) Aspects of the life history of the salema, *Sarpa salpa* (Pisces, Sparidae), off the Canarian archipelago (central-east Atlantic). Environ Biol Fish 63:183–192

Westneat MW, Alfaro ME (2005) Phylogenentic relationships and evolutionary history of the reef fish family Labridae. Mol Phylogenet Evol 36:370–390

White DB, Palmer SM (2004) Age, growth, and reproduction of the red snapper, *Lutjanus campechanus*, from the Atlantic waters of the southeastern US. B Mar Sci 75:335–360

Williams AJ, Davies CR, Mapstone BD, Russ GR (2003) Scales of spatial variation in demography of a large coral-reef fish – an exception to the typical model? Fish Bull 101:673–683

Wilson CA, Nieland DL (2001) Age and growth of red snapper, *Lutjanus campechanus*, from the northern Gulf of Mexico off Louisiana. Fish Bull 99:653–664

Wing SR, Wing ES (2001) Prehistoric fisheries in the Caribbean. Coral Reefs 20:1–8

Woodbury D (1999) Reduction of growth in otoliths of widow and yellowtail rockfish (*Sebastes entomelas* and *S.flavidus*) during the 1983 El Nino. Fish Bull 97:680–689

Wyanski DM, White DB, Barans CA (2000) Growth, population age structure, and aspects of the reproductive biology of snowy grouper, *Epinephelus niveatus*, off North Carolina and South Carolina. Fish Bull 98:199–218

3. AGE IN YEARS FROM OTOLITHS OF ADULT TROPICAL FISH

A.J. FOWLER

South Australian Research and Development Institute, PO Box 120, Henley Beach 5022, South Australia

1 Introduction

Otoliths of fish are hard, calcified internal structures that assist in orientation and sound perception (Popper et al. 2005, Green et al., Chapter 1, this volume). As otoliths grow inside each fish's head they record an extraordinary amount of information about the life of that fish and the environments that it experiences (Begg et al. 2005). The challenge to scientists is to retrospectively access this information, to interpret it accurately in terms of the biology and life history of the fish, and to use the data to appropriately understand and manage the natural resources of the fish population and the broader aquatic ecosystem. The greatest application of otoliths to date has been in providing information on the age and growth of fish in years, which has ultimately been used in the management of associated fisheries (Campana & Thorrold 2001, Campana 2005). Prior to the 1980s, and stretching back to the late 19th Century, most such ageing work was done in the temperate regions of the world and contributed to managing the enormous fisheries in the oceans of these regions (Beamish 1992, Beamish & McFarlane 1995). There was little attention paid to ageing fish from tropical regions (Longhurst & Pauly 1987). This imbalance has been redressed to some extent over the past couple of decades, however, as scientific effort in the tropics has increased, resulting in a substantial increase in the number of studies and publications on age-related subjects for tropical fishes (Morales-Nin & Panfili 2005).

This chapter is concerned with the use of growth increments that form each year (annuli) in otoliths to age tropical fish to a resolution of years. The information presented is based largely on the numerous population biology studies that have been done since the early 1990s. The first section of the chapter considers historic, geographic and taxonomic issues, followed by a discussion on the nature and characteristics of the otoliths of tropical fishes. Some of the population characteristics of tropical fish that have become apparent through ageing studies are examined, particularly with respect to their life histories, demography and ecology. Issues and uncertainties that remain about using otoliths for ageing tropical fish are also presented. The second section of the chapter is primarily concerned with the practical issues of working with and interpreting otoliths to provide estimates of fish age in years. A summary of how the methodological approach to ageing work has changed through the 20th Century is followed by a description of the sequential steps involved in a fish ageing study once a validated ageing protocol for that species has been established.

B.S. Green et al. (eds.), *Tropical Fish Otoliths: Information for Assessment, Management and Ecology*, Reviews: Methods and Technologies in Fish Biology and Fisheries 11,
DOI 10.1007/978-1-4020-5775-5_3,

1.1 HISTORIC, GEOGRAPHIC AND TAXONOMIC CONSIDERATIONS

1.1.1 Development of annual ageing for tropical fish using otoliths

The determination of fish ages in laboratories around the world for fishery stock assessment purposes constitutes the most significant application of direct ageing and age-structured analyses in any biological application (Campana & Thorrold 2001). Approximately 2 million fish are aged each year as part of this process. The resulting estimates of age provide the fundamental information for understanding the demography and life history of each fish species and for assessing the productivity of its fishery. These estimates of fish age are interpreted to provide insights into the population-specific characteristics of growth, longevity and mortality and to relate to other age-related population processes, such as reproduction and movement.

Otoliths are the preferred anatomical structure for direct ageing of many species of fish because they provide the most accurate and precise estimates of age over the broadest age ranges whilst being relatively easy to prepare and interpret (Secor et al. 1995, Campana & Thorrold 2001). The remarkable success of otoliths as chronometers relates to a number of their characteristics. Firstly, otoliths continue to grow, at least in some dimensions, throughout the lives of the fish, as daily increments are added to the growing surface of the otolith (Campana & Neilson 1985, Morales-Nin 2000). Secondly, otolith growth varies seasonally, which manifests optically as the banded macrostructure that is interpreted as annual increments to estimate fish age. Finally, otoliths are not subject to resorption and reworking, as happens with skeletal bone, and so represent a complete chronological record of the life of the fish (Campana 1999).

Ageing studies based on annuli in otoliths have been most successful to date for temperate species of fish. Ageing applications for tropical species have never attained the same level of significance for several reasons. Firstly, the much smaller and less valuable fisheries of the tropics have never attracted the level of research effort that has been directed towards the enormous temperate water fisheries (Munro & Williams 1985, Longhurst & Pauly 1987). This has been exacerbated by the limited research capability in many tropical countries (Longhurst & Pauly 1987). Secondly, there was an historical perspective that originated as early as the 1920s (Longhurst & Pauly 1987) that tropical fish would not manifest an incremental structure in their otoliths (Pannella 1974). This dogma was based on the conceptual impression of the tropical environment as one of constancy, i.e., lacking in seasonality, to which the fish supposedly responded by growing and spawning continuously throughout the year without the physiological variation in growth that was believed to be central to the formation of seasonal patterns in otolith structure (Longhurst & Pauly 1987). Accordingly, the potential usefulness of otoliths for ageing tropical fishes remained unexplored for most of the 20th Century.

In the late 1970s and early 1980s it was discovered that the otoliths from the adults of some tropical fish did in fact display a macrostructure of banding that may represent an annual pattern of growth increments (Loubens 1978, Brothers 1987, Brothers & Mathews 1987, Manooch 1987, Samuel et al. 1987), in addition to the microstructural banding that was hypothesised to represent daily increments. These early studies considered the generality of such macrostructure amongst different taxa and explored methods for preparing and examining the otoliths. There was a considerable focus on understanding the nature of the otolith macrostructure by exploring its relationship with the microstructure. This was done by counting the number of daily increments that constituted a complete sequence of an opaque and a translucent band in the macrostructure (Brothers & Mathews 1987, Hill & Radtke 1988, Morales-Nin 1989).

There were several focussed population studies in the early 1990s that were the first attempts to validate the periodicity of macroincrement formation for tropical species based on treating fish with tetracycline (Fowler 1990, Fowler & Doherty 1992). There has been considerable expansion in the number of age-based studies on the population biology of tropical species since the mid-1990s that is likely to be related, at least in part, to the considerable expansion in production of the tropical marine and inland fisheries that has taken place since 1990 (Morales-Nin & Panfili 2005). These studies generally have heeded the warnings from the 1980s about the requirement to validate the periodicity of increment formation and otolith interpretation (Beamish & McFarlane 1983). They have undertaken validation studies, developed ageing protocols and then provided age-related information such as age structures, estimates of longevity, and rates of growth and mortality. Such biological studies have become more sophisticated since 2000, describing spatial, temporal and age-based differences in population characteristics (Adams & Williams 2001, Gust et al. 2002, Kritzer 2002, Williams et al. 2003, Choat et al. 2003).

1.1.2 Geographic and taxonomic issues

The tropical and sub-tropical regions of the earth support an extraordinary diversity of aquatic environments that can be broadly classified as inland, estuarine and marine. Each broad category is represented by many different types of habitats ranging in diversity, for example, from high altitude mountain ponds, enormous rivers to the brackish waters where the rivers meet the sea and the great depths of the ocean. Such habitats would clearly provide an immense diversity of physicochemical environments. Not surprisingly, there is an extraordinary diversity of ichthyofauna distributed throughout these many tropical habitats (Sale 1980, Blaber 2000).

The formation of an annual macrostructure within an otolith reflects seasonal growth of the otolith, presumably related to seasonal variation in the life cycle and physiology of the fish in response to environmental change throughout the year (Wright et al. 2002). The two forcing physical environmental variables that would most likely influence the lives of tropical fishes are water temperature and hydrology (Morales-Nin & Panfili 2005). Tropical sea surface temperatures can show considerable seasonal variation that can even exceed that of some temperate oceanic waters (Morales-Nin & Panfili 2005). The forcing variable that is most likely to influence the biological cycles of living organisms in inland systems is seasonal rainfall, such as the monsoons of South-East Asia. Temperature variations can be superimposed on these wet and dry seasons and result in highly seasonal environmental conditions. The flood regimes associated with seasonal rains must also impose considerable seasonality on the physicochemical regimes and the natural trophic systems of estuarine environments. Many tropical fishes are adapted to exploit this seasonal variation in conditions, as manifested through their trophic relations, cycles of reproduction, growth and condition. Such seasonality may well also result in differential growth of fish otoliths, which becomes evident in the otolith macrostructure (Jepsen et al. 1999, Faunce et al. 2002, Panfili et al. 2004, Morales-Nin & Panfili 2005).

There is an uneven distribution of otolith-based ageing studies from the different tropical environments. The greatest success so far appears to have been for the demersal species that occupy the marine neritic zone, particularly those that are associated with tropical reefs and inter-reefal areas (Fowler 1995). These fishes come from the Order Perciformes, Suborder Percoidei and include families such as the Acanthuridae, Chaetodontidae, Haemulidae, Labridae, Lethrinidae, Lutjanidae, Scaridae, Scianidae and Serranidae. Some success has been achieved also for a number of pelagic, open water species from the Suborder Scombroidei, Family Scombridae, primarily the mackerel species (McPherson 1992, Begg & Sellin 1998, Tobin & Mapleston 2004). There has been considerable ageing work on the Clupeiforme fishes that have a pelagic existence in the tropical, neritic and oceanic waters. Such fishes have generally proven to be relatively short-lived, however, living only up to a maximum of about 2 years and so the ageing work for these species has generally been based on daily increments in their otoliths (Struhsaker & Uchiyama 1976, Gjosaeter et al. 1984, Milton et al. 1993, Hoedt 2002, Sponaugle, Chapter 4, this volume). There have been far fewer ageing studies reported for fish from tropical inland systems. Some notable success has been achieved for the family Cichlidae, for which studies have been done in West Africa (Panfili et al. 2004), the rivers of Venezuela (Jepsen et al. 1999), and in south-east Florida (Faunce et al. 2002). The lack of diversity for the taxa considered from tropical inland systems may reflect that the production of fisheries from this environment is approximately only one sixth of that from tropical marine environments (Morales-Nin & Panfili 2005).

1.2 CHARACTERISTICS OF OTOLITHS OF TROPICAL FISH

1.2.1 The nature of otolith structure

For the otoliths of tropical fish to be useful in providing estimates of fish age in years they must fulfil three criteria:

1. they must display an internal structure of increments that can be quantitatively resolved, optically or otherwise;
2. the formation of increments must conform to a regular and determinable time scale; and
3. the otoliths must continue to grow throughout the life span of the individual fish (Fowler & Doherty 1992).

It was assumed initially that the first of these criteria was the one that the otoliths of tropical species would fail. Once researchers began to examine such otoliths, however, they realised that a macrostructure often was apparent (Loubens 1978, Manooch 1987). The second criterion has also been fulfilled by many tropical species. Early attempts to validate the periodicity of increment formation involved determining the relationship between the macrostructure and microstructure of the otoliths by enumerating the number of daily increments that constituted a complete macro-increment (= one opaque and translucent band) using scanning electron microscopy (Brothers & Mathews 1987, Hill & Radtke 1988, Morales-Nin 1989, Morales-Nin & Ralston 1990). Most attempts since 1990 to determine the periodicity of increment formation have involved the use of oxytetracycline tagging (Fowler 1990, Newman et al. 1996, Cappo et al. 2000, Williams et al. 2005) or marginal increment analysis (Begg & Sellin 1998, Brown & Sumpton 1998, Pilling et al. 2000, Faunce et al. 2002, Panfili et al. 2004). The third criterion above generally has been assessed through considering the relationship between otolith weight and fish age. Such relationships often have proven to be approximately linear, suggesting that relatively consistent amounts of otolith material are added to the otolith surface annually, throughout each fish's life (Ferreira & Russ 1994, Worthington et al. 1995, Newman et al. 1996, Craig et al. 1997, Lou et al. 2005).

The general conclusion from the numerous validation studies to date is that the otoliths of many tropical species do display a macrostructure that can be displayed optically. These increments consist of either opaque or translucent bands. The appearance of these bands depends on the method of illumination. Opaque bands are denser than the translucent bands and are relatively dark under transmitted light as they transmit less light. They appear lighter under reflected light because they are more reflective than the translucent bands (see Figure 1 in Green et al., Chaper 1, this volume). This is further described under Section 2.2.3 of this chapter, *Interpretation of otolith macrostructure to estimate fish age*. The terminology of opaque and translucent bands used in this chapter assumes transmitted light.

The periodicity of formation of the macrostructure has generally proven to be annual, regardless of the environment that the different species occupied (Fowler 1995, Choat & Robertson 2002, Faunce et al. 2002). One annulus involves the combination of one opaque band and its adjacent translucent band. For numerous marine species, the opaque band forms at some time throughout the spring and early summer (Fowler 1995, Choat & Axe 1996, Begg & Sellin 1998) but for inland fishes its formation depends on the timing of the local wet and dry seasons (Jepsen et al. 1999, Panfili et al. 2004). Furthermore, a relatively consistent mass appears to be deposited onto the growing surface of the otoliths during each year throughout the fish's life (Worthington et al. 1995, Lou et al. 2005). These latter two points indicate that the otoliths demonstrate a systematic pattern of growth that is manifested as an optical pattern that can provide an accurate estimate of age in years. That is, the macrostructure of otoliths of many tropical fishes so far examined do represent chronometers of the fishes' lives.

1.2.2 Optical characteristics

A short review in 1995 summarised the information that was available to that time on the ageing of coral reef fish species (Fowler 1995). Fowler had examined the sagittae, i.e. the largest pair of otoliths, from 28 species of percoid fishes from eight reef fish families and identified that the otoliths of most species had some characteristics in common. There have been many more studies on tropical percoid species since then and the characteristics of the otoliths, based on their descriptions and photographs in the literature, appear to be very consistent with the earlier descriptions. These characteristics are summarised below.

1. The whole otoliths of the percoid species have a glassy appearance and when immersed in oil transmit some light, which means that, to some extent, they are translucent, contrasting with the otoliths from some high latitude species whose otoliths are so opaque they do not transmit light.

2. Transverse sections of otoliths display a macrostructure consisting of narrow opaque and broad translucent bands (Figures 1a, b and 2). This also contrasts with the otoliths of some high latitude species, in which both bands are relatively similar in width (Blacker 1974, Williams & Bedford 1974, Beckman & Wilson 1995, Campana 2001, Campana & Thorrold 2001). This may reflect that the period of the year during which the opaque band is deposited is shorter for tropical than for temperate species, possibly accounting also for their differences in opacity.
3. The alternation between the opaque and translucent bands can be abrupt and distinct and therefore may not represent a gradient of opacity (Figure 1b).
4. The opaque band may be represented by a narrow, distinct check, or can be broader consisting of numerous closely spaced discontinuities (compare Figure 1b, c).
5. The different opaque bands within the same otolith are not necessarily of the same optical density. Specifically, the first one or two closest to the nucleus may be relatively broader but more diffuse and less distinct than those that are further from the nucleus in older fish (Figure 1a, b). Because of this, it can be easier procedurally to determine the ages of older fish than younger ones whose otoliths display only one or two diffuse opaque bands.
6. The opaque band has been related to a short period of fast growth during the year, and the translucent band associated with a longer period of slow growth (Brothers & Mathews 1987, Morales-Nin 1989, Morales-Nin & Ralston 1990). The evidence for this conclusion, however, was not made apparent in these earlier publications and the relationship between opaque and translucent bands and specific growth patterns remains equivocal for coral reef fish species. This is considered further below in Section 1.4.5.

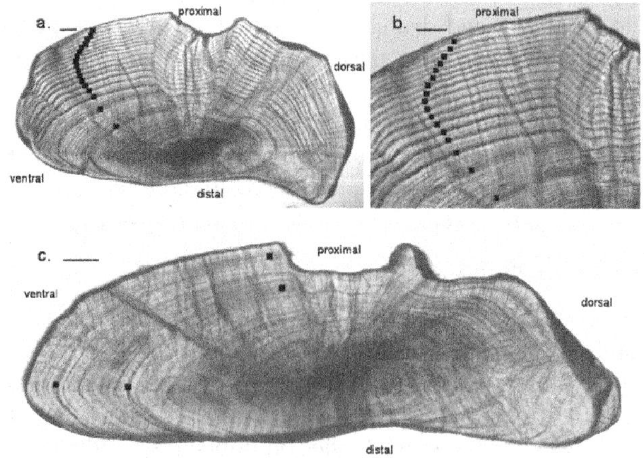

Figure 1. Transverse sections of sagittae from *Pomacentrus moluccensis* from the Great Barrier Reef. (**a**) View of transverse section with 17 opaque bands. (**b**) Higher magnified view of the same otolith showing the macrostructure of the ventral side of the sulcus. (**c**) View of a transverse section from an otolith with two relatively poorly defined opaque bands. *Dark dots* indicate the opaque bands on each image when viewed with transmitted light. Scale bars = 100 μm.

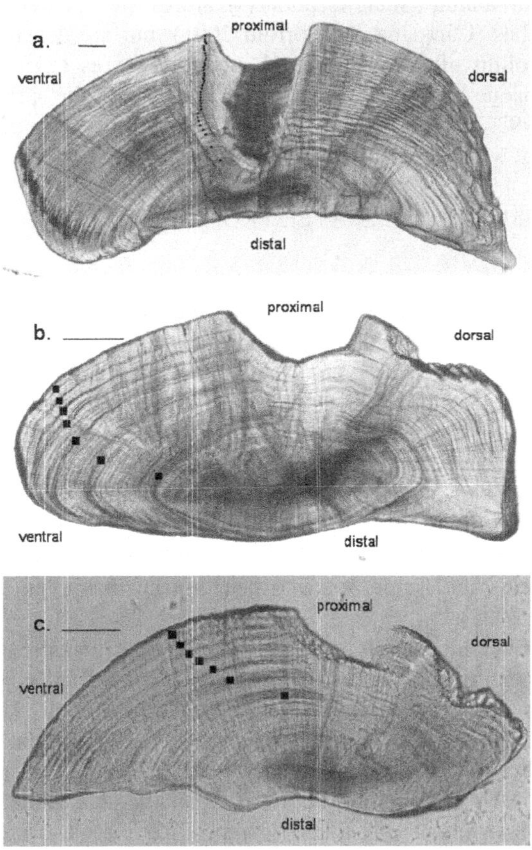

Figure 2. Transverse sections of sagittae from adult individuals from each of three species of fish from the central Great Barrier Reef. (**a**) *Acanthurus lineatus* with 37 opaque bands. (**b**) *Pomacentrus moluccensis* with seven opaque bands. (**c**) *Chaetodon aureofasciatus* with seven opaque bands. *Dark dots* indicate opaque bands when viewed with transmitted light. Scale bars = 100 μm.

The otoliths of cichlid species from inland systems display a number of the characteristics described above for marine reef fish (Jepsen et al. 1999, Faunce et al. 2002). The otoliths of scombrid fishes, however, show some interesting differences. Such otoliths are elongate and laterally compressed, which means that the translucent and opaque bands can be relatively clear in the whole otoliths (McPherson 1992, Begg & Sellin 1998, Tobin & Mapleston 2004). The first several translucent bands are characterised by radial striae that refract light and give the otoliths a crazed appearance (McPherson 1992, Buckworth 1998, Lewis & Mackie 2002). The first translucent band is broad, whilst subsequent ones become increasingly narrow. The first opaque band can be quite indistinct, after which the subsequent ones become more distinct. The later-formed opaque bands are relatively broader than the alternating translucent ones. This can make it easier to interpret the otoliths of older fish relative to younger ones. The interpretation of otolith structure also can be complicated by the presence of secondary opaque bands or sub-annuli. Such secondary features are apparent in the otoliths of some temperate fishes (Campana & Thorrold 2001), but are generally not significant structures in the otoliths of tropical percoid species (Figures 1, 2; Fowler 1995). The formation of the opaque band is more often associated with the spring–summer period (Schmidt et al. 1993, De Vries & Grimes 1997, Begg & Sellin 1998, Tobin & Mapleston 2004) but winter formation was recorded in one case (McPherson 1992).

1.3 USES OF FISH AGE ESTIMATED FROM ANNUAL INCREMENTS

Fish populations are dynamic biological entities that vary in size over time as influenced by several input and output processes (Hilborn & Walters 1992, King 1995). Population abundance and biomass increase through the natural processes of reproduction and recruitment of new individuals. The total biomass also increases through the accumulation of mass by individual fish as they grow over time. Simultaneously, the abundance and biomass are reduced as individuals die through predation, starvation and disease and, in some cases, harvest. Developing an understanding of these dynamic processes provides fundamental insights into the demography and population dynamics of a species. Understanding population characteristics is fundamental to the stock assessment procedures to determine stock status of fished species and to provide advice about potential yields and appropriate fishery management strategies (Hilborn & Walters 1992, King 1995).

Developing an understanding of the population dynamics and the controlling influences for non-exploited fish populations is also important. This indicates how natural populations are influenced by environmental variation, independent of the confounding influence of fishing mortality (Hilborn & Walters 1992). This may contribute to the better management of exploited fish stocks as well as providing valuable information for formulating ecosystem management plans. For example, the establishment and assessment of aquatic reserves and marine protected areas would benefit from detailed knowledge for key indicator species or rare species with high conservation value (Russ 2002).

It is essential to obtain estimates of growth, mortality and recruitment rates to understand the demographics and population dynamics of fish populations. Estimates of fish age provide the measures of elapsed time that are required for estimating these rates (Campana 2001). Fish ageing underpins the stock assessment procedures for many species of fishes from temperate regions of the world, which accounts for why approximately 2 million fish are aged annually around the world (Campana & Thorrold 2001). The earlier perception that direct ageing procedures would not work for fishes in tropical areas, however, (Longhurst & Pauly 1987) meant that such parameters were estimated using alternative methods such as length-based procedures (King 1995), whose applicability for tropical species is particularly concerning (Section 1.4.1). Consequently, the recent successes in using otoliths for direct ageing of tropical fish have been particularly revealing about the demography and life histories of such species and for addressing hypotheses about the factors that limit their abundances and influence their size and age structures (Thorrold & Hare 2002). These advances are evident in the following summary of some of the recent findings.

1.3.1 Demographic and life history characteristics

Longevity

Perhaps one of the biggest surprises from studies to date is that the estimates of life spans of a broad range of tropical fish have been in decades rather than just a few years. This finding is contrary to the notion that tropical systems are characterised by species with fast growth, high productivity and high turn-over (i.e., short lives) (Pannella 1974, Sale 1980). In fact, many species have maximum ages that exceed 30 years of age. Such unexpected longevity is true not only for some species that attain a large maximum size for which it might be expected that older individuals attain some protection from predation because of their size, but also applies to the small reef-associated species. One species of pomacentrid on the Great Barrier Reef, for example, had maximum ages of approximately 20 years (Doherty & Fowler 1994a). Several other species from the Galapagos Islands even reached >30 years of age, despite barely attaining 100 mm in length (Meekan et al. 2001). Several species of acanthurids had some of the highest estimates of maximum age with *Acanthurus lineatus* living up to 42 years, despite attaining only 183 mm in length (Choat & Axe 1996, Choat & Robertson 2002). Similarly, the small serranid *Cephalopholis cyanostigma* that grows to 350 mm in length reached up to 46 years of age (Mosse et al. 2002), whilst the tropical snapper *Lutjanus bohar* attained a maximum age of 54 years (Marriott 2002). Indeed, there is no apparent relationship between maximum age and asymptotic size when numerous tropical percoid species from several families are considered, since even relatively small species attain considerable ages (Figure 3a). Alternatively, there is a significant linear relationship between maximum age and asymptotic size for the scarids, indicating that those species that live longer generally attain larger sizes than the shorter lived scarid species (Figure 3b). Most estimates of maximum age for the scombrids exceeded 10 years, with the highest estimate attained so far being 22 years for the Spanish mackerel (*Scomberomorus commerson*) (Mackie et al. 2003).

Growth

The best way to determine the pattern of growth of fish is to sample a broad size range of individuals from the population, measure their size and obtain an estimate of age in years from their otolith macrostructure. This approach has been applied in numerous recent studies for tropical fishes, which have described growth in terms of the von Bertalanffy equation, i.e., $L_t = L_\infty[1 - e^{-K(t - t_0)}]$, where L_t = length at age t, L_∞ = theoretical maximum or asymptotic length that fish would reach if they lived indefinitely, K = rate at which maximum size is reached, and t_0 = the theoretical age at zero length (King 1995). This equation to describe growth has a strong physiological basis (Longhurst & Pauly 1987) and has become a standard in fishery studies.

The uniform approach of describing growth using the von Bertalanffy equation has provided a basis for comparing growth characteristics amongst species and between populations of the same species from different places (e.g. Newman et al. 2000a,b; Kritzer 2002). This has revealed that tropical species display different types of growth patterns, which relate, to some extent, to their longevities. Some species display the typical indeterminate growth pattern of fish from higher latitudes, where somatic growth is continuous throughout the life of the fish, albeit at a decreasing rate, ultimately resulting in a relationship between fish size and age (Figure 4a). Such growth has been described for species of coral trout (*Plectropomus leopardus* and *P. maculatus*) from the Great Barrier Reef (Ferreira & Russ 1992, Russ et al. 1996), as well as some of the species of the Scaridae family from the same region (Figure 4a) (Choat et al. 1996). Both the spotted mackerel (*Scomberomorus munroi*) (Begg & Sellin 1998) and Spanish mackerel (*Scomberomorus commerson*) (Tobin & Mapleston 2004), which occur in the offshore, open, tropical waters of Queensland, also exhibit this type of growth. Studies of cichlid species from different inland systems have also demonstrated such continuous growth throughout cichlids' relatively short lives (Jepsen et al. 1999, Faunce et al. 2002, Panfili et al. 2004).

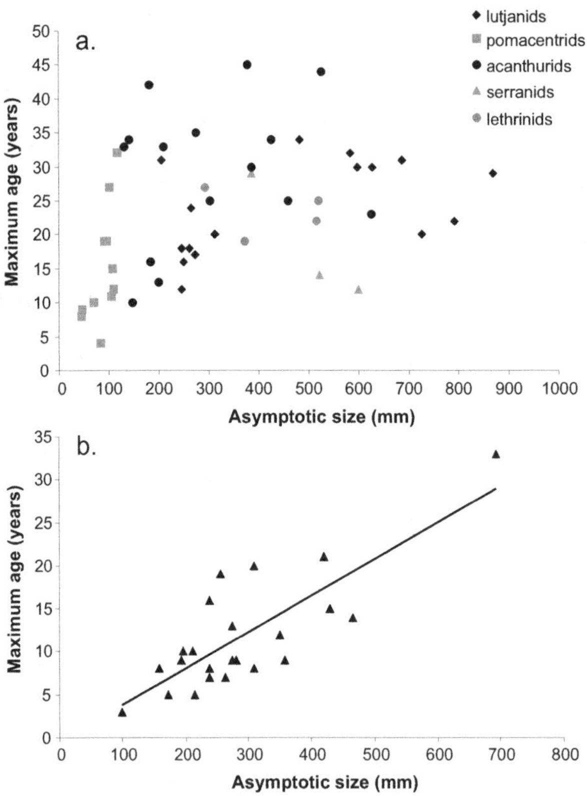

Figure 3. Relationships between maximum recorded age in years and the asymptotic size. (**a**) Data from several species from each of five reef fish families; (**b**) Data for a number of species from the Scaridae family with the line of best fit also indicated. Data extracted from the following sources: Fowler (1990), Ferreira and Russ (1992), Fowler and Doherty (1992), Ferreira and Russ (1994), Choat et al. (1996), Newman et al. (1996), Brown and Sumpton (1998), Newman et al. (2000a,b), Meekan et al. (2001), Burton (2002), Choat and Robertson (2002), Grandcourt (2002), Gust et al. (2002), Kritzer (2002), Newman (2002), Newman and Dunk (2002), Schwamborn and Ferreira (2002), Choat et al. (2003), Newman and Dunk (2003).

Other tropical fishes grow very fast over a relatively short part of the early life and quickly reach asymptotic size, after which there is virtually no further growth in length over the remainder of a fish's life. Such determinate growth results in a characteristic flat-topped growth curve (Figure 4b). Some species from the lutjanid, acanthurid, pomacentrid and serranid families demonstrate this type of growth. For example, *Acanthurus lineatus* from the Great Barrier Reef, which has a maximum age of 42 years, achieves 95% of its asymptotic size in only 6 years (Choat & Robertson 2002), meaning that individuals might experience no further growth in length for up to 85% of their life span (Figure 4b). The pomacentrid *Stegastes acapulcoensis* from the Galapagos Islands lives for up to 32 years but requires only 1 year to attain 80% and 3 years to reach 95% of its asymptotic size (Meekan et al. 2001). Thus, individuals can remain at approximately the same size for up to 29 years.

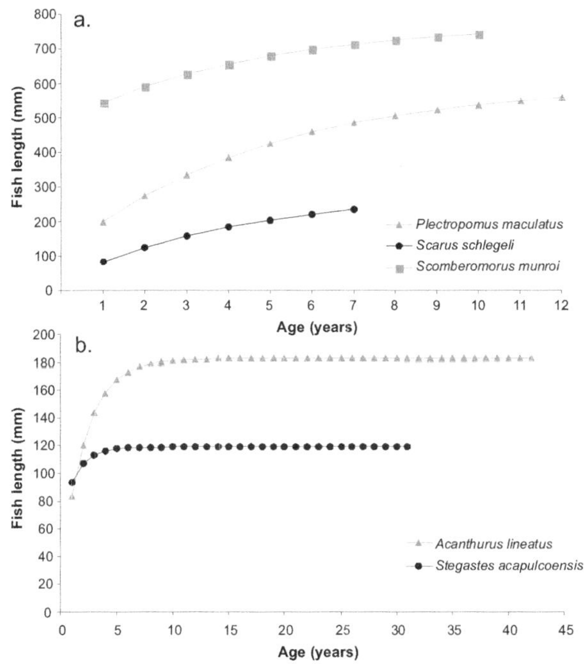

Figure 4. Von Bertalanffy growth curves for a number of tropical species. (**a**) Three species of fish with indeterminate growth, where size is related to age; (**b**) Two species for which growth is determinate and size and age are decoupled through much of a fish's life. Data extracted from Ferreira and Russ (1994), Begg and Sellin (1998), Meekan et al. (2001), Choat and Robertson (2002).

The species of reef fish with indeterminate growth tend to be relatively short-lived, whilst many long-lived species conform to the flat-topped growth curve. Thus, tropical species differ with respect to the extent to which size is related to age. For example, the largest size class of many long-lived species will be composed of many age classes, which means that for most of their lives their size and age will be decoupled (Choat & Robertson 2002). Some calculations were done to estimate what proportion of the maximum age would be required to attain 80% of the asymptotic size based on the von Bertalanffy growth parameters that are reported in the literature (Figure 5). These estimates demonstrated a clear decreasing spread of datapoints as maximum age increased. Some short-lived species, mainly the scarids, required a greater proportion of their lives to approach their asymptotic size, indicating that they conformed to the indeterminate growth pattern. The lutjanid species generally required 10–40% of their lifespan to attain 80% of the asymptotic size, whilst most acanthurid species required <20% to achieve this size. The pomacentrids had the fastest relative growth in this sense, generally requiring <10% of their life span to attain 80% of their asymptotic size.

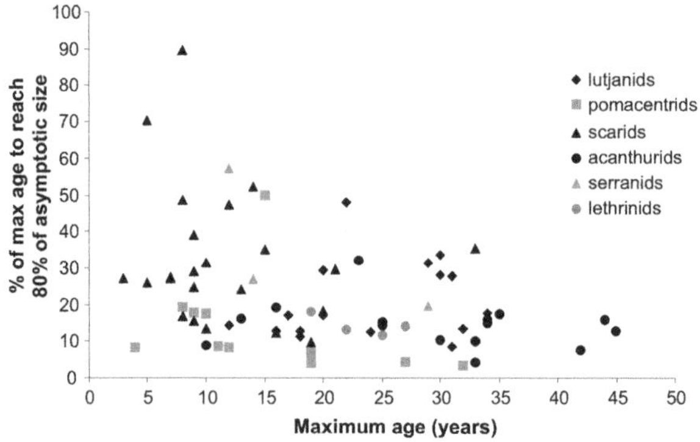

Figure 5. The percentage of maximum age required to attain 80% of the asymptotic size plotted against maximum age for several species from each of six reef fish families. Data were calculated from von Bertalanffy growth parameters provided in references listed for Figure 3.

Clearly, there are differences in growth patterns amongst the different fish taxa related to their longevities but such differences appear to not be reflected in the von Bertalanffy growth parameters of K, L_∞ and t_0 (Choat & Robertson 2002). In general, the reef fishes tend to lie somewhere along a continuum of small fish with fast growth rates and large fish with slower rates of growth (Figure 6). The species differ not so much in how they grow but rather in the number of years they persist without further growth. Such information is not well encapsulated in the von Bertalanffy parameters of K and L_∞ (Choat & Robertson 2002), indicating that it is necessary to simultaneously report the von Bertalanffy growth parameters along with estimates of maximum age to describe adequately the pattern of growth of such species.

Some understanding of the factors that influence the growth patterns of tropical fish have emerged from numerous population studies. For example, there were sex-based differences in growth for some species such as *Lutjanus sebae* (Newman & Dunk 2002) and *Lutjanus carponotatus* (Newman et al. 2000b), where in each case the males attained a larger L_∞ but grew at a slower rate than the females. Furthermore, size-at-age differed with sexual identity for sex-changing species such as the scarids *Scarus schlegeli* and *S. rivulatus*, where the terminal phase males were larger than similar-aged initial phase males and females (Choat et al. 1996). Males of the protogynous coral trout *Plectropomus maculatus* had larger mean size at age throughout life compared to individuals that remained females (Adams & Williams 2001).

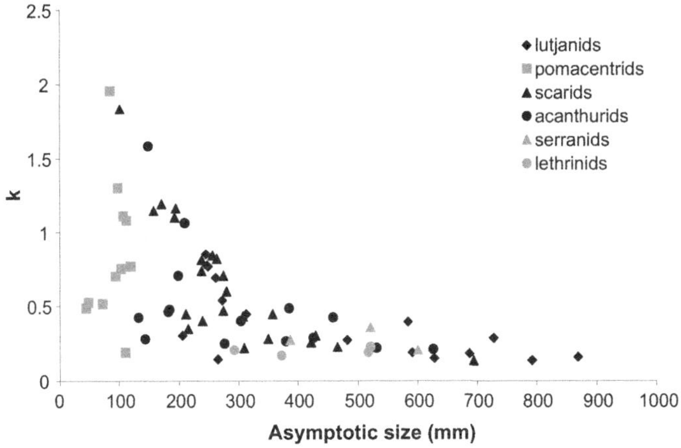

Figure 6. The relationship between the von Bertalanffy growth parameters 'k' and 'L_∞' (asymptotic size) for several species from each of six reef fish families. Data from the references listed for Figure 3.

Some studies have used hierarchical sampling designs and begun to explore the issues of spatial variation in growth under different environmental conditions, and identified considerable variation at different spatial scales (Kritzer 2002, Gust et al. 2002, Williams et al. 2003). There is a trend for populations to have a shorter longevity, reduced asymptotic size but faster growth rate as mean water temperature increases over a large geographic scale (Choat & Robertson 2002). This may be reflected as a latitudinal effect, but local environmental influences can also complicate such broad-scale patterns. For example, several species of scarids and acanthurids on the northern Great Barrier Reef showed variations in growth across the continental shelf at the same latitude over distances of 10s of kilometres. The patterns were consistent amongst several species and were thought to relate to different resource levels and population abundances among the cross-shelf locations (Gust et al. 2002). Significant small-scale variation in growth patterns has been identified for *Lutjanus carponotatus* amongst small, localised reefs (Kritzer 2002). Growth of *Lethrinus miniatus* was relatively uniform amongst reefs separated by distances of up to 10s of kilometres but differed significantly among regions separated by 100s of kilometres (Williams et al. 2003). The pattern of variation was not consistent with latitude, however, which suggests that water temperature was not the single controlling factor. Again, patterns of growth may have been influenced by the distribution and abundance of food resources.

Mortality. The construction of age structure(s) for a population, based on ageing individual fish, provides the basic data from which an estimate of total mortality rate (Z) can be derived. This 'catch curve' analysis can be done either with a single sample of fish and based on the relative abundances of the different age classes collected at the same time, or with several age structures collected in different years, allowing cohorts of individuals to be followed through time (King 1995). Estimates of mortality rates in most tropical studies to date have been based on a single age structure. The samples of fish for several such studies on important fishery species were collected from broad areas across wide ranges in latitude, in each case producing a single age frequency distribution from the numerous estimates of age (e.g. Newman & Dunk 2002, Grandcourt 2002). The slope of the line of best fit through the natural logarithm-transformed frequencies for the different age classes from the modal class to the oldest age recorded provides the estimate of mortality (Z). This can then be easily converted to the % survival (S) of each age class, where $S = 100 \times e^{-z}$ and to the % mortality for each age class per year, where $M = 100(1 - e^{-Z})$.

The historic perception of tropical fish having high rates of turn-over and poor survivorship (Pannella 1974, Sale 1980) reflects a preconception that natural mortality rates must be very high. In contrast, however, the surprisingly high estimates of longevity for some species, and the fact that some older age classes are well represented in the age structures, indicate that the rates of mortality for many tropical species are not as high as previously expected (Doherty & Fowler 1994a,b, Meekan et al. 2001, Patterson et al. 2001, Newman & Dunk 2002). The estimates of total annual mortality for a diversity of taxa presented in the literature were in the range of 0.1–0.5, which translate into relatively high estimates of annual survivorship (90.5–60.7% respectively). There was a negative curvilinear relationship between maximum age and mortality rate, where those species with the low mortality rates of around 0.1 had the highest estimates of maximum age (Figure 7). Several outlier points are evident on Figure 7, which relate to several species of lethrinid, lutjanid and serranid that had high rates of fishing mortality. Concern was expressed for several of these that the estimated levels of fishing mortality were excessive, and that some management action was required.

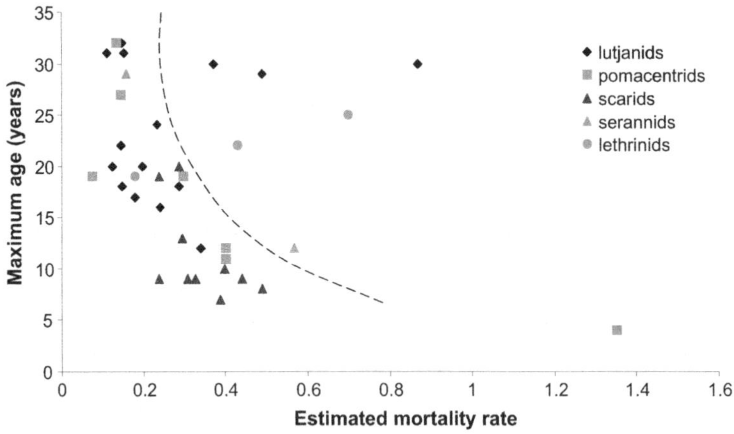

Figure 7. Relationship between maximum age and mortality rate for a variety of species from five families of reef fish species. The high estimates of mortality for the species to the right of the dashed line were related to high levels of fishing mortality. Data extracted from Ferreira and Russ (1992), Newman et al. (1996), Brown and Sumpton (1998), Newman et al. (2000a,b), Meekan et al. (2001), Burton (2002), Choat and Robertson (2002), Grandcourt (2002), Gust et al. (2002), Kritzer (2002), Newman (2002), Newman and Dunk (2002), Choat et al. (2003), Newman and Dunk (2003).

1.3.2 Ecological processes

The ability to determine the ages of adult tropical fish through interpretation of the macrostructure of otoliths has played an important role in understanding the factors that influence population abundance and structure. A significant example was in assessment of the influence of temporally variable recruitment on populations (Doherty & Fowler 1994a,b) and the debate surrounding the relative significance of pre- and post-settlement processes for the ecology of coral reef fishes (Doherty & Williams 1988, Mapstone & Fowler 1988). The issue was whether variable recruitment of new individuals to coral reef fish populations was the dominant influence over post-settlement population dynamics and demography, or whether post-recruitment events substantially modified the patterns of distribution and abundance established at the time of settlement. These alternative hypotheses have very different implications for assemblage structure and fisheries management (Doherty & Williams 1988, Mapstone & Fowler 1988). If post-recruitment events dominated, then the assemblage would likely be resource-limited and stable in structure with strong influences of intra-specific competition and density-dependent mortality. The alternative hypothesis was that reef fish populations were constrained mainly by limited numbers of larvae available to recruit to the post-settlement population, meaning that recruitment was inadequate to bring populations up to levels where compensatory, density dependent processes become important.

The study of Doherty and Fowler (1994a,b) was based on a time-series of recruitment rates for different species of fish that had been collected annually for the 9 year period of 1981–1989 for seven coral reefs at the southern Great Barrier Reef, Australia. A validated ageing protocol had been established by treatment of fish with tetracycline for two species of fish, *Pomacentrus moluccensis* and *P. wardi* (Fowler 1990, Fowler & Doherty 1992). In September 1989, the complete assemblages of fish associated with numerous patch reefs from within the lagoons of the seven coral reefs were collected, including the populations of *P. moluccensis* and *P. wardi*. Samples for these two species were aged from their sectioned otoliths and the resulting population age structures were compared with their 9-year recruitment history. The average annual rates of recruitment were good predictors of the differences in mean abundance amongst the seven reefs. Furthermore, the population age structures preserved the patterns of year-to-year variation in recruitment, with particularly strong age classes being relatable to years of exceptional settlement. It was thus apparent that the abundances and population structures for the two small species of fish were the consequence of variable recruitment and different rates of density independent mortality (Doherty & Fowler 1994a,b). This important ecological work was only possible due to the ability to age the adult fish in years based on interpreting the macrostructure of their otoliths using a validated ageing protocol.

A further conclusion can now be drawn from the work of Doherty and Fowler (1994a,b), taking into consideration the summary of demographic characteristics presented in Section 1.3.1 above. It is now reasonable to conclude that at the southern end of the Great Barrier Reef, where the Doherty and Fowler study was done, both *Pomacentrus moluccensis* and *P. wardi* were slower growing, attained larger maximum sizes and on average lived longer than the populations at lower latitudes further north on the Great Barrier Reef. The estimates of longevity for *P. moluccensis* of 17 years and 9 years for the southern and central parts of the Great Barrier Reef respectively, are consistent with this inference (Doherty & Fowler 1994a,b, Fowler 1990). Thus, such geographic variation in demographics for individual species along a latitudinal gradient cautions against applying the conclusion of recruitment limitation to the populations outside those that were directly considered.

Another age-based study that was done on the central Great Barrier Reef (18°30′S), and which involved the important fishery species the coral trout (*Plectropomus leopardus*), also identified the likely significance of recruitment variation to population dynamics and structure (Russ et al. 1996). Population age structures were developed in this study from samples of fish collected from two reefs that had been closed to fishing for a number of years. These age structures indicated the existence of a single strong year class of fish that had settled to the two reefs in early 1984. This strong year class was then tracked through the age structures from samples that were collected annually through the period of 1990–1993, which was the first time that temporal tracking of a cohort of fish had been achieved for a species of coral reef fish. It also provided the rare opportunity to estimate the rate of natural mortality for such an important fishery species (Russ et al. 1998). The most tractable explanation for the existence of this strong year class was inter-annual variation in recruitment, which ultimately influenced population abundances and age structures more than post-settlement, density-compensating processes. This significant finding was made only due to the availability of population age structures, determined from the macrostructure of the otoliths of the samples collected (Ferreira & Russ 1994).

1.4 ISSUES AND UNCERTAINTIES

1.4.1 Length-based methods for parameter estimation

Estimation of demographic parameters for stock assessment work on tropical fish was done largely using length-based analytical procedures prior to the relatively recent discoveries about the usefulness of otoliths for ageing tropical fish. The conceptual basis here was that discrete spawning events produced different size classes of fish that formed a polymodal length frequency distribution, where the different modes represented the normally distributed sizes of fish derived from the different spawning events. The different modes were relatable to time by making assumptions about the age of the first mode and the time periods between subsequent modes, thus making it possible to estimate parameters for growth and mortality (King 1995). There are two main analytical procedures for achieving this, based on single-sample and multiple-sample methods. In the former case, a single length frequency distribution is divided into separate modes. In the latter, multiple length distributions that are collected over time are arranged sequentially and the modes of the cohorts are traced as the fish increase in size over time.

There are several serious problems with using length-based methods of analysis, regardless of whether for tropical or temperate species. The first problem is that attaining length samples of fish that are representative of wild populations is extremely difficult and costly (Hilborn & Walters 1992). One reason for this is that fishing gear is always selective of fish size to some extent. Size selectivity is an interactive process between the method of capture and the behaviour of fish of different sizes. Furthermore, the life histories of many species of fish include ontogenetic migration amongst habitats and locations. As such, sampling the sizes of fish at particular places must select for particular size classes, which means that it would be necessary to sample the different size classes in different places, possibly using different fishing gears, in order to reconstruct representative size distributions for such populations. Finally, Hilborn and Walters (1992) also indicate that it is difficult to define a field-sampling regime and to train field workers to properly take representative samples of the size distributions from fishery catches.

There are also problems with length-based analyses that relate to the biology of fish species. Clearly, length-based methods will work best for those species that have discrete modes in their size distributions. Such species would be likely to have a discrete spawning season that leads to recruitment over a relatively short time period, after which growth is relatively fast and continuous. Alternatively, if the spawning period is extended or growth is slow or ceases at some point in life, then the normally-distributed sizes of fish originating from the different spawning events will overlap to such a degree that it will not be possible to distinguish separate modes. Furthermore, the rate of growth of fish slows as they get bigger, which means that the older age classes will ‘bunch’ together and form a mode in which the numerous age classes are indistinguishable.

The account of the demographics of tropical fish species presented above suggests that many species would be unsuitable for analysis using length-based analytical techniques. Many species are relatively long-lived, with relatively low rates of mortality, meaning it is likely that their size distributions will not display discrete modes that relate to individual age classes. This is likely to be the case for those species with indeterminate growth, but will be exacerbated for those species with determinate growth. The tropical snapper *Lutjanus carponotatus* lives for up to 18 years on the coral reefs of the central Great Barrier Reef but the size distributions on five different reefs were normally distributed, not displaying modes that were relatable to age classes (Kritzer 2002). Similarly, the males and females of *Lutjanus sebae* from North-western Australia, that live to 30 and 34 years respectively, each had length frequency distributions that were normally distributed without strong modes being apparent (Newman & Dunk 2002). Even the size distribution of the first 6 year classes for the black grouper, *Mycteroperca bonaci*, from Florida waters, which can live to 32 years of age, did not show modes that were relatable to age classes (Crabtree & Bullock 1998). Finally, despite the presence of a very strong year class in the age structures of coral trout (*Plectropomus leopardus*) collected annually over 4 years, there was no obvious modal progression in length distributions from year-to-year and the modal lengths of catches for the 4 years overlapped substantially, reflecting the large variability in size-at-age for this species (Russ et al. 1996).

1.4.2 Sampling for fish ages

Some consideration must be given to the process of sample collection and to the number of fish to be aged to develop a population age structure. There are two general ways to sample fishes for developing age structures. The simplest method is simple random sampling where a sample of fish is removed from a population and all or as many as possible are aged using a validated ageing protocol. It is assumed that the fish are sampled independently and have the same probability of being sampled, meaning that cohorts will be represented in the sample in proportion to their abundances in the wild population. Such sampling is not usually characteristic of fishery studies (for example the sequence of individual fish captures from line fishing is almost certainly non-independent), and many tropical studies have sought to address this problem. For example, researchers have produced combined samples of fish accessed from multiple sources, such as commercial and recreational fishers, and research sources. Such sampling also has been used in ecological studies involving hierarchical sampling. For example, Kritzer (2002) collected samples of *Lutjanus carponotatus* by spear fishing, where fish were targeted as sighted, without preference based on body size, so as to ensure that the sample was as representative as possible. The coral trout *Plectropomus leopardus* were sampled on different coral reefs and occasions using a standardised approach using recreational line fishers (Russ et al. 1996). Doherty and Fowler (1994a,b) used rotenone to collect all individuals of *Pomacentrus moluccensis* and *P. wardi* from patch reefs in the lagoons of coral reefs. In each of these examples, as many fish as possible from the samples were aged for development of age structures. The study of Kritzer et al. (2001) is an example of investigating the optimal sample numbers required to estimate demographic parameters for populations of tropical reef fish.

An alternative form of sampling has been called two-stage random sampling (Quinn & Deriso 1999). This method, which appears to not have been commonly applied in tropical applications, is more likely to be used in fishery applications where there is greater accessibility to samples. Here, the first stage of sampling is where a simple random sample is removed from the catch and each individual is measured for generation of a size frequency distribution. The second stage of sampling is where a sub-sample of the first sample is then selected for ageing work based on the interpretation of otolith structure. The age frequencies for each length class of fish are used to generate an age-length key, which is then used to convert the larger length sample into estimates of age. Clearly the main efficiency here is that not all fish are aged, thus representing a cost saving in labour time and material costs.

There are two well-known methods for selecting those fish to be aged from the length sample. The first is called 'fixed allocation', where a constant number of fish are aged from each length class. The alternative is 'proportional allocation' where the number in the age sample is proportional to the number of fish in the different length categories. Such proportional allocation of samples results in the self-weighting of data across these length categories and is likely to constitute the best default allocation in any new application (Kimura 1977, Quinn & Deriso 1999).

The application of the age-length key provides an age structure based on the frequencies of age classes in each length category. The estimated variance for each age class comes from the theory of two-stage sampling and involves both a within-length and between-length component of variance (Quinn & Deriso 1999). The numbers of fish that are required for measurement and ageing can be determined statistically in order to achieve a pre-determined level of precision in the age structure.

1.4.3 Accuracy and precision of fish ageing

The process of determining fish age in years from the structure of otoliths involves several possible sources of error that can result in erroneous estimates of fish age. This is true regardless of whether it is a tropical or temperate species and is independent of the size or significance of the ageing study. Such errors, if undetected or uncorrected, would propagate through the subsequent quantitative processes, resulting in incorrect frequency distributions, erroneous estimates of growth rates, mortality rates, yield estimates and the misunderstanding of the potential productivity of the fishery (Lai & Gunderson 1987, McFarlane & Beamish 1995, Eklund et al. 2000). There are two major sources of such error: the first has been called 'process' error and relates directly to the otolith structure being interpreted; the second relates to errors in the interpretation of the otolith structure (Campana 2001, Morison et al. 2005).

As indicated earlier, otoliths must satisfy three criteria to be useful for ageing work (Section 1.2.1, Fowler & Doherty 1992). Otoliths used to estimate fish age when any of the three criteria is not satisfied will result in counts that do not accurately represent the fish age in years. This is the 'process' error identified above. Such errors were common in many studies on temperate fish species done prior to the 1980s based on interpreting either the structure of fish scales or whole otoliths. The errors resulted from using procedures that had not been validated to estimate fish age. A landmark publication in 1983 highlighted the lack of attention that scientists had paid, to that time, to appropriate validation procedures (Beamish & McFarlane 1983). The authors also pointed out that such inattention was likely to have resulted in serious misunderstandings of the biology and population dynamics of the important commercial fish species that were involved, with serious economic consequences.

One advantage of the relatively late start to tropical fish ageing is that researchers have learned from the earlier mistakes and heeded warnings from their temperate fish counterparts to ensure that fish ageing methods are appropriately validated. There are a number of methods that can be used for such validation work that vary in their scientific value, as well as their applicability, advantages and disadvantages (Campana 2001). Tropical fish researchers to date have paid considerable attention to ensuring that their ageing protocols have been validated appropriately (see also Choat et al. in Chapter 2 of this volume).

It is inevitable that some ageing errors related to the preparation and interpretation of the incremental structure of the otoliths will occur even when a validated ageing protocol is used (Morison et al. 2005). The structure of otoliths is complex, displaying a variety of increments and discontinuities (Pannella 1974). The annual macrostructure can vary considerably in clarity and interpretability both amongst individuals from the same population and among populations from different places (Fowler 1995). Reading and interpreting such complex structures, therefore, must itself be a complex process. Correctly interpreting otolith structure is an analytical process that must be learned and for which skill level increases with practise and experience, but also declines with lack of practise (Morison et al. 2005). Not surprisingly, there is considerable variation in the aptitude of different personnel for this task.

Differences in aptitude of fish agers and variation over time in their skill levels should be taken into consideration in ageing studies. This requires the implementation of appropriate quality assurance and control procedures. Many fish ageing laboratories around the world have established quality control protocols that are aimed at detecting and redressing errors in otolith interpretation to minimise the influence of incorrect age estimates on stock assessments (Campana 2001, Kimura & Anderl 2005, Morison et al. 2005). Such quality assurance and control measures are based on monitoring the consistency of the ageing work over time, particularly between years, to ensure that the way the age readers interpret the otoliths does not drift through time. 'Drift' would introduce bias to the estimates relative to those from earlier age determinations. Using appropriate quality measures also helps to ensure that age interpretations are comparable between readers. This process will detect systematic or biased errors, which are of greatest concern because of their consequences for the calculation of age-based parameters that relate to the population dynamics (Campana 2001). Thus, using appropriate quality control monitoring is an important part of the data collection process that underpins stock assessment and fishery management procedures. The appropriate implementation of such procedures will likely provide confidence to the otolith readers, as well as those who use the age data in quantitative processing, and to the fishery managers whose decision-making depends on the outcomes of the stock assessments.

1.4.4 Geographic variation in otolith clarity and interpretability

A further issue for tropical fish otoliths, compared to those from temperate fish, is that there is considerable variation in the clarity and interpretability of the otolith structure. Such variation is apparent amongst individuals within the same population, but is more apparent among populations of the same species from different places (Fowler 1995). There can be considerable differences in the usefulness of otoliths among confamilial and even congeneric species (Fowler 1995). These taxonomic and geographic differences have been demonstrated most clearly for species from the pomacentrid family of coral reef fishes. There were notable differences in the clarity of the incremental macrostructure of otoliths from each of *Pomacentrus moluccensis* and *P. wardi* between locations on the central and southern Great Barrier Reef, which differed in latitude between 19° and 23°S (Fowler & Doherty 1992). Furthermore, the otoliths from the former species were easier to interpret than those from the latter at each location. A similar phenomenon was described for the congeneric species *Stegastes planifrons* and *S. partitus* collected from five locations throughout the tropical western Atlantic, whose latitudes ranged from about 9° to 28°N (Caldow & Wellington 2003). In both examples, the fish collected from the higher latitudes had the clearer otoliths. Those locations with broader temperature ranges had clearer otoliths, but an incremental structure was still apparent in otoliths of fish from locations with a seasonal temperature range of only 3°C in the latter study. This suggests that seasonal water temperature variation is to some extent implicated in otolith increment formation in tropical marine fishes. The situation is more complex than this, however, since there were clarity differences between otoliths of two *Stegastes* species collected from places that experienced similar seasonal temperature regimes (Caldow & Wellington 2003).

There was also considerable geographic variation in the clarity of the macrostructure of the opaque and translucent bands for several species of peacock cichlids sampled from a number of tropical rivers and reservoirs in Venezuela (Jepsen et al. 1999). Otolith clarity was more pronounced for fishes from fluvial ecosystems, whilst those from reservoirs were mostly diffuse and difficult to interpret. The formation of opaque bands in the clear cichlid otoliths corresponded with periods of slow somatic growth that were associated with the heaviest rainfall period of the year.

Thus, there was a correlation between hydrological seasonality and otolith macrostructure. Alternatively, the populations living in the reservoir environment, which is a relatively benign and aseasonal environment compared to that of the rivers, had otoliths that were so diffuse and unclear that their macrostructure could not be interpreted with confidence.

1.4.5 Otolith macrostructure – causes and need for validation

It is now apparent for a range of taxa from a diversity of tropical systems around the world that fish otoliths have proven extraordinarily useful for ageing. Nevertheless, there remains a lack of understanding of the process of otolith growth and the nature of what is perceived optically as an otolith macrostructure. In fact, there remains a general lack of fundamental understanding of the relationship between otolith macrostructure, the physiological processes of the fish and the environmental conditions in which the fish has grown (Wright et al. 2002). This may not necessarily impede the use of otoliths for direct ageing, but it does mean that there must remain a level of caution and some limitations to their use. For example, there remains the requirement to undertake validation studies in new applications of otoliths for ageing work for tropical fish species. This requirement might not be so stringent, however, if our fundamental understanding about otolith macrostructure was more advanced.

There are two issues related to understanding the formation of annuli in fish otoliths: understanding the basis of what is perceived as an otolith macrostructure – i.e., the sequence of bands of alternating optical density; and identifying the processes of physiological control over otolith formation and how this is influenced by exogenous factors (Fowler 1995). There was little understanding of either of these factors in 1995 and it was recommended that this be a key focus for future research. This challenge appears to not have been met in the interim (Wright et al. 2002). Otolith-based studies in the tropics since then have focussed primarily on the use of otoliths as chronometers, with little attention to the nature of otoliths themselves. Few papers have provided comprehensive descriptions of otolith structure and there has been a lack of consideration of the physiology of macrostructure formation.

The studies from the 1980s that related otolith macrostructure to the incremental microstructure were not all in agreement with regard to the nature of the variation in optical density at the macroscopic level. Some suggested that the opaque bands reflected wide daily increments indicative of fast growth in spring and early summer, and the translucent bands related to thin daily increments (Brothers & Mathews 1987, Morales-Nin 1989). Alternatively, a study of a pomacentrid species from Midway Island in the Pacific Ocean suggested that the opaque band reflected a period of slow growth (Hill & Radtke 1988). This latter finding is similar to the more recent findings for the cichlid species for which it has been determined that the opaque bands correspond to periods of slow somatic growth in response to environmental stress (Jepsen et al. 1999, Panfili et al. 2004).

There has been little recent work on tropical species to resolve the ambiguity about the nature of the macrostructure of the otoliths. Some work for a suite of South African species from the Sparidae family from the latitude range of 33–34°S (Mann-Lang & Buxton 1996), however, may be useful here because sparid otoliths share a number of characteristics with those described earlier for tropical percoid fishes. Their otoliths have a glassy appearance and, when sectioned, show an alternating sequence of narrow opaque and broad translucent bands, which occasionally display discontinuities (Mann-Lang & Buxton 1996, Fowler personal observation). The microstructure of the different bands in the otoliths of these species was examined using scanning electron microscopy (Mann-Lang & Buxton 1996). The translucent bands were found to be composed of daily increments each with a wide calcium carbonate incremental phase, which meant that the proteinaceous discontinuities were broadly spaced. Conversely, the opaque bands were characterised by narrow incremental structures, which meant that the proteinaceous discontinuities of daily increments were closely spaced. This finding is intuitively appealing as it sensibly accounts for the optical characteristics of the macrostructure, i.e., that the opaque band is optically dense because it consists of numerous thin increments with closely aligned proteinaceous discontinuities, whilst the translucent band transmits light because of the broad crystalline incremental part of the daily microincrements. This suggests that the opaque bands represent slow otolith growth and the translucent bands relate to periods of fast growth. This finding is somewhat counterintuitive since for 9 of the 12 species of sparids the opaque bands formed in spring–summer and the translucent bands related more to autumn–winter, indicating that otolith growth was faster during the cooler part of the year and slower during the warmer part of the year. The thin opaque band also forms during spring–summer in numerous species of tropical reef fish (Fowler 1995). Maybe for these fish the opaque band also represents a period of slow otolith growth, and the broader translucent band is from the faster growth period of the year. This suggestion is contrary to the findings of early studies that explored this issue (Brothers & Mathews 1987, Morales-Nin 1989, Morales-Nin & Ralston 1990).

The sparid study also developed a model that related fish physiology to otolith growth (Mann-Lang & Buxton 1996). The timing of formation of opaque bands, which generally incorporated spring and summer, corresponded with the time of reproduction for most of the 12 species considered. Thus, the period of slow otolith growth at the warm time of the year may reflect the impact of calcium metabolism in relation to reproductive activity. This implies that otolith growth in mature fish is strongly influenced by endogenous processes, rather than being directly or predominantly driven by exogenous factors such as water temperature. Mann-Lang and Buxton (1996) proposed that water temperature had a stronger influence on otolith growth in immature fish. This model is complicated because it suggests that otolith growth and macrostructure formation are controlled by a combination of endogenous and exogenous factors throughout the lifetime of the fish, whose influence varies at different ages and life history stages. This model remains equivocal because it has been suggested that although reproducing females have elevated plasma calcium concentrations, this takes the form of protein-bound calcium, which might not affect the calcium levels in the endolymph (Wright et al. 2002). Nevertheless, if the model is correct, it suggests that there could be a shift in the timing of formation of opaque bands with the onset of maturity, from temperature driven periods of slow otolith growth in winter to physiologically driven periods of slow growth in spring–summer.

The poor understanding of the formation of otolith macrostructure for tropical fish has consequences for the on-going requirement for validation processes. Two main procedures have been used for validation in the tropical studies since 1990: the treatment of fish with oxytetracycline to form a time marker in the otolith structure; and marginal increment analysis. Both validation procedures can be demanding of time, labour and resources, particularly if validation is to be achieved for the total range of age classes (Beamish & McFarlane 1983), which can be up to approximately 50 years for some tropical species (Section 1.3.1). Because of the poor understanding of the nature of otolith growth and the controlling influences, however, such validation procedures remain necessary in new applications. This may be particularly so for the scombrid species that may display a secondary macrostructure that is formed sub-annually (Lewis & Mackie 2002). It might not remain essential to complete validation for each new ageing application if in the future there is improvement in our understanding of the relationships amongst otolith growth, environmental seasonality and physiological processes.

2 Methodological approaches to fish ageing studies

2.1 HISTORICAL APPROACHES FOR ESTIMATING FISH AGE FROM OTOLITHS

2.1.1 Whole versus sectioned otoliths prior to 1980

Many ageing studies prior to the 1980s, even those related to the huge fisheries in the cold water regions of the world, were based on interpreting the structure of whole otoliths (Beamish 1992). It was finally realised, however, that it may be impossible to obtain accurate estimates of age from the optical characteristics of whole otoliths for the entire size and age range of fish in a population because of the way otoliths grow in three dimensions throughout the lives of the fish. Otoliths generally do not grow consistently in length, breadth and thickness throughout life, with growth in length and breadth slowing and, in many cases, ceasing whilst they continue to accumulate thickness. This results in the 'stacking' of growth increments in the sagittal plane between the otolith core and the proximal surface, which is evident in Figures 1 and 2 (Beamish 1979a, Mann-Lang & Buxton 1996). Such a pattern of otolith growth clearly means that the alternating sequence of opaque and translucent bands cannot be observed when whole otoliths are examined from either the proximal or distal surfaces, but most otoliths are too thick for light to pass through sufficiently to show the incremental structure when examined side-on.

It became apparent that it was possible to determine ages accurately only by exposing the incremental macrostructure within the otoliths of older fish, which required exposing the otolith structure perpendicular to the plane of maximum growth. This generally has been achieved by either: (a) removing a transverse section through the otolith centre using a diamond saw; or (b) breaking the otolith through the centre to expose the transverse face and then burning the otolith to help differentiate the internal bands. Burning highlights the different bands within each annual growth increment because of their relative differences in protein (which discolours with burning) and calcium carbonate (which does not) (Christensen 1964, McFarlane & Beamish 1995, Fowler & Short 1998). It has become apparent from these more recent methods that many ageing studies from the first 80 years of the 20th century, even for some species that supported huge and highly valuable fisheries, produced estimates of age that consistently underestimated the true age of the fish. For example, the sablefish (*Anoplopomia fimbria*) was considered to be a fast-growing, short-lived species with longevity of up to 12 years until early results from broken and burnt otoliths indicated that some fish in fact lived longer than 60 years (Beamish & Chilton 1982, McFarlane & Beamish 1995). Similarly, estimates of age from sectioned otoliths of walleye Pollock (*Theragra chalcogramma*), which supports the largest fishery in the world, indicated that longevity was twice that indicated by counts from whole otoliths (Beamish 1992). Similar differences in age estimates between whole and sectioned otoliths have been identified for other important fishery species, such as the Pacific Ocean perch (*Sebastes alutus*) and Pacific hake (*Merluccius productus*), with some estimates of longevity differing by up to a factor of three (Beamish 1979a,b).

The examples presented above indicate that during the latter part of the 20th Century it was realised that many species of fish had been inaccurately aged from their scales and whole otoliths, because of insufficient attention to validation procedures (Beamish & McFarlane 1983). Such errors have highly significant implications for the estimates of population parameters such as growth and mortality rates, and associated estimates of fishery productivity and sustainable rates of harvest (Beamish & McFarlane 1995). More accurate estimates of age highlighted that many species of fish grew more slowly, lived longer and had lower rates of natural mortality than was originally estimated. It was concluded that these species of fish were far less productive and more vulnerable to overfishing than previously thought.

2.1.2 Age based on otolith measurements

Some studies in the latter part of the 20th Century explored the possibility that otolith size and weight might provide sufficiently accurate and precise estimates of fish age to by-pass the requirement for the visual interpretation of otolith structure. This was considered beneficial as it might provide savings in time and labour costs associated with preparing and interpreting the transverse sections of otoliths. Boehlert (1985) used multiple regression models to integrate data on otolith weight, length and width from individual fish to estimate their age for two species of rockfish, *Sebastes diploproa* and *S. pinniger*, from the west coast of the USA. The models typically accounted for 70–92% of the variability in age, depending on species, sex and method of age analysis.

A study of the pilchard (*Sardinops neopilchardus*) used otolith weight as an indicator of fish age, with the resulting age estimates being used to develop age frequency distributions from which population parameters were calculated (Fletcher 1991, 1995). A similar study done in the tropics assessed the utility of otolith weight to estimate fish age, and identified two sources of error in the estimates of age (Worthington et al. 1995). Firstly, the considerable overlap in otolith weight amongst fish of different age classes resulted in normally distributed errors in the estimates of age from otolith weights. Secondly, there was spatial variation in the otolith weight–age relationships that could bias subsequent age determinations. It was nevertheless concluded that otolith weight could provide an objective and economic method for age determination with the same reliability as counting annuli in sectioned otoliths provided the relationship between otolith weight and age was frequently calibrated.

2.1.3 *Transverse sections of otoliths from tropical species*

The majority of population studies for tropical fish species that have been based on direct ageing have used the transverse sections of otoliths and have enumerated the opaque and translucent bands towards the proximal surface in the vicinity of the sulcus. There appears to have been little consideration paid to the supposed potentially labour-saving methods of estimating fish age from otolith weight or size. Perhaps this preference for using transverse sections reflects that the labour involved in producing such preparations is less onerous than was thought would be the case. It has probably become evident to researchers that the laboratory work involved in preparing and interpreting transverse sections of otoliths is far lower and less demanding of resources than the field collection of the specimens to be aged. Using the preparation methods described in technical Box 2 (later in this chapter), it is quite tractable for a technician to produce several hundred transverse sections of sagittal otoliths within a week. Lou et al. 2004 have made a comparison of efficiencies of otolith processing of fishes from both tropical and temperate marine environments. With the notable exception of the Effects of Line Fishing experiment on the Great Barrier Reef (Mapstone et al. 2004), many tropical fish studies to date have been based on sample sizes of only hundreds rather than thousands of fish and the laboratory time and cost of producing and interpreting the otolith sections have been relatively low.

Researchers generally have been wary of using whole otoliths, due to the temperate fish experience, but several studies have combined the use of transverse sections with reading of whole otoliths. This introduces efficiencies into the protocol as it avoids having to section the otoliths of young fish that are effectively interpreted whole. For example, Ferreira and Russ (1994) first read whole all the otoliths of the common coral trout (*Plectropomus leopardus*) and when the increment count exceeded six or the otolith was considered difficult to interpret, it was sectioned and then reconsidered. For *Lutjanus carponotatus*, the first annulus in sectioned otoliths was often faint and difficult to discern, which meant that the counts deviated between whole and sectioned otoliths for the older age classes. Consequently, whole otoliths were used to age fish that were up to 10 years old, whilst sectioned otoliths were required for the older age classes from 10 to 18 years of age (Kritzer 2002). Williams et al. (2003, 2005) found for *Lethrinus miniatus* that fish of all ages up to the maximum of 19 years could be aged reliably from whole otoliths following a thorough comparison of readings between whole and sectioned otoliths.

2.2 PRACTICAL STEPS IN A FISH AGEING STUDY

There are a number of sequential steps involved in completing a fish ageing study once the validated ageing protocol to provide estimates of fish age has been established. These steps include: (1) collecting the specimens; (2) removing and preparing the otoliths; (3) interpreting the otolith structure to provide the count of the macro-increments that is used in deriving the estimate of fish age; and, finally, (4) completing some quality control assessment to ensure that the age estimates have an acceptably low error rate. Some notes on these steps are provided in the following text.

2.2.1 Collecting samples

The sampling strategy used for accessing samples of fish will be determined by the nature, aims and funds available to the study. There are two broad categories of sampling methods: fishery dependent and fishery independent. Clearly, the first of these relates to sub-sampling the catch of the fishery to obtain biological samples and size information. Such sampling may be through an on-board observer program or subsequently on land through port or market sampling. This approach has the advantage of being relatively cheap but can result in problems in achieving spatial coverage of the target population since fishers choose where to fish based on their knowledge, experience, and habit. Thus, samples may be dominated by those from accessible areas or places where the fish are more abundant, which could introduce biases into the estimates of age-based parameters. The spatial distribution over which the samples are collected and combined will determine the applicability and variability in the estimates of population parameters. A study of several important reef fish species in the artisinal fishery of the Republic of the Seychelles used traditional fish traps, spears and handlines to access the fish samples (Grandcourt 2002). Newman and Dunk (2002) accessed landings from the commercial Northern Demersal Scalefish Fishery of North-western Australia for a study on *Lutjanus sebae*, where samples were collected between July 1995 and December 1999 from latitudes 12º–20ºS using fish traps in water depths of 60–150 m. Begg and Sellin (1998) collected samples of school mackerel (*Scomberomorus queenslandicus*) and spotted mackerel (*S. munroi*) from commercial net fishers and recreational anglers from June 1992 and January 1995, which provided 1,172 otoliths of the former species and 1,385 from the latter.

The alternative approach is to undertake fishery independent sampling, giving the scientist more control over the spatial and temporal aspects of the sampling and allowing standardisation of sampling equipment and techniques. It is possible to obtain further information that may not be possible from landed fishes, such as tissue samples for genetic studies and stomach contents for information about the interactions amongst species. The disadvantages of fishery independent sampling are the expense, particularly if a charter or research vessel is involved, and lower sample sizes than are accessible through the fishery. Fishery independent sampling has been used in numerous ecological studies on the Great Barrier Reef (notably Mapstone et al. 2004), which have aimed to provide a more refined understanding of population characteristics at several spatial scales. In each case, hierarchical sampling regimes were used to target particular species, which provided representative samples of fish from relatively small areas and occasions (Kritzer 2002, Gust et al. 2002, Williams et al. 2003), or over a specific geographic range (Williams et al. 2003, Lou et al. 2005). Such sampling clearly allows for the comparison of population characteristics over a range of spatial scales. Targeted, regular monthly sampling using dropnets, cast nets and line fishing was used to access samples of cichlids in South-eastern United States of America and Western Africa (Faunce et al. 2002, Panfili et al. 2004).

2.2.2 Preparation of otoliths for interpretation

After collection, fish are dissected and their otoliths (most commonly the sagittae) are removed. The otoliths are then cleaned to remove remnant tissue, blood and endolymph by bathing them in alcohol or dilute bleach, or by wiping them on paper towel. They are then dried, stored and labelled. Note that some cleaning procedures may preclude use of the sample in future studies such as otolith chemistry analysis (see Chapter 8 by Thorrold & Swearer in this volume) and population genetic analysis that rely on biological material adhering to archival otolith samples. Otoliths should be weighed after cleaning and prior to further processing if otolith weight might be used as a proxy for age in later work.

There are several methods by which the otoliths can be prepared for examination, depending on the chosen ageing protocol. Some researchers have examined some or all otoliths whole (Ferreira & Russ 1994, Kritzer 2002, Williams et al. 2003, 2005). This technique would now only be considered acceptable if it had been demonstrated previously that whole otoliths provided similar counts to those from sectioned otoliths (Fowler & Short 1998, Williams et al. 2003). Whole otoliths are immersed in a liquid or oil to clear them so that they transmit light, which more clearly exposes their internal structure (Morales-Nin & Panfili 2002b). Such liquids and oils include water, alcohol, glycerine, immersion oil, clove oil or aniseed oil. The otoliths are then examined using either transmitted light through a transparent container or reflected light against a black background.

It has been necessary in most studies of longer-lived tropical fish species to expose the otolith structure perpendicular to the plane of the growth layers to provide accurate counts of the annuli. There are two approaches to achieve this: by 'breaking and burning' the otoliths; or by preparing transverse sections through the otolith centres. These methods are broadly described here and detailed instructions are provided in Boxes 1 and 2 respectively.

In the 'breaking and burning' process (Box 1), the otolith is snapped in two, perpendicular to the long axis through the centre, thus providing two halves each with a freshly exposed transverse face. One half is then slowly heated in a flame, which gradually cooks the half otolith, turning it brown as the protein is denatured, and accentuating the difference between the opaque and translucent bands. The cooked half is then mounted so that the transverse face can be smeared with immersion oil and examined using reflected light and low power microscopy.

Box 1. Recipe for breaking and burning an otolith

1. Dissect fish and remove otoliths.
2. Clean otoliths.
3. Weigh one otolith.
4. Examine the otolith using a dissecting microscope and transmitted light – examine the distal surface with the otolith resting on its proximal surface.
5. Locate the core region of the otolith and mark with a single line across the otolith between the two edges through the centre, using a sharp pencil.
6. Orientate the otolith with the distal surface facing upwards, push downwards on the two ends of the otolith using the thumb and index finger of one hand, whilst simultaneously scoring the otolith across the pencil line with a scalpel. This should cause the otolith to snap into two pieces approximately through the centre to expose the transverse faces of both halves.
7. Hold one half of the otolith with a pair of forceps, heat it slowly in an alcohol flame or the blue flame of a Bunsen burner. Orientate the otolith piece so that its bottom end receives the heat of the flame and the otolith gradually cooks and turns brown from the end towards the centre. It might be necessary to repeatedly move the otolith through the flame for a number of seconds to control the rate of heating.
8. Allow the cooked otolith half to cool and then mount it in plasticine with the transverse face orientated upwards and ready for examination.
9. Smear the transverse face with a little immersion oil, illuminate with light from above, examine the surface and interpret the alternating sequence of thin, whitish, opaque bands and broader, brown translucent bands.

The preparation of transverse sections of otoliths (Box 2) is the most used method in studies ageing tropical percoid fishes to date, regardless of their taxonomic classification or the environment they inhabited. There have been many methods developed for preparing transverse sections but there are several fundamental requirements that must be taken into consideration. The section must be cut at the appropriate angle, ideally on the plane normal to the growing surface, to achieve maximum discrimination between the different bands. The section must also be sufficiently thin to allow enough light to pass through to highlight the optical characteristics of the bands, whilst minimising artefacts that relate to viewing too thick a section. There must also be appropriate preparation of the surface of the section to minimise the interference effects of scratches that may result from the cutting process that would affect the resolution of the internal structure of the otolith.

The clean, dry otoliths should be embedded in resin prior to being sectioned, to protect them during the sectioning process. The resin block encasing the otolith is then cut using a low speed diamond saw to produce sections of 100–500 μm thick, depending on the species. The section might then be polished using either diamond paste or lapping film, and smeared with immersion oil to clear any remaining surface scratches. Alternatively, appropriate gluing of the section to a glass slide may remove any need for polishing. The section can then be examined under low power microscopy using either transmitted or reflected light. Histological staining, although not often needed, may improve contrast between opaque and translucent bands in the section, which apparently reveals similar chromophilic increments to the burning process

(McCurdy et al. 2002). Good results have been obtained using a sequence of acid etching followed by staining with aniline blue and toluidine blue (Richter & McDermott 1990), whilst the use of just toluidine blue after acid etching has also been successful (McCurdy et al. 2002).

Box 2. Protocol for preparation of a transverse section of an otolith

1. Dissect fish and remove otoliths.
2. Clean and weigh the otoliths.
3. Embed one otolith in resin using an appropriately sized mould so that the whole structure is encased in resin. First set some resin in the mould, then place the otolith on the cured resin before topping up to completely encase it. It is possible to embed a number of samples per resin block, but special care must be taken so that otoliths are aligned for simultaneous cutting, and to preserve sample identity.
4. Examine the resin block containing the otolith under a dissecting microscope illuminated with transmitted light. Score the block with a scalpel to identify where to cut the block to provide the section of desired thickness.
5. Securely clamp the resin block containing the otolith in the cutting saw jig and align the block with the saw to produce the desired cut.
6. Use a diamond saw to remove one end of the resin block and otolith, thus exposing the transverse face of the otolith.
7. Make a second cut to remove the slice of otolith of the desired thickness. Several slices of the otolith can be removed in this way to obtain multiple sections. This is particularly advisable when multiple otoliths have been set in the same block because of the difficulties in aligning multiple otoliths. Note that if a double-bladed saw is being used then steps 6 & 7 represent a single step.
8. Retrieve the slice(s) of resin that incorporates the desired otolith section(s), dry, and mount on a glass slide by carefully lowering onto a drop of liquid glue such as pre-heated Crystalbond. Since this glue has approximately the same refractive properties as the resin, any scratches in the cut surface will be filled and there will be no requirement for time-consuming polishing.
9. Examine the section and check for appropriate thickness based on the clarity of the increment structure. If the section is too think it can be ground using wet and dry paper and polished with lapping film.
10. Smear the otolith surface with immersion oil and then examine under a dissecting microscope using either reflected or transmitted light.

2.2.3 Interpretation of otolith macrostructure to estimate fish age

The whole otolith, broken and burnt half otolith or the transverse section is then examined under low power microscopy. The structure is illuminated either by transmitted or reflected light. Transmitted light is directed from below through the slide or container and through the otolith. Reflected light is provided from a cold light source, with the light directed onto the otolith plane from above and oblique to the otolith. The lighting affects the appearance of the incremental structure and thus its interpretability, and some experimentation is required to achieve the optimal arrangement that maximises the contrast between the opaque and translucent bands. The opaque bands are more optically dense than the translucent ones and thus more effectively block the transmittance of light. Therefore, under illumination by transmitted light the opaque bands appear dark, whilst the translucent ones are light. Alternatively, under reflected light the opaque bands appear bright whilst the translucent bands are dull (see also Figure 1 in Chapter 1 of this volume).

The otolith structure is then interpreted by counting either the opaque or translucent bands. This is the difficult interpretative process that was discussed in Section 1.4.3, which the age reader must learn and practise to achieve proficiency. The interpretation of the structure is complicated by such difficulties as: the need to identify the first increment; recognising annuli that may vary in optical density across the otolith; distinguishing true annuli from false checks and discontinuities; and distinguishing the type of increment on the otolith edge and the relative width of this marginal increment.

The resulting count of the number of annuli does not necessarily translate into a direct estimate of fish age. Rather, this count must then be interpreted using other information to provide the best estimate of the fish's age. This further required information is rather complex and depends on having at least a basic understanding of the life history of the species and the growth pattern of the otolith. It is necessary to know the reproductive season of the species as this is used to nominate the standard birth date to be assigned to every fish (Morison et al. 2005). It is also necessary to know the time of year when the opaque and translucent bands form in the otolith. Finally, it is also necessary to know the time of year that the fish was captured. Then, the number of annuli in the otolith, the standard birth date, the time of year of capture and the interpretation of the marginal increment, with respect to the timing of increment formation, can be combined to provide the estimate of fish age (Francis et al. 1992, Fowler & Short 1998, Morales-Nin & Panfili 2002a).

2.2.4 Quality control and precision of age estimates

Clearly, attaining accurate estimates of fish age depends on achieving accurate counts of the annuli in the otoliths. As discussed in Sections 1.4.3 and 2.2.3, however, the interpretation process is complex, involving decisions on interpretation where many factors can influence the judgement call of the reader (Campana 2001, Morison et al. 2005). Ultimately, it is inevitable that there will be some errors in the otolith interpretation that lead to incorrect estimates of age. Thus, the implementation of an appropriate quality control and assurance program is important to detect inconsistencies in the ageing process and ageing errors and to ensure consistent and accurate interpretation of otolith structure. Such a protocol would assess both the relative accuracy of the counts and their precision. Whilst identifying ageing errors is important in any ageing study, such quality control monitoring is most relevant in high-volume, production ageing where consistency in otolith interpretation from year-to-year is essential (Campana 2001).

There is considerable variation in quality assurance and control processes amongst ageing labs around the world (Morison et al. 2005). They differ with respect to: whether reference collections of otoliths are retained and how such collections are used; whether the practise of assigning common birth dates for particular species is used; the number of otoliths that should be read to appropriately train an otolith reader for a new species; the level of information about a fish that is provided to the reader when the otolith is being interpreted; and the method used for comparing repeated age estimates from the same sample of fish (Morison et al. 2005). Some laboratories have a long history of quality control and have established sophisticated, computerised routines for estimating precision and identifying outliers in the age estimates (Kimura & Anderl 2005). The appropriate use of well-designed databases as well as digital images of otoliths from image analysis systems can contribute significantly to effective quality assurance and control in age estimation (Campana 2001, Morison et al. 2005).

Campana (2001) proposed a generic quality control protocol that is recognised as having considerable merit (Morison et al. 2005). The first step in this protocol is to develop a reference collection of otoliths for the species of interest. Ideally, this collection would involve otoliths from known-age fish or by second choice some for which there is agreement on the ages by expert readers. This reference collection provides the stable base for assessing the performance of fish agers over time. The recommended way to do this is to take a random, blindly-labelled sub-sample of otoliths from the reference collection, inter-mix these otoliths with a sub-sample of recently aged ones, and read this mixed set of otoliths. Then, the new counts for the otoliths from the reference collection can be compared and assessed against the original accepted ages. This is done using age bias graphs to test for ageing consistency and estimates of coefficient of variation (CV) to quantitatively describe the precision of the estimates (Campana et al. 1995, Campana 2001). Accurate counts from the otoliths in the reference collection will strongly suggest that counts from the recently aged sample of fish will also be accurate. Alternatively, evidence of bias between the earlier and recent counts for otoliths from the reference collection would suggest that the recently aged fish will be biased, indicating that some remedial action is required to address those errors. This may require some retraining for the otolith reader and possibly re-reading all otoliths from the recent sample once the reader was (re)trained sufficiently to be unbiased.

The discussion above has been concerned primarily with minimising ageing errors to ensure that estimates of age are as accurate as possible. There is also some value in considering the precision or repeatability of age estimates. Measuring precision provides a means of assessing the relative ease of determining the age from the otoliths, for assessing the reproducibility of an individual reader's determinations and comparing the counts between different agers. The percent agreement statistic has been used often as a measure of precision in such applications, but it is not particularly robust. Alternatively, both the index of average percent error (Beamish & Fournier 1981) and coefficient of variation (CV=SD/mean) (Campana et al. 1995) have been demonstrated to be robust measures of precision. Furthermore, it has been shown that they are functionally equivalent and it is possible to convert between the two measures (Campana 2001). These measures can be used to estimate precision at different hierarchical levels, e.g., between counts by an individual reader, between readers, and between laboratories where multiple readers have been used at each lab.

The equations presented by Beamish and Fournier (1981) to calculate average percent error are as follows:

$$X_j = \frac{1}{R}\sum_{i=1}^{R} X_{ij}$$

where X_j is the average age calculated for the jth fish, X_{ij} is the ith age determination of the jth fish and R is the number of times that each fish is aged;

$$\frac{1}{R}\sum_{i=1}^{R}\frac{|X_{ij} - X_j|}{X_j}$$

is the average error in ageing the jth fish; and

$$\frac{1}{N}\sum_{j=1}^{N}\left[\frac{1}{R}\sum_{i=1}^{R}\frac{|X_{ij} - X_j|}{X_j}\right]$$

is the index of average error which, when multiplied by 100, becomes the average percent error.

3 Conclusion

The dogma about the inability to age tropical fish from their otoliths was rejected in the late 1970s and early 1980s. Discovery of the usefulness of otoliths for ageing tropical fish meant that it was possible, for the first time, to collect samples of fish from tropical species and to use their otoliths to retrospectively reconstruct how their populations worked by providing insights into their demography, life history and population dynamics. There was a considerable period, however, before any real momentum developed in establishing new research programs based on ageing tropical fish from their otoliths. In fact, it wasn't until the 1990s and early 2000s that there was a significant increase in such studies.

The findings from the early ageing studies were profound. Many tropical species were found to be remarkably long-lived, contrary to previous belief that tropical fish species were short-lived. Mortality rates were found to be lower than expected, leading to the rejection of the paradigm about tropical species having fast turnover rates. Some species demonstrated growth patterns where initial growth was fast over a relatively short period, after which there was no further growth throughout the remainder of the relatively long life. Clearly, such new knowledge revolutionised the understanding of the functionality of these populations. Otolith ageing has now been established as a fundamental and essential technique for tropical fish scientists that will continue to provide invaluable insights into the biology of species at particular places and times. The on-going development of tropical fisheries will rely on the data from such studies, whilst the process of ecosystem management will also benefit.

The call for fundamental research into the nature of tropical fish otoliths that was made by Fowler (1995) is repeated here. It is true that otoliths can be used in ageing applications, so long as appropriate validation procedures have been completed, despite the lack of understanding about the relationship between otolith macrostructure, fish physiology and environmental seasonality. It is difficult, however, to ignore the implications of this poor understanding. There may be further valuable life history information in the otoliths that is currently being missed because of such poor understanding. There is a need to explore the relationships between otolith growth at the microscopic level, the calcium metabolism of the fish, particularly in relation to reproductive activity, and somatic growth, all of which are influenced by environmental seasonality. The scientific challenge here is significant, but the rewards for assisting in otolith interpretation in future studies may make facing the challenge worthwhile.

Acknowledgements

I wish to thank several anonymous reviewers whose comments significantly helped to improve the first draft of this chapter.

References

Adams S, Williams AJ (2001) A preliminary test of the transitional growth spurt hypothesis using the protogynous coral trout *Plectropomus maculatus*. J Fish Biol 59:183–185

Beamish RJ (1979a) Differences in the age of Pacific Hake (*Merluccius productus*) using whole otoliths and sections of otoliths. J Fish Res Board Can 36:141–151

Beamish RJ (1979b) New information on the longevity of Pacific Ocean perch (*Sebastes alutus*). J Fish Res Board Can 36:1395–1400

Beamish RJ (1992) The importance of accurate ages in fisheries science. In: Hancock DA (Ed) Proceedings of the Australian Society for Fish Biology workshop on the measurement of age and growth in fish and shellfish No. 12. Bureau of Rural Resources, Australian Government Publishing Service, Canberra, Australia pp 8–22

Beamish RJ, Chilton DE (1982) A preliminary evaluation of a method to determine the age of sablefish (*Anoplopoma fimbria*). Can J Fish Aquat Sci 39:277–287

Beamish RJ, Fournier DA (1981) A method for comparing the precision of a set of age determinations. Can J Fish Aquat Sci 38:982–983

Beamish RJ, McFarlane GA (1983) The forgotten requirement for age validation in fisheries biology. Trans Am Fish Soc 112:735–743

Beamish RJ, McFarlane GA (1995) A discussion of the importance of ageing errors, and an application to walleye Pollock: the world's largest fishery. In: Secor DH, Dean JM, Campana SE (Eds) Recent developments in fish otolith research. University of South Carolina Press, Columbia, pp 545–565

Beckman DW, Wilson CA (1995) Seasonal timing of opaque zone formation in fish otoliths. In: Secor DH, Dean JM, Campana SE (Eds) Recent developments in fish otolith research. University of South Carolina Press, Columbia, pp 27–43

Begg GA, Campana SE, Fowler AJ, Suthers IM (2005) Otolith research and application: current directions in innovation and implementation. Mar Freshwater Res 56:477–483

Begg GA, Sellin MJ (1998) Age and growth of school mackerel (*Scomberomorus queenslandicus*) and spotted mackerel (*S. munroi*) in Queensland east-coast waters with implications for stock structure. Mar Freshwater Res 49:109–120

Blaber SJM (2000) Tropical estuarine fishes. Ecology, exploitation and conservation. Blackwell Science Ltd, Carlton Victoria, Australia

Blacker RW (1974) The ICNAF cod otolith exchange scheme. In: Bagenal TB (Ed) The ageing of fish. Unwin Brothers, Old Woking, England, pp 108–113

Boehlert GW (1985) Using objective criteria and multiple regression models for age determination in fishes. Fish Bull 83:103–117

Brothers EB (1987) Methodological approaches to the examination of otoliths in aging studies. In: Summerfelt RC, Hall GE (Eds) The age and growth of fish. The Iowa State University Press, Ames, Iowa, pp 319–330

Brothers EB, Mathews CP (1987) Application of otolith microstructural studies to age determination of some commercially valuable fish of the Arabian Gulf. Kuwait Bull Mar Sci 9:127–157

Brown IW, Sumpton WD (1998) Age, growth and mortality of red throat emperor *Lethrinus miniatus* (Pisces: Lethrinidae) from the southern Great Barrier Reef, Queensland, Australia. Bull Mar Sci 62:905–917

Buckworth RC (1998) Age structure of the commercial catch of Northern Territory narrow-barred Spanish mackerel. Final report to FRDC for project T94/015, Fisheries Research and Development Corporation, Canberra, Australia

Burton ML (2002) Age, growth and mortality of mutton snapper, *Lutjanus analis*, from the east coast of Florida, with a brief discussion of management implications. Fish Res 59:31–41

Caldow C, Wellington GM (2003) Patterns of annual increment formation in otoliths of pomacentrids in the tropical western Atlantic: implications for population age-structure examination. Mar Ecol Prog Ser 265:185–195

Campana SE (1999) Chemistry and composition of fish otoliths: pathways, mechanisms and applications. Mar Ecol Prog Ser 188:263–297

Campana SE (2001) Accuracy, precision, and quality control in age determination, including a review of the use and abuse of age validation methods. J Fish Biol 59:197–242

Campana SE (2005) Otolith science entering the 21st Century. Mar Freshwater Res 56:485–495

Campana SE, Annand MC, McMillan JI (1995) Graphical and statistical methods for determining the consistency of age determinations. Tran Am Fish Soc 124:131–138

Campana SE, Neilson JD (1985) Microstructure of fish otoliths. Can J Fish Aquat Sci 42:1014–1032

Campana SE, Thorrold SR (2001) Otoliths, increments, and elements: keys to a comprehensive understanding of fish populations. Can J Fish Aquat Sci 58:30–38

Cappo M, Eden P, Newman SJ, Robertson S (2000) A new approach to validation of periodicity and timing of opaque zone formation in the otoliths of eleven species of *Lutjanus* from the central Great Barrier Reef. Fish Bull 98:474–488

Choat JH, Axe LM (1996) Growth and longevity in acanthurid fishes: an analysis of otolith increments. Mar Ecol Prog Ser 134:15–26

Choat JH, Axe LM, Lou DC (1996) Growth and longevity in fishes of the family Scaridae. Mar Ecol Prog Ser 145:33–41

Choat JH, Robertson DR (2002) Age-based studies. In: Sale PF (Ed) Coral reef fishes – dynamics and diversity in a complex ecosystem. Academic Press, San Diego California, pp 57–88

Choat JH, Robertson DR, Ackerman JL, Posada JM (2003) An age-based demographic analysis of the Caribbean stoplight parrotfish *Sparisoma viride*. Mar Ecol Prog Ser 246:265–277

Christensen JM (1964) Burning of otoliths, a technique for age determination of soles and other fish. J Cons Perm Int Explor Mer 29:73–81

Crabtree RE, Bullock LH (1998) Age, growth, and reproduction of black grouper, *Mycteroperca bonaci*, in Florida waters. Fush Bull 96:735–753

Craig PC, Choat JH, Axe LM, Saucerman S (1997) Population biology and harvest of the coral reef surgeonfish *Acanthurus lineatus* in American Samoa. Fish Bull 95:680–693

De Vries DA, Grimes CB (1997) Spatial and temporal variation in age and growth of king mackerel, *Scomberomorus cavalla*, 1977–1992. Fish Bull 95:694–708

Doherty PJ, Fowler AJ (1994a) An empirical test of recruitment limitation in a coral reef fish. Science 263:935–939

Doherty PJ, Fowler AJ (1994b) Demographic consequences of variable recruitment to coral reef fish populations: a congeneric comparison of two damselfishes. Bull Mar Sci 54:297–313

Doherty PJ, Williams DMcB (1988) The replenishment of coral reef fish populations. Ocean Mar Biol Ann Rev 26:487–551

Eklund J, Parmanne R, Aneer G (2000) Between-reader variation in herring otolith ages and effects on estimated population parameters. Fish Res 46:147–154

Faunce CH, Patterson HM, Lorenz JJ (2002) Age, growth, and mortality of the Mayan cichlid (*Cichlasoma urophthalmus*) from the southeast Everglades. Fish Bull 100:42–50

Ferreira BP, Russ GR (1992) Age, growth and mortality of the inshore coral trout *Plectropomus maculatus* (Pisces: Serranidae) from the Central Great Barrier Reef, Australia. Aust J Mar Freshwater Res 43:1301–1312

Ferreira BP, Russ GR (1994) Age validation and estimation of growth rate of the coral trout, *Plectropomus leopardus*, (Lacepede 1802) from Lizard Island, Northern Great Barrier Reef. Fish Bull 92:46–57

Fletcher WJ (1991) A test of the relationship between otolith weight and age for the pilchard *Sardinops neopilchardus*. Can J Fish Aquat Sci 48:35–38

Fletcher WJ (1995) Application of otolith weight-age relationship for the pilchard, *Sardinops sagax neopilchardus*. Can J Fish Aquat Sci 52:657–664

Fowler AJ (1990) Validation of annual growth increments in the otoliths of a small coral reef fish. Mar Ecol Prog Ser 64:39–53

Fowler AJ (1995) Annulus formation in otoliths of coral reef fish – a review. In: Secor DH, Dean JM, Campana SE (Eds) Recent developments in fish otolith research. University of South Carolina Press, Columbia, pp 45–63

Fowler AJ, Doherty PJ (1992) Validation of annual growth increments in the otoliths of two damselfishes from the southern Great Barrier Reef. Aust J Mar Freshwater Res 43:1057–1068

Fowler AJ, Short DA (1998) Validation of age determination from otoliths of the King George whiting *Sillaginodes punctata* (Perciformes). Mar Biol 130:577–587

Francis RICC, Paul LJ, Mulligan KP (1992) Ageing of adult snapper (*Pagrus auratus*) from otolith ring counts: validation by tagging and oxytetracycline injection. Aust J Mar Freshwater Res 43:1069–1089

Gjosaeter J, Dayaratne P, Bergstad OA, Gjosaeter H, Sousa MI, Beck IM (1984) Ageing tropical fish by growth rings in the otoliths. FAO Fish Circ 776:54

Grandcourt EM (2002) Demographic characteristics of a selection of exploited reef fish from the Seychelles: preliminary study. Mar Freshwater Res 53:123–130

Gust N, Choat JH, Ackerman JL (2002) Demographic plasticity in tropical reef fishes. Mar Biol 140:1039–1051

Hilborn R, Walters CJ (1992) Quantitative fisheries stock assessment. Choice, dynamics and uncertainty. Chapman and Hall, New York

Hill KT, Radtke RL (1988) Gerontological studies of the damselfish *Dascyllus albisella*. Bull Mar Sci 42:424–434

Hoedt FE (2002) Growth in eight species of tropical anchovy determined from primary otolith increments. Mar Freshwater Res 53:859–867

Jepsen DB, Winemiller KO, Taphorn DC, Rodriguez Olarte D (1999) Age structure and growth of peacock cichlids from rivers and reservoirs of Venezuela. J Fish Biol 55:433–450

Kimura DK (1977) Statistical assessment of the age-length key. J Fish Res Board Can 34:317–324

Kimura DK, Anderl DM (2005) Quality control of age data at the Alaska Fisheries Centre. Mar Freshwater Res 56:783–789

King M (1995) Fisheries biology, assessment and management. Blackwell Science Ltd, Oxford

Kritzer JP (2002) Variation in the population biology of stripey bass *Lutjanus carponotatus* within and between two island groups on the Great Barrier Reef. Mar Ecol Prog Ser 243:191–207

Kritzer JP, Davies CR, Mapstone BD (2001) Characterizing fish populations: effects of sample size and population structure on the precision of demographic parameter estimates. Can J Fish Aquat Sci 58:1557–1568

Lai HL, Gunderson DR (1987) Effects of ageing errors on estimates of growth, mortality and yield per recruit for walleye Pollock (*Theragra chalcogramma*). Fish Res 5:287–302

Lewis PD, Mackie M (2002) Methods in the collection, preparation and interpretation of narrow-barred Spanish mackerel (*Scomberomorus commerson*) otoliths for a study of age and growth in Western Australia. Fisheries Research Report No 143, Department of Fisheries, Western Australia, 23p

Longhurst AR, Pauly D (1987) Ecology of tropical oceans. Academic Press Inc, London, England

Lou D, Begg GA, Petursdottir G, Mapstone BD (2004) A comparative study of otolith processing for tropical and temperate marine fishes. Third International Symposium of Fish Otolith Research and Application, Townsville, QLD

Lou DC, Mapstone BD, Russ GR, Davies CR, Begg GA (2005) Using otolith weight-age relationships to predict age-based metrics of coral reef fish populations at different spatial scales. Fish Res 71:279–294

Loubens G (1978) Biologie de quelque especes de poisson du lagon Neo-Caledonien. I. Determination de l'age (otolithometrie). Cahiers ORSTOM, Serie Oceanographie 16:263–283

Mackie M, Lewis PD, Gaughan DJ, Buckworth R (2003) Stock assessment of the narrow-barred Spanish mackerel (*Scomberomorus commerson*) in Western Australia. Final Report to FRDC for project 1999/151, Fisheries Research and Development Corporation, Canberra, Australia

Mann-Lang JB, Buxton CD (1996) Growth characteristics in the otoliths of selected South African sparid fish. S Afr J Mar Sci 17:205–216

Manooch CS (1987) Age and growth of snappers and groupers. In: Polovina JJ, Ralston S (Eds) Tropical snappers and groupers: biology and fisheries management. Westview Press Inc, Boulder Colorado, pp 329–363

Mapstone BD, Davies CR, Little LR, Punt AE, Smith ADM, Pantus F, Lou DC, Williams AJ, Jones A, Ayling AM, Russ GR, McDonald AD (2004) The effects of line fishing on the Great Barrier Reef and evaluations of alternative potential management strategies. CRC Reef Research Centre Technical Report No 52, CRC Reef Research Centre, Townsville, Australia

Mapstone BD, Fowler AJ (1988) Recruitment and the structure of assemblages of fish on coral reefs. Trend Ecol Evol 3:72–76

Marriott R (2002) Age, growth and sex structure of red sea bass populations on the Great Barrier Reef. In: Williams AJ, Welch DJ, Muldoon G, Marriott R, Kritzer JP, Adams S (Eds) Bridging the gap: a workshop linking student research with fisheries stakeholders. CRC Reef Research Centre Technical Report No. 48, CRC Reef Research Centre, Townsville, pp 85–93

McCurdy WJ, Panfili J, Meunier FJ, Geffen AJ, de Pontual H (2002) Preparation and observation techniques. Preparation of calcified structures. In: Panfili J, de Pontual H, Troadec H, Wright PJ (Eds) Manual of fish sclerochronology. Ifremer-IRD coedition, Brest, France, pp 331–357

McFarlane GA, Beamish RJ (1995) Validation of the otolith cross-section method of age determination for sablefish (*Anoplopoma fimbria*) using oxytetracycline. In: Secor DH, Dean JM, Campana SE (Eds) Recent developments in fish otolith research. University of South Carolina Press, Columbia, pp 319–329

McPherson GR (1992) Age and growth of the narrow-barred Spanish mackerel (*Scomberomorus commerson* Lacepede, 1800) in north–eastern Queensland waters. Aust J Mar Freshwater Res 43:1269–1282

Meekan MG, Ackerman JL, Wellington GM (2001) Demography and age structures of coral reef damselfishes in the tropical eastern Pacific Ocean. Mar Ecol Prog Ser 212:223–232

Milton DA, Blaber SJM, Rawlinson NJF (1993) Age and growth of three species of clupeids from Kiribati, tropical central south Pacific. J Fish Biol 43:89–108

Morales-Nin B (1989) Growth determination of tropical marine fishes by means of otolith interpretation and length frequency analysis. Aq Living Res 2:241–253

Morales-Nin B (2000) Review of the growth regulation processes of otolith daily increment formation. Fish Res 46:53–67

Morales-Nin B, Panfili J (2002a) Sclerochronological studies. Age estimation. In: Panfili J, de Pontual H, Troadec H, Wright PJ (Eds) Manual of fish sclerochronology. Ifremer-IRD coedition, Brest, France, pp 91–97

Morales-Nin B, Panfili J (2002b) Preparation and observation techniques. Observation. In: Panfili J, de Pontual H, Troadec H, Wright PJ (Eds) Manual of fish sclerochronology. Ifremer-IRD coedition, Brest, France, pp 358–369

Morales-Nin B, Panfili J (2005) Seasonality in the deep sea and tropics revisited: what can otoliths tell us? Mar Freshwater Res 56:585–598

Morales-Nin B, Ralston S (1990) Age and growth of *Lutjanus kasmira* (Forskal) in Hawaiian waters. J Fish Biol 36:191–203

Morison AK, Burnett J, McCurdy WJ, Moksness E (2005) Quality issues in the use of otoliths for fish age estimation. Mar Freshwater Res 56:773–782

Mosse JW, Adams S, Welch DJ (2002) Bommie cod, (*Cephalopholis cyanostigma*): a big surprise from a little fish. In: Williams AJ, Welch DJ, Muldoon G, Marriott R, Kritzer JP, Adams S (Eds) Bridging the gap: a workshop linking student research with fisheries stakeholders. CRC Reef Research Centre Technical Report No. 48, CRC Reef Research Centre, Townsville, pp 94–107

Munro JL, Williams DMcB (1985) Assessment and management of coral reef fisheries: biological, environmental and socio-economic aspects. Proceedings 5th International Coral Reef Congress, Tahiti 4:544–581

Newman SJ (2002) Growth rate, age determination, natural mortality and production potential of the scarlet seaperch, *Lutjanus malabaricus* Schneider 1801, off the Pilbara coast of north-western Australia. Fish Res 58:215–225

Newman SJ, Cappo M, Williams DMcB (2000a) Age, growth, mortality rates and corresponding yield estimates using otoliths of the tropical red snappers, *Lutjanus erythropterus, L. malabaricus and L. sebae*, from the central Great Barrier Reef. Fish Res 48:1–14

Newman SJ, Cappo M, Williams DMcB (2000b) Age, growth, and mortality of the stripey, *Lutjanus carponotatus* (Richardson) and the brown-stripe snapper, *L. vitta* (Quoy and Gaimard) from the central Great Barrier Reef, Australia. Fish Res 48:1–14

Newman SJ, Dunk IJ (2002) Growth, age validation, mortality and other population characteristics of the red emperor snapper, *Lutjanus sebae* (Cuvier, 1828), off the Kimberley coast of north-western Australia. Est Coast Shelf Sci 55:67–80

Newman SJ, Dunk IJ (2003) Age validation, growth, and additional population parameters of the goldband snapper (*Pristipomoides multidens*) off the Kimberley coast of northwestern Australia. Fish Bull 101:116–128

Newman SJ, Williams DMcB, Russ GR (1996) Age validation, growth and mortality rates of the tropical snappers (Pisces: Lutjanidae) *Lutjanus adetii* (Catelnau, 1873) and *L. quinquelineatus* (Bloch, 1790) from the central Great Barrier Reef, Australia. Mar Freshwater Res 47:575–584

Panfili J, Mbow A, Durand J, Diop K, Diouf K, Thior D, Ndiaye P, Lae R (2004) Influence of salinity on the life-history traits of the West African black-chinned tilapia (*Sarotherodon melanotheron*). comparison between the Gambia and Saloum estuaries. Aquat Living Res 17:65–74

Pannella G (1974) Otolith growth patterns: an aid in age determination in temperate and tropical fishes. In: Bagenal TB (Ed) The ageing of fish. Unwin Brothers Ltd, London, pp 28–39

Patterson WF, Cowan JH, Wilson CA, Shipp RL (2001) Age and growth of red snapper, *Lutjanus campechanus*, from an artificial reef area off Alabama in the northern Gulf of Mexico. Fish Bull 99:617–627

Pilling GM, Millner RS, Easey MW, Mees CC, Rathacharen S, Azemia R (2000) Validation of annual growth increments in the otoliths of the lethrinid *Lethrinus mahsena* and the lutjanid *Aprion virescens* from sites in the tropical Indian Ocean, with notes on the nature of growth increments in *Pristipomoides filamentosus*. Fish Bull 98:600–611

Popper AN, Ramcharitar J, Campana SE (2005) Why otoliths? Insights from inner ear physiology and fisheries biology. Mar Freshwater Res 56:97–504

Quinn TJ, Deriso RB (1999) Quantitative fish dynamics. Oxford University Press Inc, New York

Richter H, McDermott JG (1990) The staining of fish otoliths for age determination. J Fish Biol 36:773–779

Russ GR (2002) Yet another review of marine reserves as reef fishery management tools. In: Sale PF (Ed) Coral reef fishes – dynamics and diversity in a complex ecosystem. Academic Press, San Diego, California, pp 421–443

Russ GR, Lou DC, Ferreira BP (1996) Temporal tracking of a strong cohort in the population of a coral reef fish, the coral trout, *Plectropomus leopardus* (Serranidae: Epinephelinae), in the central Great Barrier Reef, Australia. Can J Fish Aquat Sci 53:2745–2751

Russ GR, Lou DC, Higgs JB, Ferreira BP (1998) Mortality rate of a cohort of the coral trout, *Plectropomus leopardus*, in zones of the Great Barrier Reef Marine Park closed to fishing. Mar Freshwater Res 49:507–511

Sale PF (1980) The ecology of fishes on coral reefs. Oceanogr Mar Biol Ann Rev 18:367–421

Samuel M, Mathews CP, Bawazeer AS (1987) Age and validation of age from otoliths for warm water fishes from the Arabian Gulf. In: Summerfelt RC, Hall GE (Eds) The age and growth of fish. The Iowa State University Press, Ames, Iowa, pp 253–267

Schmidt DJ, Collins MR, Wyanski DM (1993) Age, growth, maturity and spawning of Spanish mackerel, *Scomberomorus maculatus* (Mitchill) from the Atlantic coast of the south eastern United States. Fish Bull 91:526–533

Schwamborn SHL, Ferreira BP (2002) Age structure and growth of the dusky damselfish, *Stegastes fuscus*, from Tamandare reefs, Pernambuco, Brazil. Env Biol Fish 63:79–88

Secor DH, Dean JM, Campana SE (1995) Recent developments in fish otolith research. University of South Carolina Press, Columbia

Struhsaker P, Uchiyama JH (1976) Age and growth of the nehu, *Stolephorus purpureus* (Pisces: Engraulidae), from the Hawaiian Islands as indicated by daily growth increments of sagittae. Fish Bull 74:9–17

Thorrold SR, Hare JA (2002) Otolith applications in reef fish ecology. In: Sale PF (Ed) Coral reef fishes – dynamics and diversity in a complex ecosystem. Academic Press, San Diego, California, pp 243–264

Tobin A, Mapleston A (2004) Exploitation dynamics and biological characteristics of the Queensland east coast Spanish mackerel (*Scomberomorus commerson*) fishery. CRC Reef Research Centre Technical Report No. 51, CRC Reef Research Centre, Townsville

Williams AJ, Davies CR, Mapstone BD (2005) Variation in the periodicity and timing of increment formation in red throat emporer (*Lethrinus miniatus*) otoliths. Mar Freshwater Res 56:529–538

Williams AJ, Davies CR, Mapstone BD, Russ GR (2003) Scales of spatial variation in demography of a large coral reef fish: an exception to the typical model? Fish Bull 101:673–683

Williams T, Bedford BC (1974) The use of otoliths for age determination. In: Bagenal TB (Ed) The ageing of fish. Unwin Brothers, Old Woking, England, pp 114–123

Worthington DG, Doherty PJ, Fowler AJ (1995) Variation in the relationship between otolith weight and age: implications for the estimation of age of two tropical damselfish (*Pomacentrus moluccensis* and *P. wardi*). Can J Fish Aquat Sci 52:233–242

Wright PJ, Panfili J, Morales-Nin B, Geffen AJ (2002) Types of calcified structures. Otoliths. In: Panfili J, de Pontual H, Troadec H, Wright PJ (Eds) Manual of fish sclerochronology. Ifremer-IRD coedition, Brest, France, pp 31–57

4. DAILY OTOLITH INCREMENTS IN THE EARLY STAGES OF TROPICAL FISH

SU SPONAUGLE

Marine Biology and Fisheries, Rosenstiel School of Marine and Atmospheric Science, University of Miami, Miami, FL, USA

1 Introduction

Our understanding of the ecological dynamics of early life stages of marine fish has frequently been referred to as the great unknown "black box" (Sale 1980). Fortunately, however, Pannella's (1971) discovery several decades ago that the otoliths of marine fish record age and growth information on a daily basis provided a critical tool for us to begin propping open the box lid. Increasing research and much technological advancement since then have enabled scientists to obtain accurate and detailed measurements of multiple early life history traits from the otoliths of young fish. Use of the wealth of data contained in the microstructure of fish otoliths is one of the most significant advantages that studies of fish populations have over similar studies of invertebrates. Age data from individuals can be used to examine patterns of larval dispersal, recruitment dynamics, early growth and survivorship, and selective mortality in addition to the more familiar age-specific growth rates. Importantly, individual-based data can be extended to examine cohort-specific and population-wide dynamics, including the relative contributions of particular spawning populations and various nursery habitats to cohort success. Understanding the mechanisms underlying cohort success is a necessary step to eventually predicting year class strength for commercially important species. Growth rates can also be included in bioenergetics models and manipulated to examine critical processes occurring during early life. Ultimately, the examination of relative growth and survival over transitions between consecutive life history stages may provide insight into the selective advantages and evolution of particular life histories.

In this chapter I first outline the general uses of individual-based early life history data that can be obtained from the otoliths of young fish (i.e., < 1 year old), including the variability over multiple spatial and temporal scales of these otolith-derived traits. Secondly, I present some of the technical background and steps necessary to gather daily age and growth data from the otoliths of young fish. Specifically, I review the basic methods used to obtain samples, process and prepare otoliths for analysis, interpret increments, and analyze data. I also identify some inherent problems and limitations in this process. Valuable daily age and growth information can be obtained for a variety of ecological studies involving young tropical fish species from a diversity of environments ranging from the pelagic realm to coral reefs, seagrass beds, and estuarine mangrove-fringe areas by following these basic steps.

B.S. Green et al. (eds.), *Tropical Fish Otoliths: Information for Assessment, Management and Ecology*, Reviews: Methods and Technologies in Fish Biology and Fisheries 11,
DOI 10.1007/978-1-4020-5775-5_4,

1.1 OTOLITH-BASED EARLY LIFE HISTORY TRAITS

Otoliths can provide data on a suite of variables or early life history (ELH) traits. Larval age, birthdate, pelagic larval duration (PLD), juvenile age, and the timing of settlement all involve the enumeration of otolith increments along an otolith axis (Table 1). Relative daily growth at any point during this time period can be obtained by examining the width between successive increments. More material is generally deposited on the otolith during periods of faster growth, analogous to tree-rings. The comparison of increment widths during a particular stage of life among individuals or cohorts provides a relative measure of somatic growth. Further, calibration between somatic growth and otolith growth can provide estimates of real somatic growth from otolith measurements. Demonstration of a significant and consistent linear relationship between body size (e.g., standard length, SL) and otolith size and age is a prerequisite to such analyses (Campana 1990, Geffen 1992, Thorrold & Hare 2002). It is also important to note that caution must be used when comparing increment widths among stages, or longitudinally, *within* an individual because the specifics of otolith increment deposition may change ontogenetically as body geometry (allometry), habitat, and food sources change. Relative size-at-age can be measured in certain circumstances from the otolith radius to the day of interest. This measurement must also be made with caution, however, because various environmental processes can influence the somatic growth – otolith growth relationship (e.g., temperature, Mosegaard et al. 1988, food level, Baumann et al. 2005) and otolith material will continue to be deposited even when somatic growth is negligible. Thus, this decoupling of somatic – otolith growth can make the estimation of relative somatic size from otolith size misleading (Mosegaard et al. 1988, Reznick et al. 1989, Secor & Dean 1992, Morales-Nin 2000). It is worth noting that such decoupling due to negligible growth under natural conditions may be rare for young fish as young fish exhibiting such poor growth are unlikely to survive to the point of being collected in the field (Folkvord & Mosegard 2002). Regardless, validation of otolith-derived estimates of relative size between slow-growing and fast-growing fish by measuring somatic length-at-age for same-age fish (e.g., Searcy & Sponaugle 2000) can make otolith size-at-age a useful relative measurement.

Life history transitions also can be recorded in otoliths, in addition to daily increments. Settlement is the primary transitional event that is frequently evident and of interest in ecological studies of juvenile fish. Settlement (and associated metamorphosis) can be recorded in very different ways. Settlement for some species is merely the point at which narrower larval increments change to wider juvenile increments (or vice versa). A wider dark band (or check) is deposited during settlement and metamorphosis in some species. In other more unusual species, such as the bluehead wrasse, *Thalassoma bifasciatum*, metamorphosis is protracted while the fish is buried under sand and coral rubble and a wide band is deposited on the otolith during this time period (Victor 1983; Figure 1). Settlement marks can be gradual or abrupt, are frequently species-specific, and can vary within species among local populations (Wilson & McCormick 1997, 1999). Settlement in some species is characterized by the formation of secondary growth centres (accessory primordia). Accessory primordial are discussed in greater detail below and by Panfili et al. in Chapter 7 of this volume (see also Geffen et al. 2002 for additional discussion) and Wright et al. (2002a) provides a compilation of other transitional events recorded in otoliths of particular species.

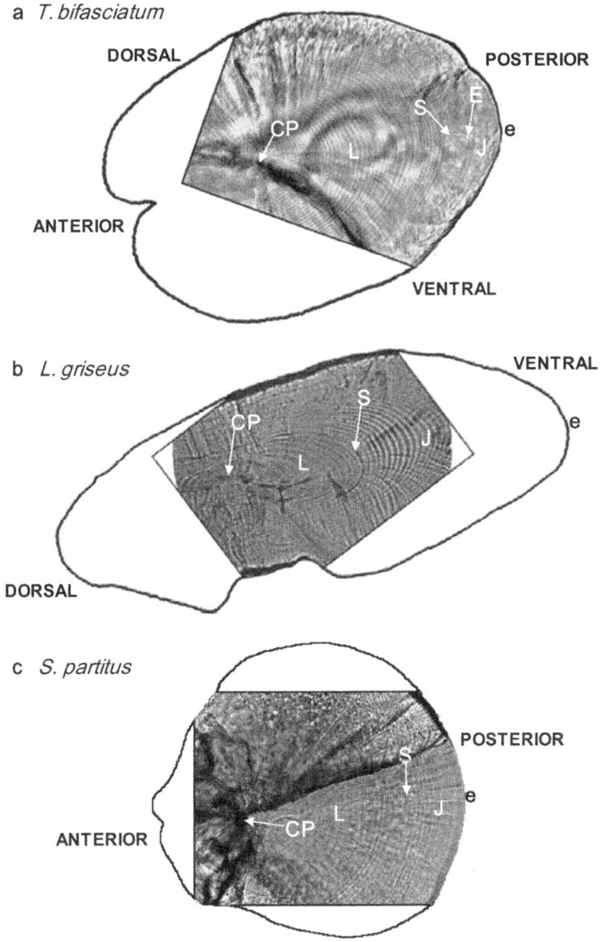

Figure 1. Features used to obtain ELH traits. (**a**) a sagitta of *Thalassoma bifasciatum*, (**b**) a transverse section of *Lutjanus griseus* sagitta, and (**c**) a lapillus of *Stegastes partitus*. CP = central primordium (core), L= larval increments, S = settlement, E = emergence, J = juvenile increments, e = otolith edge.

Differences in otolith shape have been used previously to distinguish fish species and stocks (e.g., Campana & Casselman 1993, Begg & Brown 2000, DeVries et al. 2002), but on a smaller scale, data on otolith morphology (shape) may be useful for ecological studies of young fish. Recent work on two young tropical fish, *Amphiprion akindynos* and *Pomacentrus amboinensis*, has shown that subtle differences in otolith shape can provide valuable information on recent (i.e., 13 days) feeding history (Gagliano & McCormick 2004). This is a promising new avenue of research that, together with traditional daily increment analysis, may provide high-resolution information for ecological studies.

Table 1. Early life history traits (LHT) that can be obtained by reading the otoliths of young fish with ecological definition and otolith metric used to measure traits in larvae, settlement-stage, or post-settlement juveniles.

Life history trait	Definition	Measurement
Larval age	Age in days of larva at the time of collection	Number of increments between the otolith primordium and edge
Birth date	Spawning day or hatching day	Subtraction of larval age (above) from day of collection. Depending on species, an additional factor of time to hatch (~ usually 2–4 days) should be factored in to this calculation.
Pelagic larval duration (PLD)	Number of days a larva spends in the plankton; time between spawning of pelagic eggs (or hatching from demersal eggs) and settlement	Number of increments between the otolith core and edge for settlement-stage larvae, or between core and distinct mark on otolith indicating settlement (settlement mark; see Figure 1)
Juvenile age	Number of days post-settlement and metamorphosis	Number of increments from the settlement mark to the otolith edge
Timing of settlement	Day larva settled from the plankton to benthos	Subtraction of juvenile age from day of collection
Daily growth	Relative growth rate each day	Distance between successive otolith increments. Analogous to tree-rings, more material is generally deposited during periods of faster growth.
Size-at-age	Relative size at a given age	Distance from the primordium to a given age (otolith radius-at- age)

The primary advantage of otolith-derived traits is that variables are recorded simultaneously so that events can be related to ages and, importantly, an actual calendar date can be associated with a particular age, event, or growth rate in the life of an individual fish. Furthermore, otoliths are acellular and not subjected to resorption. This allows an historical analysis of events in the early life of a fish, such as a close coupling between the tidal cycle and the transport of larvae over daily scales. Daily ageing is critical to population replenishment studies aimed at identifying patterns of settlement, recruitment, and larval transport mechanisms. When spawning grounds are unknown (frequently the case for pelagic species), source populations can be identified using larval age data in conjunction with hydrographic and current data (Hare & Cowen 1991, 1996, Fowler et al. 2000, Jenkins et al. 2000, Hare et al. 2002, Govoni et al. 2003, Serafy et al. 2003). The role and relative value of different juvenile habitats can be compared by examining relative growth of fish collected from those habitats (e.g., Ross

2003). Otolith-derived early life history traits can also be compared among initial larvae or settlers and survivors to examine patterns of selective mortality (Meekan & Fortier 1996, Hare & Cowen 1997, Sogard 1997, Searcy & Sponaugle 2001, Bergenius et al. 2002, Vigliola & Meekan 2002, Wilson & Meekan 2002, Sponaugle & Grorud-Colvert 2006, Gagliano et al. 2007, Hamilton et al. 2008). The collection of such data from individuals allows a fine resolution of scales of variability and provides data for "individual based models" (IBM). These relatively new techniques can be a powerful means of defining relationships among variables and eventually will improve our understanding of the ecology of populations by revealing important ecological processes that affect the early life histories of fish (e.g., Allain et al. 2003).

1.2 VARIABILITY IN EARLY LIFE HISTORY TRAITS

Studies of otolith-derived early life history traits frequently involve looking beyond mean measurements to the variability in these traits expressed over multiple spatial and temporal scales. Several examples of how variability in these traits contributes to our examination of ecological processes are presented below.

1.2.1 Among-taxa and geographic variation in early life history traits

Pelagic larval duration leads the list of ELH traits that have and continue to receive substantial attention in ecological studies. The amount of time a larva requires to develop into a juvenile not only plays a central role in the evolution of its life history strategy, but likely influences population connectivity on ecological scales. PLD in fish varies substantially among taxa, which may influence the degree to which populations are ecologically related (see Sponaugle et al. 2002 for discussion). Early work was designed to examine this variability among and within taxa over geographic scales (Victor 1986, Thresher et al. 1989, Wellington & Victor 1989, Victor & Wellington 2000, Zapata & Herron 2002). More recent studies have examined genetic relatedness of populations of fish with different PLDs (Shulman & Bermingham 1995, Riginos & Victor 2001, Taylor & Hellberg 2003).

1.2.2 Within-taxa variation in early life history traits

The degree of within-species variation in PLD differs among taxa. Particular taxa have relatively invariant PLDs (e.g., pomacentrids) while others exhibit a wider range in PLD (some labrids, gobies, mullids, acanthurids; Victor 1986, Cowen 1991, McCormick 1994, Sponaugle & Cowen 1994, 1997). Such within-taxa variation can be used to examine the role of ecological processes such as food availability and water temperature over shorter temporal scales. For example, among-cohort variation in PLD is evident in monthly cohorts of settlers of the bluehead wrasse, *Thalassoma bifasciatum*, at Barbados: fish with higher larval growth rates have shorter PLDs (Searcy & Sponaugle 2000). Lower growth and longer PLDs have been attributed to reduced food availability and larval encounter with low-salinity physical oceanographic features moving through the system (North Brazil Current rings; Searcy & Sponaugle 2000, Sponaugle & Pinkard 2004a). Where water temperatures are more seasonally variable (i.e., Florida Keys), cohorts of *T. bifasciatum* settling during warm months of the year have higher larval growth rates and thus shorter PLDs than cohorts settling during cold months

(Sponaugle et al. 2006). Larvae that grow and develop more quickly and settle sooner may have higher survivorship rates because mortality in the plankton is typically high and size-dependent (stage duration hypothesis: Anderson 1988). For example, variation within species in larval growth at Panama, accounted for a significant portion of settlement variability for two pomacentrids (Wilson & Meekan 2002, Meekan et al. 2003) and an acanthurid (Bergenius et al. 2002).

1.2.3 Within-cohort variation

Within-cohort variation in traits such as growth and PLD can be used to explore the dynamics of early survivorship. Repeated collections over time of the same cohort (i.e., "cohort-tracking") allows comparisons of the traits of survivors (those collected most recently) with those of the settling population (earliest collection) to examine whether survivorship is related to early life history traits such as larval or juvenile growth or the duration of the larval period (Meekan & Fortier 1996, Hare & Cowen 1997, Sogard 1997, Searcy & Sponaugle 2001, Bergenius et al. 2002, Vigliola & Meekan 2002, Wilson & Meekan 2002, Sponaugle & Grorud-Colvert 2006, Gagliano et al. 2007, Hamilton et al. 2008). This approach is explored in greater detail by Vigliola and Meekan in Chapter 6 of this volume.

2 Methods

2.1 SAMPLE SIZES

An understanding of the various scales of variation in ELH traits is a prerequisite for designing otolith-based studies to ensure that sufficient samples (individual fish) are collected for otolith interpretation to answer the main questions of interest. Sufficient numbers are needed to accurately estimate mean values for a species from a particular locality and the sampling design of comparative studies must be carefully planned to encompass other scales of variation. For example, if a geographic analysis is planned, sample collections from different localities over different months of a year may introduce variation that is seasonal rather than, or in addition to, geographic. Furthermore, sampling must factor in known difficulties with otolith processing and interpretation (discussed in greater detail below). Cohort-tracking studies to examine how the distributions of early life history traits changes over time must sample an excess of individuals thought to be from the same cohort because specific cohort membership can only be determined once otolith ageing is complete, typically resulting in the elimination of a portion of each sample. Adequate supply of larvae, settlers, or new and older recruits is needed for any otolith-based study of young fish. It is clearly helpful to have knowledge of the species' natural history, including its settlement habitat to secure this supply.

2.2 SAMPLE COLLECTION

Larvae and young recruits can be collected using a number of sampling techniques. Young larvae are typically collected by towing various types of nets from medium to large vessels. The smaller the net opening and mesh size, and the slower the net is towed, the greater the likelihood that larger, older larvae will be able to evade the sampling gear. Older larvae must be collected with nets with large openings and a large mesh size (>1 m^2 opening and 1 mm mesh). These larger nets can and should be towed more rapidly to increase the likelihood of collecting actively-swimming older larvae, although rapid towing also may increase larval extrusion through the net, resulting in damaged specimens that cannot be identified. Tow speed must be chosen to balance these opposing constraints since accurate identification of larvae is a prerequisite to further analysis.

State of the art ichthyoplankton sampling gear includes the Multiple Opening and Closing Environmental Sampling System (MOCNESS; 1–4 m^2 opening; 0.3–1 mm mesh), and a coupled MOCNESS system with two different nets and mesh sizes (Guigand et al. 2005) allows simultaneous sampling of larvae and their prey. This gear can collect samples from discrete depths and measure hydrographic variables simultaneously, but it must be fished from a large, dedicated research vessel. MOCNESS gears have been successfully used in a variety of tropical and subtropical locations (e.g., Barbados, Cowen & Castro 1994, Paris & Cowen 2004; Straits of Florida, Llopiz & Cowen 2008), though the high cost of operating large research vessels capable of towing the gear has limited their use somewhat. Rapid sample decay may limit tow duration in tropical waters (> 25°C), though one solution is to immediately ice the samples upon retrieval. A less technical, but still effective, net for collecting vertically discrete samples from smaller vessels is the standard opening-closing Tucker trawl (see Pepin & Shears 1997 for comparison of Tucker trawl and bongo nets).

Late, or settlement stage larvae are frequently targeted with larval light traps. These traps are deployed immediately offshore of coral reefs to intercept late stage larvae as they return to reefs to settle. Numerous designs have been used in a diversity of geographic settings (Doherty 1987, Brogan 1994, Sponaugle & Cowen 1996a,b, Valles et al. 2001, Wilson & Meekan 2002), all of which are selective for particular species (Doherty 1987). Light traps have also been arranged vertically to evaluate vertical positioning of larvae in the water column (Fisher & Bellwood 2002). Channel nets (Shenker et al. 1993) and crest nets (Dufour & Galzin 1993, Doherty & McIlwain 1996) can be effective at collecting larvae moving through reef channels or over the reef crest to settle in lagoons, but require particular environmental conditions (strong, unidirectional flow) for deployment. Larval purse seines are a cumbersome, but often productive means of collecting larvae, particularly older "pre-settlement" or late stage larvae (Kingsford & Choat 1985, Choat et al. 1993).

Young recruits can be collected from the benthic habitat (e.g., reef, seagrass bed) by slurp guns or hand nets, with or without the use of anesthetizing chemicals such as clove oil or quinaldine. Clearly, knowledge of species-specific settlement habitats and the behaviour of young fish is helpful. For example, the very youngest reef settlers are often under boulders or hidden in macroalgae (e.g., Sale et al. 1984, Tolimieri 1995, Paddack & Sponaugle 2008). Careful investigation of these habitats is necessary. Older recruits can be similarly collected by hand or by spear-fishing with small hand spears, by herding into drop nets, or by benthic trawling. Beach seines are a common technique for collecting juveniles in nearshore, estuarine areas, including seagrass beds. Sampling among mangrove prop roots can be challenging, and methods include pop nets (Hindell & Jenkins 2005) or hand nets with anesthetizing chemicals.

Specimens should be preserved in $\geq$ 75% ethanol, regardless of the means by which fish are collected. Freezing is also an acceptable means of short-term storage of specimens. Some researchers have had success with drying as a means of preservation but this can be risky and other preservatives should be tested for their effects on otoliths prior to use for important samples. For example, the use of certain preservatives will damage the otoliths (e.g., formaldehyde or formalin, which is mildly acidic).

2.3 DAILY DEPOSITION VALIDATION

Otoliths are valuable tools for obtaining important ecological data for short-lived or young fish provided increments are deposited on a consistent, daily basis. Most reef fish have otoliths with increments that are deposited daily, although a few cases exist where daily deposition is not evident or consistent. A necessary first step in using otoliths, therefore, is validating the existence of daily deposition of increments. This can be done in a direct manner involving marking or rearing larvae (Wright et al. 2002b) or through the use of indirect or statistical techniques (Morales-Nin & Panfili 2002a). Daily deposition rates have been validated for larvae or juveniles of a number of species from different families (see Table 2 for examples). The means by which daily deposition rates can be validated include the following.

1. *Chemical marking.* Chemically marking otoliths is coupled with maintaining marked fish in the laboratory for short periods (7–10 days; Figure 2) to identify structures formed in the otoliths between the time of marking and sacrifice of the fish. Otoliths are generally marked by immersion of the live fish in a chemical bath for 24 h or injection of the fish with elevated concentrations of the chemical. Chemicals that are taken up readily by the otoliths include tetracycline, oxytetracycline, calcein fluorescent green, alizarin complexone, and strontium chloride (Tsukamoto 1988, Brothers 1990, Geffen 1992, Wilson 1995, see Wright et al. 2002b for summary table of effective chemical concentrations).
2. *Stress marking.* This method involves inducing an atypical mark in the otoliths by exposing fish to elevated temperatures, abbreviated photoperiods, or reduced food, maintaining the fish for additional days, and comparing the number of increments deposited after the stress mark to days held (see review in Wright et al. 2002b). Otolith increments deposited by fish maintained under laboratory conditions frequently do not resemble increments deposited by fish in the wild (Figure 2) but the periodicity of deposition should remain constant, allowing validation of the rate of increment formation. Chemically or stress-tagged fish can also be released into the wild and re-captured. This method results in the most natural appearance of increments but recapture rates can be low. Maintenance of tagged individuals in field enclosures or outdoor mesocosms is another option.
3. *Collections around settlement.* Collection of late-stage larvae that are in the process of settling allows comparison of the otoliths of those individuals with the otoliths of settled juveniles collected a short period of time after the arrival of late-stage larvae. This method not only validates the daily deposition of juvenile increments but also validates the settlement mark (i.e., late stage larvae should not have such a mark; Raventos & MacPherson 2001).

4. *Laboratory rearing.* Hatching a batch of eggs in the lab and examining the otoliths of the larvae after several days allows validation of the periodicity of increments at the beginning of larval life. Difficulties in rearing many species through the larval phase make it difficult to validate periodicity right through to settlement, though improving rearing technologies are making this approach increasingly viable.

5. *Sequential field collections of larvae.* Periodicity of increments in otoliths can be established by re-sampling a water mass (containing a single cohort of larvae) over a short period (~7 days) and comparing the age and length frequency distributions of larvae among collections. This method assumes that the same cohort is re-sampled over time, that there are few new fish entering the water mass and that fish have a constant, predictable mortality schedule. This technique will not work well in heterogeneous environments with substantial water column mixing, but can be applied in certain circumstances, as well as to settled juveniles. The comparison of growth rates calculated via differences in length frequency over time can be compared to growth rates calculated from otolith increment analysis (size-at-age) as a means of corroborating otolith increment interpretation (e.g., Morales-Nin & Aldebert 1997).

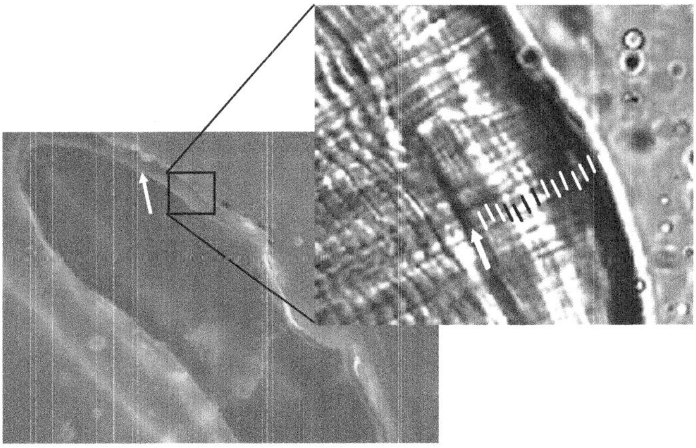

Figure 2. Validation of daily deposition in the sagitta of the gray snapper, *Lutjanus griseus*. Freshly caught juveniles were placed in an aerated 10 L saltwater aquarium with a 200 mg/L alizarin red concentration for 24 h. The aquarium was kept in the dark due to the photosensitivity of alizarin red. After 24 h, the fish were removed and placed in flow-through seawater tanks and fed until they were euthanized 14 days later and the otoliths prepared for viewing. Marked otoliths were viewed through an inverted microscope with UV light at 25× to mark the increment and then viewed again under standard light microscopy to enumerate the number of increments (*Black and white lines* in enlarged box) beyond the fluorescent mark (*white arrow* in each image). It is common for fish maintained under laboratory conditions to have increments that do not resemble the appearance of increments deposited in the wild but validation of the increment periodicity is still possible.

Table 2. Selected studies that validated the daily deposition of otolith increments in tropical marine fishes. Studies were compiled from published sources in the primary literature only. More validation studies likely exist in the gray literature.
Legend: Area = Broad Distribution, T= tropical, ST= subtropical; *Region = Specific distribution,* COSMO = cosmopolitan, AT= Atlantic, WA = Western Atlantic, EA = Eastern Atlantic, GOM = Gulf of Mexico, C = Caribbean, P = Pacific, EP = Eastern Pacific, NEP = North East Pacific, SEP = South East Pacific, WP = Western Pacific, SWP = South West Pacific, IWP = Indo-West Pacific, NWP = North Western Pacific, ECP = East Central Pacific, EA = East Asia, M= Mediterranean Sea, AS = Asia, IO = Indian Ocean, RS = Red Sea, AF = Africa; *Validation = Validation Technique*, TC = tetracycline, OTC = oxytetracycline, ALC = alizarin complex one, SC = Strontium Chloride, C = Calcein, MIC = marginal increment analysis, RL = reared larvae, SM = stress marking, F = followed larvae.

Species	Family	Area	Region	Validation	References
Megalops atlanticus	Elopidae	T	WA, GOM, E,P	OTC	Zerbi et al. (2001)
Megalops cyprinoides	Elopidae	T	IP	ALC	Tsukamoto and Okiyama (1993)
Anguilla celebesensis	Anguillidae	T	WP	ALC	Arai et al. (2000)
Anguilla japonica	Anguillidae	T	AS	ALC	Tsukamoto (1989)
Anguilla marmorata	Anguillidae	T	IP	ALC	Sugeha Hagi et al. (2001)
Anguilla rostrata	Anguillidae	ST	C,AT	TC	Martin (1995)
Herklotsichthys castelnaui	Clupeidae	T	SWP	TC	Thorrold (1989)
Spratelloides delicatulus	Clupeidae	T	IP	TC	Schmitt (1984)
Anchoa mitchilli	Engraulidae	ST	WA, GOM	RL	Fives et al. (1986)
Encrasicholina devisi	Engraulidae	T	IP	OTC	Hoedt (2002)
Engraulis encrasicolus	Engraulidae	ST	EA, M, W, IO	OTC	Cermeno et al. (2003)
Stolephorus carpentariae	Engraulidae	T	WP	OTC	Hoedt (2002)
Stolephorus nelsoni	Engraulidae	T	WP	OTC	Hoedt (2002)
Stolephorus purpureus	Engraulidae	T	P (Hawaii only)	F	Struhsaker and Uchiyama (1976)
Thryssa aestuaria	Engraulidae	T	WP	OTC	Hoedt (2002)
Chanos chanos	Chanidae	T	IP, EP	OTC	Tzeng and Yu (1989)

Species	Family	Area	Region	Validation	References
Benthosema suborbitale	Myctophidae	T	WA, EA, P, IO	MIC	Gartner (1991)
Diaphus dumerilii	Myctophidae	T	EA, WA	MIC	Gartner (1991)
Lepidophanes guentheri	Myctophidae	T	EA, WA	MIC	Gartner (1991)
Mugil cephalus	Mugilidae	ST	WA, EA, M, GOM	RL	Radtke (1984a)
Atherina presbyter	Atherinidae	ST	EA, M	SC	Moreno and Morales-Nin (2003)
Hypoatherina tropicalis	Atherinidae	T	WP	TC	Schmitt (1984)
Hirundichthys affinis	Exocoetidae	T	IP, EA, WA	TC	Oxenford et al. (1994)
Centopomus undecimalis	Centropomidae	T	WA, C, GOM	RL	Tucker and Warlen (1986)
Coryphaena hippurus	Coryphaenidae	T/ST	AT, IO, P	RL	Massuti et al. (1999)
Trachurus japonicus	Carangidae	ST	WP, NWP	ALC	Xie et al. (2005)
Lutjanus griseus	Lutjanidae	T	WA, C, GOM	ALC	Ahrenholz (2000)
Lutjanus synagris	Lutjanidae	T	WA, C, GOM	ALC	Ahrenholz (2000)
Pristipomoides filamentosus	Lutjanidae	T	IP	TC	Ralston and Miyamoto (1983)
Pomadasys kaakan	Haemulidae	T	RS, IWP	ALC	Husaini et al. (2001)
Diplodus puntazzo	Sparidae	ST	EA, AF	ALC	Villanueva and Moli (1997)
Diplodus vulgaris	Sparidae	ST	EA, AF	ALC	Villanueva and Moli (1997)
Pagrus major	Sparidae	ST	NWP	RL	Tsuji and Aoyama (1982)
Sparus aurata	Sparidae	ST	EA, M	RL	Morales-Nin et al. (1995)
Archosargus probatocephalus	Sparidae	ST	WA, GOM	OTC	Parsons and Peters (1989)
Bairdiella chrysoura	Sciaenidae	ST	WA, GOM	C	Hales and Hurley (1991)
Cynoscion nebulosus	Sciaenidae	T/ST	AT	ALC	Powell et al. (2000)
Sciaenops ocelatus	Sciaenidae	ST	WA, GOM	RL	Peters and McMichael (1987)
Chaetodon plebius	Chaetodontidae	T	IWP	TC	Fowler (1989)
Chaetodon rainfordi	Chaetodontidae	T	WP-GBR only	TC	Fowler (1989)

Species	Family	Area	Region	Validation	References
Chelmon rostratus	Chaetodontidae	T	IWP	TC	Fowler (1989)
Sarotherodon melanotheron	Cichlidae	T	AF	RL	Ekau and Blay (2000)
Tilapia mariae	Cichlidae	T	AF	RL	Rosa and Re (1985)
Pomacentrus amboinensis	Pomacentridae	T	WP	TC	Pitcher (1988)
Pomacentrus melonochir	Pomacentridae	T	IP	TC	Pitcher (1988)
Pomacentrus wardi	Pomacentridae	T	WP	TC	Pitcher (1988)
Stegastes partitus	Pomacentridae	T	WA, GOM	SM	Robertson et al. (1988)
Halichoeres bivittatus	Labridae	T	WA, GOM	TC	Victor (1982)
Thalassoma bifasciatum	Labridae	T	WA, GOM	TC	Victor (1982)
Parapercis cylindrica	Pinguipedidae	T	WP	TC	Walker and McCormick (2004)
Amblygobius bynoensis	Gobiidae	T	IP	TC, SC, C	Hernaman et al. (2000)
Amblygobius phalaena	Gobiidae	T	IP	TC, SC, C	Hernaman et al. (2000)
Aphia minuta	Gobiidae	ST	AT, M	SC	Iglesias et al. (1997)
Asterropteryx semipunctatus	Gobiidae	T	IP	TC, ALC, SC	Hernaman et al. (2000)
Favonigobius reichei	Gobiidae	T	IWP	TC, SC, C	Hernaman et al. (2000)
Glossogobius biocellatus	Gobiidae	T	IP	TC, SC, C	Hernaman et al. (2000)
Istigobius goldmanni	Gobiidae	T	WP	TC, SC, C	Hernaman et al. (2000)
Valenciennea muralis	Gobiidae	T	IWP	TC, SC, C	Hernaman et al. (2000)
Acanthurus chirurgus	Acanthuridae	T	WA	OTC	Bergenius et al. (2002)
Euthynnus pelamis	Scombridae	T	COSMO (ex M)	RL	Radtke (1984b)
Thunnus albacares	Scombridae	T	COSMO (ex M)	RL	Uchiyama and Strusaker (1981)
Parophrys vetulus	Pleuronectidae	ST	NEP (into Baja)	RL	Laroche et al. (1982)
Stephanolepis hispidus	Monacanthidae	T	WA, C, EA	ALC	Rogers et al. (2001)

Marginal increment analysis, which is frequently used in validating annual increment deposition of older fish, is not generally effective for validating daily increments given the significantly shorter time periods involved and limitations on microscope and observer resolution (but see Ré 1984, Jenkins & Davis 1990, Gartner 1991). Campana (2001) provides a detailed comparison of the strengths and weaknesses of various validation techniques.

The second process requiring validation for different species is the timing of first increment deposition. Use of a mean time of first increment deposition is sufficient for some purposes, but it is necessary to validate this for the species of interest to answer more precise questions such as hind-casting larval transport or spawning areas. The first increment is typically deposited sometime after hatching (Campana & Neilson 1985), usually 2–3 days after spawning of pelagic eggs. Passive drift must be factored in to account for this early pelagic period when estimating adult spawning locations from daily increments in larval otoliths of such species. A longer time lag between spawning and first increment deposition occurs for larvae hatching from demersally-spawned eggs, but these eggs generally are not subjected to drift prior to hatching.

2.4 SUB-DAILY INCREMENTS

Validation of daily deposition rates can be useful or even necessary for distinguishing sub-daily increments, i.e., increments that are deposited more than once per day. Such marks are not uncommon in tropical fishes and their inclusion in readings results in over-estimation of age and under-estimation of growth rates. Sub-daily increments are most frequently slightly lighter in colour and less distinct in contrast than daily increments (e.g., bluehead wrasse), although sub-daily increments also can appear in some species as darker banding that can be confused with daily increments (e.g., barracuda). Adjusting the magnification of the microscope to where increments are slightly blurry can help distinguish sub-daily from daily increments because sub-daily increments tend to disappear when the image is blurred (Figure 3). Sub-daily increments, when apparent, may be difficult to distinguish from daily increments only over certain regions of the otolith. It is important to note, however, that in these regions sub-daily increments should occur repeatedly in groups. They do not generally occur singly. Furthermore, sub-daily increments should be approximately half the width of neighbouring daily increments where they do become apparent. Sub-daily increments may be a particular problem when fish are young because it may be more difficult to distinguish sub-daily from daily increments without the benefit of many increments to compare and it is more difficult to use microscope adjustments to resolve differences when only few increments are present. Validation by marking and holding fish for 7–10 days can help resolve which marks are daily and which are sub-daily. The primary reason for identifying sub-daily increments is so that they are *not* counted or measured. There is not yet any compelling reason to compare specific patterns of sub-daily increments in tropical reef fish.

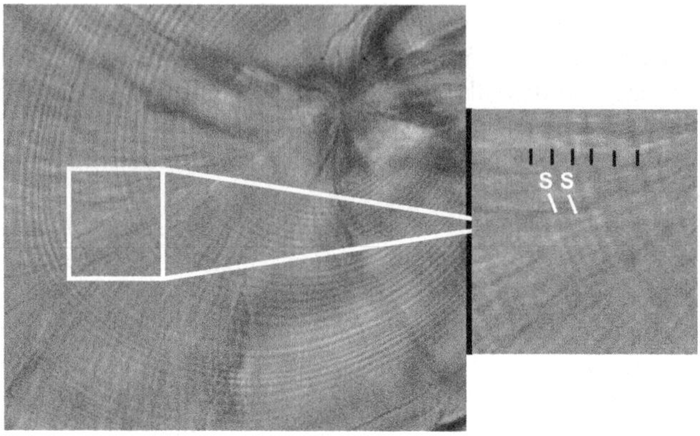

Figure 3. Daily (*vertical black lines*) and sub-daily (*white S*) increments in a sagitta of *Thalassoma bifasciatum*. Sub-daily increments can be resolved by blurring the microscope image slightly if necessary: sub-daily increments will disappear; daily increments will remain.

2.5 DISSECTION AND OTOLITH PREPARATION TECHNIQUES

Retrieving the otoliths from small fish may seem like a daunting task initially but with the right tools and techniques, it is actually a reasonably straightforward process. Note that all morphometric data should be obtained before beginning any dissection because the fish body will be damaged or destroyed during dissection. For example, each fish should be assigned a sample number, weighed (wet weight) and its appropriate length (standard, total, or fork length) measured. The details of otolith dissection and preparation depend on the species and life stage of interest. Consequently, the following discussion of techniques is based on a "case study" approach, although the techniques presented should be applied easily to a diversity of other species. Secor et al. (1992) provide a compilation of techniques used for other species. A valuable manual on otolith preparation (Secor et al. 1991) is also available online at http://cbl.umces.edu/~secor/otolith-manual.html

2.5.1 Larvae (e.g., larval bluelip parrotfish, Cryptotomus roseus)

Most larvae are sufficiently small that opaqueness due to preservation can be effectively reversed by immersing larvae in medium viscosity immersion oil for 1–30 days. The pairs of otoliths can be viewed easily through the intact body under polarization and microscope magnification once the body is transparent (Figure 4). Otoliths in very young fish sometimes can be read in situ directly through the body after dissolving the flesh with bleach or enzymes (Secor & Dean 1992, Green et al. 2002) and staining the otolith using the Von Kossa staining method (Green et al. 2002). In most situations, however, the otoliths should be removed from the body and all tissue remnants cleared away. Removal of the otoliths involves gentle tearing of the tissue and prying out of the otoliths using very fine forceps or dissecting or insect pins (Box 1). Cleaning off tissue can be done manually by scraping with dissecting or insect pins or electrostatically-sharpened tungsten needles or by flushing the otolith with a light bleach solution (Secor & Dean 1992).

Box 1. Extracting and preparing larval otoliths

Supplies: Two fine forceps, two dissecting or insect pins, glass slides, medium viscosity immersion oil with dropper, disposable wipes, dissecting microscope.

Protocol

1. After taking morphological measurements, place the larva on a clean, labelled glass slide in a small pool of immersion oil (multiple larvae can be cleared simultaneously by placement in cell culture tray with multiple wells).
2. Let sit undisturbed in a protected location until larval body is transparent (generally 2–3 days). Note that the first two steps can be bypassed for many species and the otoliths simply found and extracted using only dissecting needles and a polarized light source (see step 3).
3. Place slide on a dissecting microscope with a polarizing filter or film under the glass protective plate covering the lower light source. Rotate the polarizing filter until the otoliths become visible (Figure 4). It is possible that the otoliths will also be visible without a polarizing filter.
4. Locate the otoliths and gently tear apart the flesh with dissecting pins and forceps to release the otoliths.
5. Using a dissecting pin, clean off tissue from the otolith surface. Move otoliths to a clean area of oil.
6. Prepare a new clean, labelled slide with one drop of medium viscosity immersion oil (e.g., Cargille Laboratories, non-drying immersion oil, type B) in the centre (or multiple drops if you wish to keep the otoliths separate).
7. Place the new slide next to the dissection slide and transfer the otoliths to the clean slide by dipping a dissecting or insect pin in the immersion oil, touching the oiled tip to an otolith and transferring it to the drop of oil on the clean slide. Repeat for each otolith.
8. Gently push the otoliths down through the oil so they lie flat on the surface of the slide. Depending on species, select and maintain a consistent orientation of the otoliths on the slide in preparation for reading.
9. Wipe excessive oil off the slide and, if otoliths are sufficiently clear to read, set aside for reading; if not, store the slides flat in a covered location to reduce dust falling into the oil while waiting for the otoliths to clear (check periodically to determine when otoliths have cleared sufficiently). Many larval otoliths can be read almost immediately. In fact, in some cases, otoliths can "over-clear" if left too long in immersion oil. Some researchers prefer to remove the oil from the otoliths after all readings are complete and store the otoliths "dry" to prevent over-clearing. The advantage to this is that dry otoliths can be stored and transported easily in small vials. If otoliths are stored in immersion oil, they should be examined frequently (e.g., weekly, then monthly) to avoid "over-clearing".

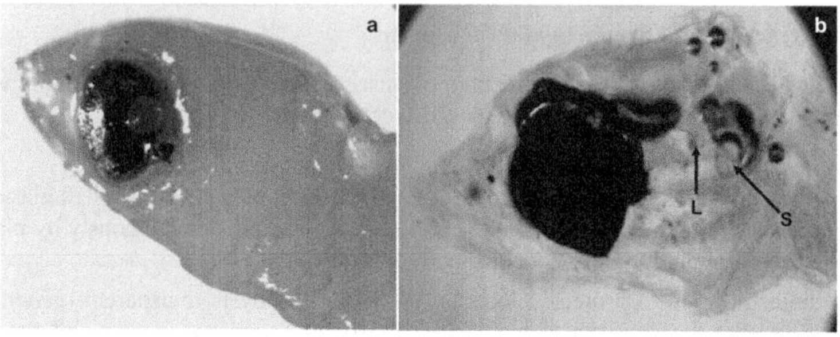

Figure 4. Clearing of larval parrotfish to reveal otoliths. (**a**) Whole larvae or heads of larvae preserved in ethanol can (**b**) be soaked in immersion oil to clear the flesh and reveal otoliths under polarized light. L = lapillus; S = sagitta.

2.5.2 Young juveniles (e.g., bluehead wrasse, Thalassoma bifasciatum)

The integument of juvenile fish is tougher than larvae and requires one of several possible dissection techniques. The use of these techniques will differ by taxa, depending on body shape, age of the fish, relative size of the otoliths, and skill of the dissector. The following protocol (Box 2) works well for many young reef fish, but techniques should be customized to the study species.

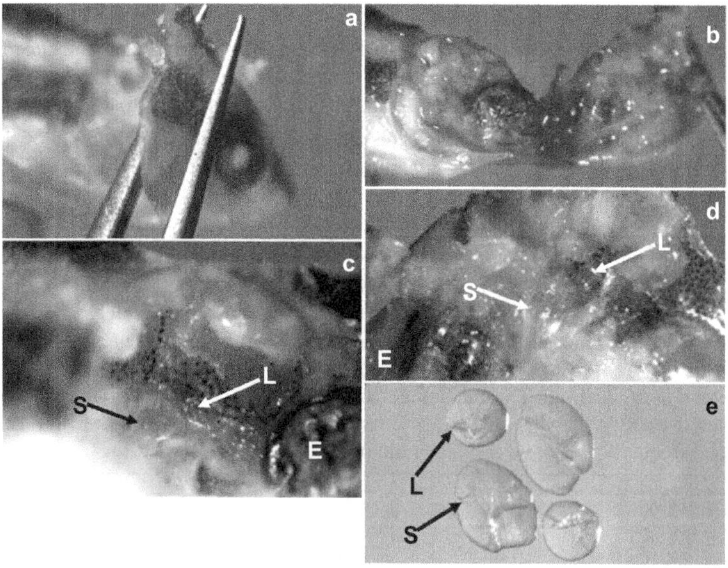

Figure 5. Dissection of a bluehead wrasse, *Thalassoma bifasciatum,* recruit. Grab the operculum (**a**) and (**b**) peel back one side of the head to (**c**) reveal the sagitta (S) and lapillus (L) on the removed flap or (**d**) in the exposed head cavity. Once the top pair of otoliths are removed, the brain mass can be pushed aside to reveal the lower pair of otoliths. Both pairs should be cleaned and (**e**) placed in immersion oil for clearing. E = eye.

Box 2. Extracting and preparing otoliths from small juvenile fish

Supplies: Two fine forceps, two dissecting or insect pins, glass slides, small Petri dish of water, immersion oil with dropper, disposable wipes, dissecting microscope.

Protocol

1. Soak fish in a Petri dish of water for a minute to reduce evaporative drying of ethanol during dissection. This step is unnecessary if fish were frozen.
2. Place the fish on a clean glass slide on one side such that the posterior of the body can be held down by the index finger of one hand.
3. Place a drop or two of water to one side of the same slide.
4. Slide one tip of the forceps under the exposed operculum and insert diagonally into the body up through the midline on the top of the head.
5. Pinch the forceps together and peel back the operculum and outer skin of the lateral (visible) side of the head (Figure 5a, b).
6. The precise position of the otoliths will be taxon-specific, but in many cases a portion of the sagitta in the upper side should be visible. The lapillus is generally more lateral and might be pulled away with the flap of skin (Figure 5c, d).
7. Retrieve the otoliths from the top side then gently push the brain mass up and retrieve the otoliths from the lower side.
8. Place the tissue-covered otoliths in the drop of water on the slide until all the otoliths are removed.
9. Return the fish to the sample vial.
10. Clean off the tissue from the otoliths by gently scraping with dissecting pins.
11. Slide the clean otoliths out of the water drop and carefully wipe dry the slide around the otoliths.
12. Prepare a new clean, labelled slide with one drop of medium viscosity immersion oil (see previous protocol for oil type) in the centre (or multiple drops if you wish to keep the otoliths separate).
13. Place the new slide next to the dissection slide and transfer the otoliths to the clean slide by dipping a dissecting or insect pin in the immersion oil, touching the oiled tip to an otolith and transferring it to the drop of oil on the clean slide. Repeat for each otolith.
14. Push the otoliths down through the oil so they lie flat on the surface of the slide (Figure 5e). Depending on species, select and maintain a consistent orientation of the otoliths on the slide in preparation for reading.
15. Wipe off the dissecting slide and continue with the next specimen. Slides with otoliths must be stored flat in a covered location to reduce dust fall into the oil.
16. Periodically examine the otoliths to determine when the otoliths have cleared sufficiently for reading. This period can range from immediately to 30 days. Frequent checking is required for some species to prevent "over-clearing" of the otoliths (see previous protocol).

2.5.3 Older juveniles (e.g., gray snapper, Lutjanus griseus)

Older juveniles or fish with tougher integuments will require more rigorous otolith extraction methods such as the technique described for this species (Figure 6, Boxes 3–6). Any number of variations on this technique are possible, including lateral cuts with a scalpel (parallel with the operculum) or dorsal head cuts with the specimen held with its ventral surface on the slide. Otoliths from older juveniles frequently require at least polishing (see alternative polishing methods below, Box 6), while others (see Table 3) may need to be embedded and sectioned for reading, as in this example. See Secor et al. (1992) and McCurdy et al. (2002) for other techniques and embedding media.

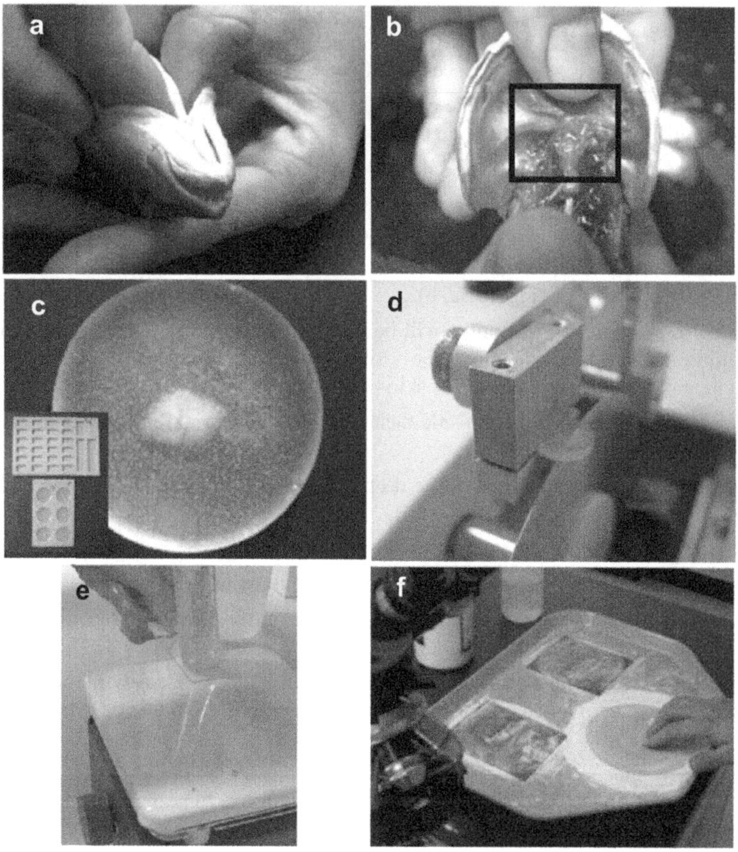

Figure 6. Dissection of juvenile *Lutjanus griseus* to (**a**) snap the isthmus, (**b**) find and remove otoliths (Box 3), (**c**) embed the otoliths in resin (Box 4), (**d**) section the otoliths with a diamond blade isomet saw (Box 5), (**e**) attach the section to a slide and then (**f**) polish the section (Box 6).

Box 3. Extracting otoliths from older juvenile fish

Supplies: Two fine forceps, two dissecting or insect pins, labelled plastic centrifuge tubes for otolith storage, microscope.

Dissection Protocol

1. Thaw frozen specimens until the body is just flexible. Specimens preserved in ethanol will need to be soaked in water first to reduce evaporative drying.
2. Using either forceps or your finger break the isthmus of the fish (Figure 6a).
3. Clear away the tissue until the spine and neurocranium (braincase) are visible.
4. Break the spine at the base of the braincase.
5. Use forceps to carefully lift open the braincase. This step will require some force, so care is needed not to crush the otoliths inside the case. Once the braincase is open the sagittae and lapilli will be clearly visible (Figure 6b).
6. Remove the otoliths and clean off any remaining tissue using forceps or dissecting pins. The cleaned otolith can now be mounted (Box 4) for sectioning (Box 5) or stored in a small dry vial until needed.

Box 4. Embedding otoliths in resin for sectioning

Supplies: Fine forceps, two dissecting or insect pins, two part epoxy resin, disposable wipes, disposable syringe, rubber moulds, microscope.

Embedding Protocol

1. Mark the core of the otolith before mounting by drawing a pencil line in the transverse plane that intersects the core of the otolith.
2. Have all the otoliths to be mounted available and marked before beginning the preparation of epoxy mixture. Follow epoxy safety procedures and work in a well-ventilated area.
3. Make clear, low viscosity epoxy (e.g., from Buehler or West System) by mixing, from separate syringes, 5 parts resin to 1 part hardener by *weight*. This ratio must be correct or the otoliths will not section properly. See McCurdy et al. (2002) for a summary of different mounting media (each requiring different ratios).
4. Mix the resin very well (for at least 2 min) before application to otoliths.
5. Use a disposable plastic syringe to pour epoxy into the wells of a rubber mould.
6. Place an otolith on the epoxy in each mould well. Orientation is not critical, but some prefer to position the otolith with the concave surface down.
7. Use a dissecting pin to manoeuvre the otolith to the middle of the block and push it down under the epoxy, being careful no air bubbles are trapped below otolith.
8. Repeat this process for each otolith to be mounted and allow the epoxy to cure overnight (Figure 6c). Three grams of epoxy will embed 12–13 otoliths in small moulds or 1–3 otoliths in larger moulds. If moulds are not available, each otolith can be left in a blob of viscous epoxy on a labelled wax sheet, with the otolith turned over and another blob of epoxy placed on the lower flat surface within 12 h.

Box 5. Sectioning embedded otoliths

Supplies: Fine forceps, glass slides, hot plate, Crystalbond thermoplastic cement, a multi-speed isomet saw equipped with a diamond blade, microscope.

Sectioning Protocol

1. Section the otoliths using a multi-speed isomet saw with a diamond blade.
2. Cut the otolith to one side and then the other of the pencil line marked on the otolith before it was mounted in resin (Box 4), so the line remains in the middle of the section. The thickness of section will vary by species and otolith size.
3. Place a slide on the hot plate and add a small amount of Crystalbond thermoplastic cement.
4. Once the cement has melted, use forceps to lower the otolith section onto the slide. Allow to cool and harden for several minutes before polishing (Box 6).

Box 6. Polishing otolith sections

Supplies: Fine forceps, sandpaper of multiple grits, polishing cloth, alumina powder, water bottle, hot plate, microscope.

Polishing

1. Attach pieces of 400 grit and 600 grit wet & dry sandpaper to a stable surface (Sandpaper grit may vary from 200 to 2,000 depending on the size of the section). Also attach a polishing cloth with a 0.3 alumina slurry on top of it. The alumina slurry is made by adding 0.3 alumina powder and water to the top of the cloth.
2. Add some water to the 400 grit sandpaper and begin sanding in a circular motion. The circular motion keeps polishing even. The time required to polish a section will vary, depending on the thickness of the section. Thicker sections can be started with 250 circles on the 400 grit sandpaper and thinner with 50–100 circles.
3. Polish the section on the cloth and inspect the slide under a dissecting microscope to observe where the central primordium, or core, is located (Figure 7a). If the primordium is not visible, continue sanding with the lower grit sandpaper and polish and re-check frequently until the primordium becomes visible.
4. Use the higher grit sandpaper and polishing cloth alternately as the primordium nears the surface. Proceed slowly until the polisher is familiar with the distance to the primordium or samples will be lost through over-polishing (Figure 7b).
5. Use the polishing cloth and slurry after sanding is complete to smooth the section surface and improve the clarity of the section.
6. Heat the slide back up and use forceps to flip the section over.
7. Repeat the polishing process on the second side until a clear section, including the primordium, is visible. Prepared sections can be stored in standard slide boxes.
8. It may take 30 min or more of sanding, polishing and checking to obtain a good section for ageing. Automated polishing machines are available but they are costly and require training to use. Distance holders can also be used where available to hold the slides and ensure that otoliths are not over-polished.

Figure 7. Polishing a transverse section of *Lutjanus griseus* to (**a**) to reveal the central primordium (*arrow*), but (**b**) not beyond. Note that the black primordium is missing in (**b**).

Alternative polishing protocol for un-sectioned otoliths

Some otoliths have surfaces that are too rough for clear viewing of the whole, un-sectioned otolith, but also are quite small, making sectioning difficult or unnecessary. Polishing the surface of these whole, un-sectioned otoliths can be done by placing the otolith directly on the fingertip and polishing the otolith (and fingertip!) in repeated circular (clockwise and counter-clockwise) motions with the alumina slurry and polishing cloth. Similarly, with much less wear and tear on the fingertip, the otolith can be glued directly to a slide with Crystalbond thermoplastic cement and polished one side at a time as described for the sections above (Box 6). The slide provides a means of holding the otolith while following the polishing steps (Secor et al. 1992).

2.5.4 Transmission and scanning electron microscopy preparation

Occasionally, the analysis of daily otolith increments is constrained by the resolution of light microscopy. Light microscopy resolution generally is limited to 0.25 μm and practically to 0.5–1.0 μm. The use of transmission electron microscopy (TEM) and scanning electron microscopy (SEM) can circumvent this problem and enable the high resolution of very fine-scale structure. The processing of samples for TEM and SEM is very specialized, labour-intensive, and expensive, however, meaning that the use of these techniques is limited to very special circumstances where light microscopy fails, sample sizes are small, and obtaining the information revealed by SEM or TEM is considered essential. Neither TEM nor SEM have been used frequently in daily ageing efforts for young fish, although SEM has been used in a few cases (e.g., Radtke & Dean 1981, Radtke 1984b,c, Prince et al. 1991, Correia et al. 2002). SEM is used more frequently in annual ageing efforts, however, and techniques and preparation of the samples are discussed in greater detail elsewhere (Panfili et al. 2002, Chapter 7, this volume, Morales-Nin & Panfili 2002b).

Table 3. Sample otolith statistics from individual tropical fishes collected from Barbados, the Florida Keys, St. Croix, Cayos Cochinos, and North Carolina. Late stage larvae (L) from light traps and juvenile (J) fish collected from reefs or mangroves were measured for standard length (SL), longest radius of the sagitta (S) and lapillus (L). The clearest otolith [sagitta (S) or lapillus (L)] was selected to obtain the pelagic larval duration (PLD), juvenile age, and mean increment width (IW, μm) of the larval and juvenile increments. For ease of comparison, all samples where the lapillus was used are highlighted. All otoliths were examined under 400× magnification.

Family	Species	Stage	SL (mm)	S-radius	L-radius	S or L	PLD	Larval IW	Juv Age	Juv IW
Holocentridae	*Holocentrus coruscus*	J	30.9	976	601	L	29	14.5	12	12.1
Scorpaenidae	*Scorpaena calcarata*	L	8.9	208	100	L	21	4.6	–	–
Serranidae	*Seranus tigrinus*	L	8.6	131	89	S	15	8.1	–	–
Apogonidae	*Phaeoptyx type A*	L	8.8	289	119	L	30	3.9	–	–
Apogonidae	*Phaeoptyx conklini*	J	14.0	632	238	L	15	6.9	14	7.8
Lutjanidae	*Lutjanus griseus*	J	31.7	377	–	S	22	8.5	29	5.5
Lutjanidae	*Lutjanus jocu*	L	14.0	336	162	L	17	8.9	–	–
Lutjanidae	*Ocyurus chrysurus*	L	12.7	147	146	L	17	8.3	–	–
Gerreidae	*Euconostomus type A*	L	8.8	182	105	L	15	6.8	–	–
Haemulidae	*Haemulon aurolineatum*	J	39.2	757	338	L	13	5.7	47	8.4
Pomacanthidae	*Holocanthus sp.*	L	10.9	380	186	L	16	11.2	–	–
Pomacanthidae	*Pomacanthus paru*	L	8.3	304	127	L	18	6.9	–	–
Pomacentridae	*Stegastes partitus*	J	12.3	438	292	L	27	7.8	12	5.4
Pomacentridae	*Abudefduf saxatilus*	J	~13	490	242	L	16	10.9	6	6.8
Pomacentridae	*Chromis multilineata*	J	17.6	518	266	L	27	7.8	5	6.7

Family	Species	Stage	SL (mm)	S-radius	L-radius	S or L	PLD	Larval IW	Juv Age	Juv IW
Sphyraenidae	*Sphyraena barracuda*	L	14.3	327	108	S	21	11.7	10	4.1
Labridae	*Halichoeres bivittatus*	J	11.8	207	170	S	22	6.3	5	9.1
Labridae	*Thalassoma bifasciatum*	J	10.3	198	182	S	31	4.4	4	7.6
Labridae	*Xyrichtys type B*	L	10.8	126	109	S	27	4.6	–	–
Scaridae	*Cryptotomus roseus*	L	8.4	93	62	S	26	3.5	–	–
Scaridae	*Sparisoma type A*	L	8.6	95	53	S	39	2.4	–	–
Scaridae	*Sparisoma viride*	J	9.5	181	93	S	45	2.7	7	7.9
Blenniidae	*Malacoctenus triangularis*	J	19.5	260	160	S	23	7.4	5	12.2
Blenniidae	*Ophioblennius atlanticus*	L	37.0	427	255	S	36	8.7	7	13.9
Gobiidae	*Coryphopterus glaucofraenum*	J	8.5	263	86	S	27	6.5	6	9.9
Gobiidae	*Coryphopterus type A*	L	9.1	195	68	S	32	6.0	–	–
Gobiidae	*Gnatholepis thompsoni*	J	10.0	321	122	L	58	3.0	7	7.1
Acanthuridae	*Acanthurus bahianus*	J	29.5	754	447	L	54	7.9	26	4.3
Bothidae	*Bothus sp.*	L	17.4	160	148	S	33	4.5	–	–
Monacanthidae	*Monocanthus sp.*	L	7.6	121	103	L	15	6.4	–	–
Monacanthidae	*Aluterus schoepfi*	L	11.1	147	116	S	22	6.5	–	–

2.6 SELECTION OF OTOLITHS TO READ

The selection of which of the otoliths to use in an ageing study depends on the study species, relative age of fish, traditional or previous use by other researchers, and available time, manpower, and sectioning equipment. The choice will be obvious for some species, whilst for others there will be no obvious advantage to using one otolith over the other. The sagittae of the bicolor damselfish, *Stegastes partitus*, for example, is quite large, even in late-stage larvae, while the whole lapillus is rather small and clearly readable using only oil immersion preparation. Selecting the easiest and quickest otolith to prepare for interpretation greatly facilitates the work and enables the examination of larger numbers of individuals. It generally is wise to select the same otolith for examination as in previous studies or work on the same species from the same or different locations. Where prior experience is not available, preliminary effort should be made to age both the sagitta and lapillus from each of a number of individuals to explore their relationship and ease of preparations and reading. One otolith may form earlier in development, so there may be a consistently lower number of increments in one otolith type, or deposition rates may change during different periods or transitions (e.g., Hare & Cowen 1994). Once this relationship is identified, either otolith can be aged and a correction factor applied if necessary. Data from only one otolith type, however, should be used for the final analysis in a given study. The sizes and choice of the "best" otoliths are provided for a selected number of young tropical fish in Table 3.

Each otolith in a pair of sagittae and lapilli should be similar and can be randomly selected for increment interpretation. Occasionally, one otolith of the pair will be abnormally shaped or fractured in such a way that accurate reading would be compromised (Figure 8). Clearly, these should not be used, but frequently the other member of the pair is normal and readable. The initial steps of an otolith study should include careful examination of all otoliths to familiarize the reader with the shape, orientation, and readability of all otoliths for all life stages of interest. Labelled sketches can be useful during this phase.

2.7 OTOLITH INCREMENT INTERPRETATION

The first essential tool in reading fish otoliths for daily increments is a good compound microscope equipped with a polarizing filter and oil immersion lenses of several magnifications (25×, 40× and/or 100×), depending on the study species and size of their otoliths (Table 3). Examination of whole otoliths is frequently improved using oil immersion, while sectioned otoliths can be read with or without oil immersion. Selecting the correct magnification for reading is a balance between being able to clearly resolve daily increments and being able to view the entire reading axis (often the longest axis) in one image. Magnification should not be so high that sub-daily increments become difficult to distinguish from daily increments nor so low that daily increments cannot be distinguished. Lower magnifications are needed for finding very small otoliths on the slide and obtaining complete images of whole otoliths. It is sometimes impossible to fit the entire reading axis in one image, meaning that two readings of the otolith must be made and the data later combined. It is necessary to identify otolith landmarks in each reading to avoid double counting increments.

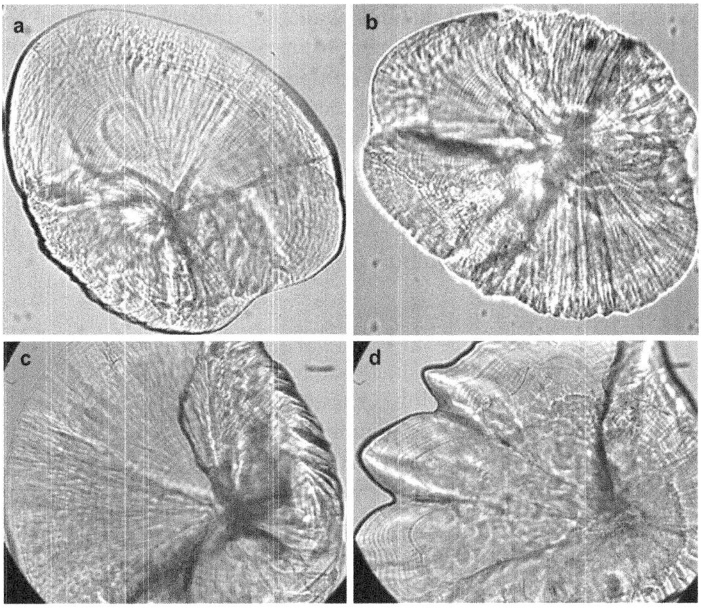

Figure 8. (**a**) Normal and (**b**) abnormally-shaped sagittae from *Thalassoma bifasciatum*, and (**c**) normal and (**d**) abnormally-shaped lapilli from *Stegastes partitus*.

Ideally, the microscope is set up with a camera that is attached to a computer equipped with image-analysis software. Image-analysis software has revolutionized the use of otoliths in daily ageing studies by improving the accuracy with which otoliths are read as well as enabling the measurement of small differences in otolith increment widths for comparing daily growth rates. This technology also greatly speeds up the process of otolith reading. Troadec and Benzinou (2002) provide a detailed discussion of visual data processing and the use of computer technology in age and growth studies. It is really only possible to obtain simple age measurements by eye through the microscope without such software. Use of an ocular micrometer is possible, but it is a very slow and relatively inaccurate method. Thus, the remainder of this discussion will assume access to such image analysis software. There are several different systems used currently in otolith ageing studies. The system upon which much of the remainder of this discussion is based is Image-Pro Plus®, version 4.5 (Media Cybernetics Inc, USA), although much of the discussion applies to other systems, such as OPTIMAS (BioScan, USA) and the free software NIH Image (http://rsb.info.nih.gov/nih-image/about.html).

2.7.1 Microscope and system calibration

The first critical step in setting up an image analysis system is to calibrate the system to the microscope, including each objective. The calibration must be performed on the size of image that will actually be read. If the still image that is captured from the live microscope image is enlarged for reading purposes, either by use of a macro (see below) or manually, the calibrations must be made on the larger sized image. Selecting an incorrect calibration greatly compounds errors and without correct calibration, measurements are not comparable and so unusable.

One of the most convenient features available in many image analysis programs is the ability to create and save measurement or mark settings ("calliper settings") to be loaded and used for each otolith read. The calliper settings should include unique edge markers ("edge detectors") for each type of increment (e.g., larval, settlement, juvenile) to be assigned. Edge detector tick marks can be assigned different colours and labels for easy visual reference. Measurements within the calliper settings can be set to include measurements of the distance from the otolith core to each increment and distances between individual increments. These measurements are all included when the data are exported to a spreadsheet (such as Microsoft Excel®).

2.7.2 Macros

Modern software is designed to incorporate programming that facilitates the interpretation and recording of otolith increments. Designing a macro to automate otolith reading in an image analysis program can greatly increase efficiency, although it is perfectly feasible to simply calibrate a microscope, create calliper settings, and proceed to reading otoliths. Macros can be written to automatically key repeated actions such as selecting and loading the correct calibration and calliper settings, capturing and sharpening the live microscope image, applying filters or other image-enhancing tools, assigning marks according to whether increments are larval or juvenile, making basic calculations from the data, exporting the data to a spreadsheet, and generating reports containing otolith images and tables of the data. The details of a macro are very specific to particular investigators, study species, and systems and so it is not feasible here to provide detailed instructions on creating macros. Nevertheless, a short excerpt from a macro created in Visual Basic for use with Image-Pro Plus® 4.5 is provided in Box 7 to demonstrate how codes can include commands to: (1) acquire a live preview of the otolith; (2) capture a still image; (3) enlarge the image; (4) select the appropriate calibration; and (5) load the appropriate calliper settings.

Box 7. Example of Visual Basic macro for image processing

```
Private Sub OtolithMeasure()
    ret = IpAcqShow(ACQ_LIVE' 1)                          'opens picture tool bar
    ret = IpDocClose()
    ret = IpAcqShow(ACQ_LIVE' 0)                          'live preview
    ret = IpAcqSnap(ACQ_CURRENT)                          'captures ('snaps') a still image
    ret = IpWsScale(948' 711' 0)                          'increases image size
    ret = IpAppSelectDoc(0)
    ret = IpDocClose()
    ret = IpAppSelectDoc(1)
    ret = IpFltSharpen(7' 10' 1)                          'applies filter to sharpen image
    ret = IpCalLoad("C:\IPWin4\Otolith.cal")              'loads calibration file
    ret = IpSCalSelect("DMLB68_40x")                      'selects calibration
    ret = IpClprShow(1)
    ret = IpClprSettings("C:\IPWin4\Otolith.cps"' 0)      'loads caliper setting file
End Sub
```

2.7.3 *Reading an otolith*

All readings must be done along a consistent reading axis, often the longest otolith radius. Consistency is key, however, and alternate axes can be selected for all otoliths if the longest axis is not the clearest axis. It is useful for many species to determine how otoliths have been read previously. Maintaining consistency among studies cannot only save time by avoiding "re-invention of the wheel", but also can enhance the general applicability and comparative use of the work. Several problems can arise, however, depending on the species and life stage, three of which are listed below.

(1) *Changing growth axes with ontogenetic transitions.* Larval increments may be deposited along one "long axis" but once metamorphosis has occurred, a new "long" axis of growth may emerge. One example of this is the Caribbean stoplight parrotfish, *Sparisoma viride* (Figure 9). Reading should be split into separate readings along each long axis to account for this change in growth axis.

(2) *Secondary growth centres (accessory primordia).* Also generally occurring at transitional times during early life (such as metamorphosis, or eye migration in flatfish), secondary growth centres can greatly complicate the interpretation of otolith increments. Once new growth centres form, growth increments are deposited around those centres instead of the original primordium (Figure 10). Readings that encompass these areas will often need to be broken into multiple readings. Again, consistency will enable comparisons to be made among individuals. Examples of secondary growth centres include bluefish (Hare & Cowen 1994), flatfish (Campana 1984, Sogard 1991), myctophids (Gartner 1991, Linkowski 1991) and sphyraenids (Figure 10).

(3) *Fractures.* The long axis can be interrupted by fractures that overshadow clear otolith increments. In this situation, it is best to continue reading along this same axis and interpolate marks based on the relative position of clear neighbouring increments. A complete read can be made along a shorter but clearer axis if necessary and this marked line kept on the computer screen as a reference during the "real" read along the long axis. Such interpolation should only be done on a few individuals. A new common reading axis should be selected if fractures along the longest reading axis are a recurring problem.

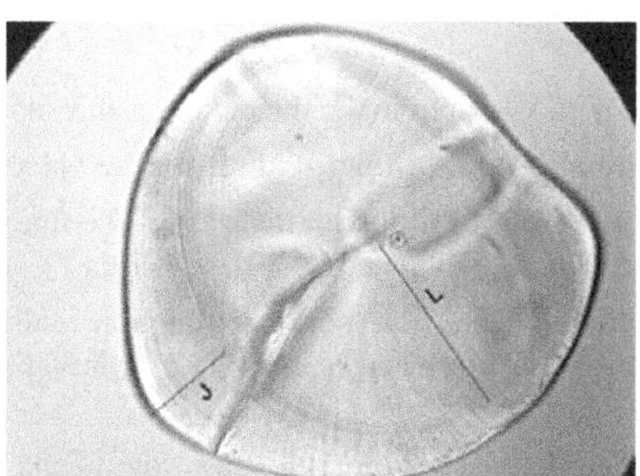

Figure 9. Changing otolith growth axes in a larval parrotfish, *Sparisom viride*. After settlement, juvenile increments (J) are deposited along an axis that is perpendicular to the larval (L) growth axis.

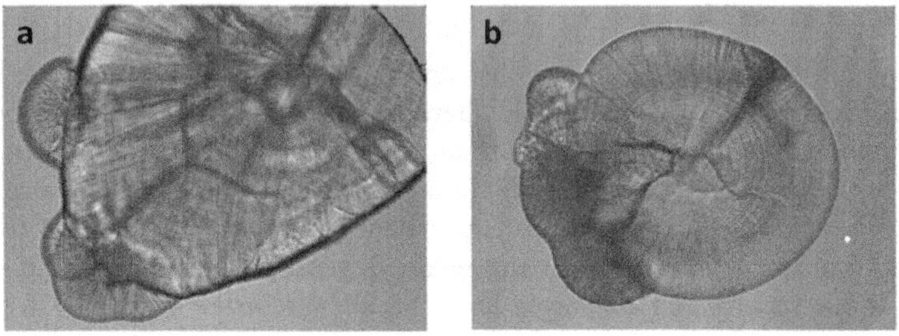

Figure 10. Formation of secondary growth centres (accessory primordia) during transitional events (e.g., metamorphosis or eye migration in bothids) for (**a**) a bothid, *Bothus sp*. and (**b**) the barracuda, *Sphyraena barracuda.*

Reading an otolith involves a continuous, dynamic analysis by the reader through the microscope and image analysis system. Continual reference through the microscope is necessary to interpret the computer image even after the "best" image is obtained for making the read and marking increments. Frequent changes of focus and polarization can help resolve difficult areas and viewing other parts of the otolith not obvious on the computer screen also can help with increment interpretation along the read axis. Regardless of otolith preparation details and reading axis, most readings will be done as outlined in Box 8 [see also Brothers (1987) for standard techniques].

Box 8. Basic steps in reading otoliths under oil immersion

Protocol

1. Turn on microscope, camera, and computer and start image analysis system.
2. Place slide on the microscope stage and adjust position until the oil drop containing the otolith is centred, lower the oil immersion lens onto the top of the oil drop, and adjust microscope until the otolith is centred in the field of view.
3. View the "live preview" on the computer screen and adjust the focus, contrast, polarization, depth of field, and light level on the microscope until the increments are visible and in focus. Obtaining the best image is critical, so patience and experimentation are needed.
4. Load a macro, if available.
5. Capture (snap) an image with which to make the readings and sharpen the image.*
6. Load the appropriate calibration and calliper settings for the objective being used.*
7. Mark the centre of the core of the otolith with the start of the line (—) function and drag the line along the longest axis (radius) to the edge of the otolith.
8. Begin marking increments with the tick mark (–) function (calliper settings can be set to create and label different types of tick marks; e.g., "L" for larval, "S" for settlement, "J" for juvenile).
9. Export the data from the completed read by selecting "Send Data" in the Input–Output page of the calliper window (ensure "DDE to Excel" option is selected.*)

* Steps 5, 6, 7, and 10 can be automated in a macro.

2.7.4 *Initiation period*

Three philosophical perspectives may be helpful during the initiation of otolith reading with a new species.

1. Otoliths often look difficult or even impossible to read during the first few readings of a new species or life stage. Otolith interpretation requires perseverance and patience. Otoliths also generally look better as more effort is put into reading, as readers become more familiar with the features of the otoliths. There will continue to be good and bad days: more than a few days of ageing work are needed to encounter a good day.
2. Otolith interpretation is a science, but increment identification and interpretation also will require judgment and a reliance on rules of parsimony.
3. It takes a period of time and experience reading the otoliths of a species before readings can be considered accurate. Interpretation of particular otolith increments and transitional marks may change over time during the initiation period, so it is important to stabilise this interpretation before using a reading as real data. It is beneficial to make a first reading of 50–100 random specimens from throughout the collection and then discard those readings and start again.

2.7.5 *Blind reading*

This topic is not an oxymoron but refers to the importance of interpreting otoliths "blind", i.e., without prior knowledge of the fish size, collection period, or any other detail of sampling which could bias the interpretation. This can be difficult and necessitates a little extra effort (such as identifying samples by number only and not other data such as fish standard length, collection location, or date). Blind reading will reduce the error introduced by readers by eliminating any subconscious bias that might be introduced during reading by knowing the identity of the sample.

2.7.6 *Multiple readers and reference training collection*

It is often ideal to have a single person responsible for otolith interpretation in a study to maintain consistency and reduce inter-reader variability, but this is not always feasible. Attempts should be made to keep the number of otolith readers on a given study to a minimum and have all readers train together, read otoliths together, and go through the initiation period together before reading samples independently when having multiple readers is unavoidable.

It is critical that readers interpret otoliths in as similar a manner as possible. It is particularly valuable to establish a reference set of otoliths for each species and stage to aid in maintaining the consistency of readings. These reference otoliths, selected from specimens of a range of known ages, can then be used to train new readers and verify that new and experienced readers are continuing to interpret otolith increments in a similar manner. It is useful to use the reference set several times during a study, not only when training new readers. Even with high consistency among readers, it is recommended that multiple readers read otoliths from a given study randomly so that any consistent differences or trends in interpretation are not superimposed on other features of the data (e.g., size categories) that could confound analyses.

Finally, at least a portion of the study otoliths should be read by all of the readers to identify any systematic differences among readers. Various statistical comparisons of reader error can be performed, with the ultimate goal of obtaining the highest quality data. Jones (2002) recommends using symmetry tests to estimate bias or imprecision in daily increment readings between multiple readers or even within a single reader. Precise readings should fall on the diagonal of a matrix of the first reading versus the second reading (see Jones 2002 for example). Reducing reader error and standardizing readings are integral parts of otolith analysis, but, ultimately, only the most objective, high quality data should be used (i.e., publication of data on reader error is not inherently useful).

2.7.7 Repeat readings and acceptable error

Even when a single reader is interpreting all the otoliths, there will be interpretation errors associated with the readings due either to human error or characteristics of the otolith. A certain proportion of all otoliths are simply not interpretable. Multiple readings of each otolith are necessary to reduce the error associated with interpretation. Several techniques have been used and these approaches have changed over time with improved technology. The most common is repeated readings until the error between or among readings is less than an acceptable maximum level, usually determined by the total number of increments being enumerated and, in part, by the question being asked. Image analysis systems have improved over time to the point where images used by one reader during increment interpretation can be saved for future checking by the same or a second reader. Using these captured images (with and without increments labelled) and reports generated by the software, multiple images (or blind readings) can be directly compared side by side. In every study, there is a number of otoliths that simply need to be thrown out because all repeated readings differ by more than the acceptable amount. It is important to keep a record of these otoliths as there may be a trend, for example, associated with some type of environmental stress. As discussed earlier, sampling design must factor in this loss.

2.8 REPORTS AND DATA HANDLING

One of the most useful advancements made to image analysis systems is the means by which data are handled. Current software enables automatic retrieval of ageing data, the creation of reports from each individual reading, and the transfer of data to a spreadsheet such as Microsoft Excel®. The specifics of how the software can be used to accomplish these tasks will vary by system, but two sample outputs are included for reference (Figure 11). Many of the steps to create these two outputs can be automated with the creation of macros.

a

Larval Fish Ecology Lab: Otolith Report

Site Code = PI
Day = 010814
Collection = 01
Sample # = 05
Species = TB
Collection # = PI010814-05
Reader: DP
Read #: 01
Date Read: 4/7/2003
Time Read: 11:29:04 AM

Comments: larval increments near the core wide. Emergence confirmed in another region of otolith.

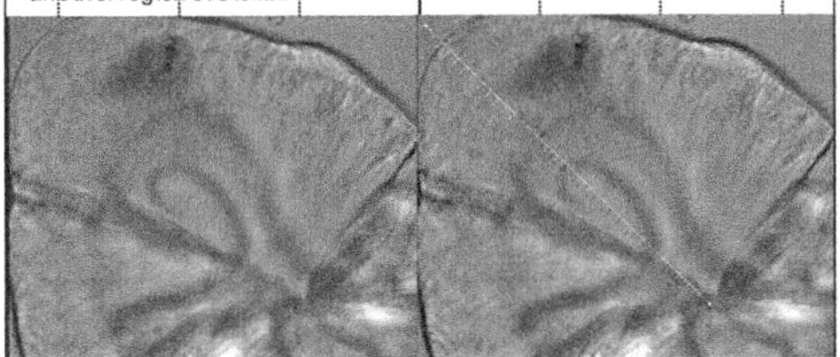

	DISTANCE FROM CORE			INCREMENT WIDTHS		
	LARVAL	SETTLEMENT	JUVENILE	LARVAL	SETTLEMENT	JUVENILE
Min	2.71	168.77	193.12	2.71	6.09	7.10
Max	163.36	187.38	264.83	5.75	6.43	8.46
Mean	75.53	178.16	228.50	4.46	6.20	7.97
StdDev	48.66	6.92	23.13	0.87	0.16	0.43
Samples	37.00	4.00	10.00	36.00	3.00	9.00
Sum	2794.73	712.63	2285.03	160.66	18.60	71.70

b

Species TB
Site PI
Day 010814
Collection01
Sample # 01
ReaderID DP
Read # 01

L = larval
S = settlement
J = juvenile
Rad. = radius
Inc. = increment

Site	Day	Collection	Sample#	Reader	L Age	S Age	J Age
PI	010814	1	1	DP	42	3	6

L Rad.	S Rad.	Otolith Rad.
209.36	228.42	263.48

Site	Day	Collection	Sample #	Inc. #	Inc. Type	Otolith Rad.	Inc. Width
PI	010814	1	1	0	L	2.72	0
PI	010814	1	1	1	L	5.11	2.39
PI	010814	1	1	2	L	7.83	2.72
PI	010814	1	1	3	L	10.55	2.72
PI	010814	1	1	4	L	13.62	3.07
PI	010814	1	1	5	L	16.68	3.06
PI	010814	1	1	6	L	19.74	3.06
PI	010814	1	1	7	L	23.49	3.75
PI	010814	1	1	8	L	27.23	3.74
PI	010814	1	1	9	L	30.98	3.75
PI	010814	1	1	10	L	34.38	3.40
PI	010814	1	1	11	L	38.13	3.75
PI	010814	1	1	12	L	42.55	4.42
PI	010814	1	1	13	L	46.64	4.09
PI	010814	1	1	14	L	51.40	4.76
PI	010814	1	1	15	L	56.17	4.77
PI	010814	1	1	16	L	59.23	3.06
PI	010814	1	1	17	L	62.64	3.41
PI	010814	1	1	18	L	66.04	3.40
PI	010814	1	1	19	L	70.81	4.77
PI	010814	1	1	20	L	75.91	5.10
PI	010814	1	1	21	L	81.36	5.45
PI	010814	1	1	22	L	86.47	5.11
PI	010814	1	1	23	L	91.23	4.76
PI	010814	1	1	24	L	97.02	5.79
PI	010814	1	1	25	L	102.81	5.79
PI	010814	1	1	26	L	108.25	5.44
PI	010814	1	1	27	L	114.04	5.79
PI	010814	1	1	28	L	119.49	5.45
PI	010814	1	1	29	L	125.95	6.46
PI	010814	1	1	30	L	131.74	5.79
PI	010814	1	1	31	L	138.89	7.15
PI	010814	1	1	32	L	144.34	5.45
PI	010814	1	1	33	L	150.46	6.12
PI	010814	1	1	34	L	155.57	5.11
PI	010814	1	1	35	L	161.02	5.45
PI	010814	1	1	36	L	167.15	6.13
PI	010814	1	1	37	L	174.29	7.14
PI	010814	1	1	38	L	182.12	7.83
PI	010814	1	1	39	L	189.61	7.49
PI	010814	1	1	40	L	197.10	7.49
PI	010814	1	1	41	L	203.57	6.47
PI	010814	1	1	42	S	209.36	5.79
PI	010814	1	1	43	S	215.83	6.47
PI	010814	1	1	44	S	222.97	7.14
PI	010814	1	1	45	J	228.42	5.45

Figure 11. Sample data outputs from ageing a sagittal otolith of *Thalassoma bifasciatum*. (**a**) Summary data and images are contained in an otolith report for each sample. Age and growth data are exported to (**b**) a spreadsheet for data analysis.

2.9 DATA ANALYSIS

There are various statistical approaches to analyzing daily otolith data which depend on the question being asked and the sampling design. In many cases, ELH traits obtained through otolith analysis can be compared among individuals, cohorts, and populations using analysis of variance (ANOVA, repeated measures, or ANCOVA tests). It is necessary, however, to distinguish between cross-sectional data where a single measure is analysed from each fish (e.g., otolith radius at settlement) and longitudinal data where multiple measures obtained over time for an individual are being analysed (e.g., increment widths; Chambers & Miller 1995). Cross-sectional data are easily compared as above, while the lack of independence among successive measurements within an individual place particular constraints on longitudinal data. The analysis of longitudinal data such as daily growth increments is discussed in detail by Vigliola and Meekan in Chapter 6 of this volume, but, generally, repeated measures ANOVA techniques (MANOVA) and other more advanced techniques are needed to account for the lack of independence between increments of individual fish (Chambers & Miller 1995, see Mosegaard et al. 2002 for summary of techniques).

Age and growth data obtained from otolith ageing have been used frequently by fisheries scientists to back-calculate individual fish sizes at earlier periods. This rather common use makes numerous assumptions and several approaches have been used in attempts to minimize inherent problems associated with the technique. Back-calculation is more frequent in annual ageing (Francis & Campana 2003, chapters in this volume) and considerable work has been done on appropriate methods for back calculation (Folkvord & Mosegard 2002, Thorrold & Hare 2002, Vigliola & Meekan in Chapter 6 of this volume). Researchers wishing to back-calculate size at age using daily age data should thoroughly review the various regression and proportional methods used in these analyses and pay particular attention to the problems associated with back-calculation. Francis and Campana (2003) provide a thorough comparison of various methods as applied to annual ageing but the limitations they discuss can also inform back-calculations based on daily increments. Back calculation and its associated errors can be bypassed entirely for many uses of daily age and growth data, however, by using relative comparisons of otolith measurements from individuals directly without back-calculating somatic size (e.g., Sponaugle & Pinkard 2004a,b).

It is sometimes desirable to obtain the ages of a large population of fish (e.g., a cohort of new recruits) to hind-cast population characteristics such as settlement patterns but processing very large numbers of otoliths is time consuming and expensive. Thus, the otoliths of a subsample of the population (cohort) sometimes are examined and the length-to-post-settlement age relationship used to calculate the age distribution of the population. Approaches to this include the use of linear least squares, geometric mean, or major axes regressions (e.g., Thorrold & Milicich 1990, Sponaugle & Cowen 1997), and non-linear models (Campana & Jones 1992, Rogers et al. 2001, Thorrold & Hare 2002). Back-calculating population age distributions can sometimes provide a coarse measure of settlement timing, despite inherent problems associated with some of these approaches. Increasing efforts to age individuals are revealing high variation in larval growth, pelagic larval duration, and size-at-settlement, however, and such back-calculations may be less useful or accurate than expected even at population level.

Data on the timing of settlement typically require analysis with time series techniques, including spectral analysis. There is a high level of inherent autocorrelation in such data because settlement is frequently cyclical. It is important to identify cyclical patterns in the data before applying statistical techniques to remove autocorrelation. Cyclic behaviour in the time domain can be identified by techniques such as the autocorrelation function (Jassby & Powell 1990). Simple techniques such as Rayleigh circular statistics (Batschelet 1981, Zar 1984) can be applied to identify the specific timing of cyclical patterns (e.g., over a lunar or tidal amplitude cycle) once the periodicity has been identified. A settlement or recruitment time series can be compared to a time series of physical (e.g., current meter) data by fitting autoregressive moving average (ARMA) models to remove autocorrelations and then cross-correlating the residuals to identify any relationships (e.g., Reyns & Sponaugle 1999). One important caveat to the hind-casting of recruitment patterns from otolith ageing of collected juveniles is that such methods rarely consider the effect of daily mortality on apparent patterns. Fishes settling closer to the time of collection will have been subjected to lower cumulative mortality than those settling earlier and so simple hind-casting of recruitment patterns will generally underestimate the contributions of older, earlier-settling fish. Thorrold and Hare (2002) recommend factoring in a mortality schedule to address this bias. Unfortunately, these are rarely available and, thus, rarely used. The collection and ageing of only the smallest (youngest) fish can minimize these problems where mortality is unknown.

3 Conclusion

Like many good ecological tools, the analysis of daily age and growth from fish otoliths requires careful use of and training in advanced techniques. The investment is worthwhile, however, given the value of the information that can be obtained from otolith work. The approach, techniques, and limitations outlined in this chapter illustrate that these techniques should not be overwhelming to the average scientist. Some investment in equipment (microscope, image analysis software, diamond blade isomet saw) is necessary in addition to the investment of time and training. Two other investments are also critical to the successful aquisiton and use of daily information from otoliths: patience and perseverance!

Acknowledgements

Kelly Denit assisted with multiple components of the chapter, including otolith protocols; images of otoliths, fish dissection, and otolith processing; table compilation; and general manuscript review. Kirsten Grorud Colvert helped in the later stages of figure construction, literature checking, and text review; and Doreen Murray quickly chased down a few lost references. The chapter benefited from the comments of Bob Cowen, the editors, and two anonymous reviewers.

References

Ahrenholz DW (2000) Periodicity of growth increment formation in otoliths of juvenile gray snapper (*Lutjanus griseus*) and lane snapper (*Lutjanus synagris*). J Elisha Mitchell Sci Soc 116:251–259

Allain G, Petitgas P, Grellier P, Lazure P (2003) The selection process from larval to juvenile stages of anchovy (*Engraulis encrasicolus*) in the Bay of Biscay investigated by Lagrangian simulations and comparative otolith growth. Fish Oceanogr 12:407–418

Anderson JT (1988) A review of size dependent survival during pre-recruit stages of fish in relation to recruitment. J NW Atl Fish Sci 8:55–66

Arai T, Limbong D, Tsukamoto K (2000) Validation of otolith daily increments in the tropical eel *Anguilla celebesensis*. Can J Zool 78:1078–1084

Batschelet E (1981) Circular statistics in biology. Academic Press, New York

Baumann H, Peck MA, Herrmann J-P (2005) Short-term decoupling of otolith and somatic growth induced by food level changes in postlarval Baltic sprat, *Sprattus sprattus*. Mar Freshwater Res 56:539–547

Begg GA, Brown RW (2000) Stock identification of haddock *Melogrammus aeglefinus* on Georges Bank based on otolith shape analysis. Tran Am Fish Soc 129:935–945

Bergenius MAJ, Meekan MG, Robertson DR, McCormick MI (2002) Larval growth predicts the recruitment success of a coral reef fish. Oecologia 131:521–525

Brogan MW (1994) Distribution and retention of larval fishes near reefs in the Gulf of California. Mar Ecol Prog Ser 115:1–13

Brothers EB (1987) Methodological approaches to the examination of otoliths in aging studies. In: Hall GE (Ed) Age and growth of fish. Iowa State University Press, Des Moines, Iowa, pp 319–330

Brothers EB (1990) Otolith marking. Am Fish Soc Symp 7:183–202

Campana SE (1984) Microstructural growth patterns in the otoliths of larval and juvenile starry flounder, *Platichthys stellatus*. Can J Zool 62:1507–1512

Campana SE (1990) How reliable are back calculations based on otoliths? Can J Fish Aquat Sci 47:2219–2227

Campana SE (2001) Accuracy, precision and quality control in age determination, including a review of the use and abuse of age validation methods. J Fish Biol 59:197–24

Campana SE, Casselman JM (1993) Stock discrimination using otolith shape analysis. Can J Fish Aquat Sci 50:1062–1083

Campana SE, Jones CM (1992) Analysis of otolith microstructure data. Can J Fish Aquat Sci 117:73–100

Campana SE, Neilson JD (1985) Microstructure of fish otoliths. Can J Fish Aquat Sci 42:1027–1029

Cermeno P, Uriarte A, de Murguia AM, Morales-Nin B (2003) Validation of daily increment formation in otoliths of juvenile and adult European anchovy. J Fish Biol 62:679–691

Chambers RC, Miller TJ (1995) Evaluating fish growth by means of otolith increment analysis: special properties of individual-level longitudinal data. In: Secor DH, Dean JM, Campana SE (Eds) Recent developments in fish otolith research. University of South Carolina Press, Columbia, pp 155–175

Choat JH, Doherty PJ, Kerrigan BA, Leis JM (1993) A comparison of towed nets, purse seine, and light-aggregation devices for sampling larvae and pelagic juveniles of coral reef fishes. Fish Bull US 91:195–209

Correia AT, Antunes C, Coimbra J (2002) Aspects of the early life history of the European conger eel (*Conger conger*) inferred from the otolith microstructure of metamorphic larvae. Mar Biol 140:165–173

Cowen RK (1991) Variation in the planktonic larval duration of the temperate wrasse *Semicossyphus pulcher*. Mar Ecol Prog Ser 69:9–15

Cowen RK, Castro LR (1994) Relation of coral-reef fish larval distributions to island scale circulation around Barbados, West Indies. Bull Mar Sci 54:228–244

DeVries DA, Grime CB, Prager MH (2002) Using otolith shape analysis to distinguish eastern Gulf of Mexico and Atlantic Ocean stocks of king mackerel. Fish Res 57:51–62

Doherty PJ (1987) Light traps: selective but useful devices for quantifying the distributions and abundance of larval fishes. Bull Mar Sci 41:423–431

Doherty PJ, McIlwain JL (1996) Monitoring larval fluxes through the surf zones of Australian coral reefs. Mar Freshwater Res 47:383–390

Dufour V, Galzin R (1993) Colonization patterns of reef fish larvae to the lagoon at Moorea Island, French Polynesia. Mar Ecol Prog Ser 102:143–152

Ekau W, Blay J (2000) Validation of daily increment deposition and early development in the otoliths of *Sarotherodon melanotheron*. J Fish Biol 57:1539–1549

Fisher R, Bellwood DR (2002) A light trap design for stratum-specific sampling of reef fish larvae. J Exp Mar Biol Ecol 269:27–37

Fives JM, Warlen SM, Hoss DE (1986) Aging and growth of larval bay anchovy *Anchoa mitchilli* from the Newport River Estuary, North Carolina, USA. Estuaries 94:362–367

Folkvord A, Mosegard H (2002) Some uses of individual age data. A. Growth and growth analysis. In: Panfili J, de Pontual H, Troadec H, Wright PJ (Eds) Manual of fish sclerochronology. Brest, France, pp 146–166

Fowler AJ (1989) Description, interpretation and use of the microstructure of otoliths from juvenile butterflyfishes (Family Chaetodontidae). Mar Biol 102:167–182

Fowler AJ, Black KP, Jenkins GP (2000) Determination of spawning areas and larval advection pathways for King George whiting in southeastern Australia using otolith microstructure and hydrodynamic modelling I. South Australia. Mar Ecol Prog Ser 199:243–254

Francis RICC, Campana SE (2003) Inferring age from otolith measurements: a review and a new approach. Can J Fish Aquat Sci 61:1269–1284

Gagliano M, McCormick MI (2004) Feeding history influences otolith shape in tropical fish. Mar Ecol Prog Ser 278:291–296

Gagliano M, McCormick M, Meekan MG (2007) Against all odds: ontogenetic changes in selective pressure mediate growth-mortality trade-offs in a marine fish Proc R Soc B 274:1575–1582

Gartner JV, Jr (1991) Life histories of three species of Lanternfishes (Pisces: Myctophidae) from the eastern Gulf of Mexico. I. Morphological and microstructural analysis of sagittal otoliths. Mar Biol 111:11–20

Geffen AJ (1992) Validation of otolith increment deposition rate. Can Spec Publ Fish Aquat Sci 117:101–113

Geffen AJ, de Pontual H, Wright PJ, Mosgaard H (2002) Sclerochronological studies B. Life history events. In: Panfili J, de Pontual H, Troadec H, Wright PJ (Eds) Manual of fish sclerochronology. Brest, France, pp 99–104

Govoni JJ, Laban EH, Hare JA (2003) The early life history of swordfish (*Xiphias gladius*) in the western North Atlantic. Fish Bull US 101:778–789

Green BS, Reilly SM, McCormick MI (2002) A cost-effective method of preparing larval fish otoliths for reading using enzyme digestion and staining. J Fish Biol 61:1600–1605

Guigand CM, Cowen RK, Llopiz JK, Richardson DE (2005) A coupled asymmetrical multiple opening closing net with environmental sampling system. Mar Tech Soc J 39:22–24

Hales LS, Jr, Hurley ADH (1991) Validation of daily increment formation in the otoliths of juvenile silver perch *Bairdiella chrysura*. Estuaries 14:199–206

Hamilton SL, Regetz J, Warner RR (2008) Postsettlement survival linked to larval life in a marine fish. Proc Natl Acad Sci 105:1561–1566

Hare JA, Cowen RK (1991) Expatriation of *Xyrichtys novacula* (Pisces: Labridae) larvae: evidence of rapid cross-slope exchange. J Mar Res 49:801–823

Hare JA, Cowen RK (1994) Ontogeny and otolith microstructure of bluefish *Pomatomus saltatrix* (Pisces: Pomatomidae). Mar Biol 118:541–550

Hare JA, Cowen RK (1996) Transport mechanisms of larval and pelagic juvenile bluefish (*Pomatomus saltatrix*) from south Atlantic Bight spawning grounds to mid–Atlantic Bight nursery habitats. Limnol Oceanogr 41:1264–1280

Hare JA, Cowen RK (1997) Size, growth, development, and survival of the planktonic larvae of *Pomatomus saltatrix* (Pisces: Pomatomidae). Ecology 78:2415–2431

Hare JA, Churchill JH, Cowen RK, Berger TJ, Cornillon PC, Dragos P, Glenn SM, Govoni JJ, Lee TN (2002) Routes and rates of larval fish transport from the southeast to the northeast United States continental shelf. Limnol Oceanogr 47:1774–1789

Hernaman V, Munday PL, Schlappy ML (2000) Validation of otolith growth-increment periodicity in tropical gobies. Mar Biol 137:715–726

Hindell JS, Jenkins GP (2005) Assessing patterns of fish zonation in temperate mangroves, with emphasis on evaluating sampling artefacts. Mar Ecol Prog Ser 290:193–205

Hoedt FE (2002) Growth in eight species of tropical anchovy determined from primary otolith increments. Mar Freshwater Res 53:859–870

Husaini M, Al-Ayoub S, Dashti J (2001) Age validation of nagroor, *Pomadasys kaakan* (Cuvier, 1830) (Family: Haemulidae) in Kuwaiti waters. Fish Res 53:71–81

Iglesias M, Brothers EB, Morales-Nin B (1997) Validation of daily increment deposition in otoliths: age and growth determination of *Aphia minuta* (Pisces: Gobiidae) from the NW Mediterranean. Mar Biol 129:279–287

Jassby AD, Powell TM (1990) Detecting changes in ecological time-series. Ecology 71:2044–2052

Jenkins GP, Black KP, Hamer PA (2000) Determination of spawning areas and larval advection pathways for King George whiting in southeastern Australia using otolith microstructure and hydrodynamic modelling I. Victoria. Mar Ecol Prog Ser 199:231–242

Jenkins GP, Davis TLO (1990) Age, growth rate, and growth trajectory determined from otolith microstructure of southern bluefin tuna *Thunnus maccoyii* larvae. Mar Ecol Prog Ser 63:93–104

Jones CM (2002) Age and growth. In: Fuiman LA, Werner RG (Eds) Fishery science: the unique contributions of early life stages. Blackwell Science Ltd, Oxford, pp 33–63

Kingsford MJ, Choat JH (1985) The fauna associated with drift algae captured with a plankton-mesh purse seine net. Limnol Oceanogr 30:618–630

Laroche JL, Richardson SL, Rosenburg AA (1982) Age and growth of a pleuronectid, *Parophrys vetulus*, during the pelagic larval period in Oregon coastal waters. Fish Bull 80:93–104

Linkowski TB (1991) Otolith microstructure and growth pattern during the early life history of lanternfishes (family Myctophidae). Can J Zool 69:1777–1792

Llopiz JK, Cowen RK (2008) Precocious, selective and successful feeding of larval billfishes in the oceanic Straits of Florida. Mar Ecol Prog Ser 358:231–244

Martin MH (1995) Validation of daily growth increments in otoliths of *Anguilla rostrata* (LeSeur) elvers. Can J Zool 73:208–211

Massuti E, Morales-Nin B, Moranta J (1999) Otolith microstructure, age, and growth patterns of dolphin, *Coryphaena hippurus*, in the western Mediterranean. Fish Bull 97:891–899

McCormick MI (1994) Variability in age and size at settlement of the tropical goatfish *Upeneus tragula* (Mullidae) in the northern Great Barrier Reef lagoon. Mar Ecol Prog Ser 103:1–15

McCurdy WJ, Panfili J, Meunier FJ, Geffen AJ, de Pontual H (2002) Preparation and observation techniques. C. Preparation of calcified structures. In: Panfili J, de Pontual H, Troadec H, Wright PJ (Eds) Manual of fish sclerochronology. Brest, France, pp 331–357

Meekan MG, Fortier L (1996) Selection for fast growth during the larval life of Atlantic cod *Gadus morhua* on the Scotian Shelf. Mar Ecol Prog Ser 137:25–37

Meekan MG, Meekan MG, Carleton JH, McKinnon AD, Flynn K, Furnas M (2003) What determines the growth of tropical reef fish larvae in the plankton: food or temperature? Mar Ecol Prog Ser 256:193–204

Morales-Nin B (2000) Review of growth regulation processes of otolith daily increment formation. Fish Res 46:53–67

Morales-Nin B, Aldebert Y (1997) Assessment of growth and hatchdate distribution of juvenile *Merluccius merluccius* in the Gulf of Lions (NW Mediterranean) determined by otolith microstructure and length frequency distribution. Fish Res 30:77–85

Morales-Nin B, Gutiérrez E, Massutí S (1995) Patterns of primary growth increments in otoliths of *Sparus aurata* larvae in relation to water temperature and food consumption. Sci Mar 59:57–64

Morales-Nin B, Panfili J (2002a) Validation and verification methods. C. Indirect validation. In: Panfili J, de Pontual H, Troadec H, Wright PJ (Eds) Manual of fish sclerochronology. Brest, France, pp 135–137

Morales-Nin B, Panfili J (2002b) Preparation and observation techniques. D. Observation. In: Panfili J, de Pontual H, Troadec H, Wright PJ (Eds) Manual of fish sclerochronology. Brest, France, pp 358–369

Moreno T, Morales-Nin B (2003) Age determination and validation on otoliths of the sand-smelt *Atherina presbyter* (Cuvier, 1829) (Pisces: Atherinidae) from the central-east Atlantic. Fish Res 62:77–87

Mosegaard H, Folkvord A, Wright PJ (2002) Some uses of individual age data. B. Ecological applications. In: Panfili J, de Pontual H, Troadec H, Wright PJ (Eds) Manual of fish sclerochronology. Brest, France, pp 167–178

Mosegaard H, Svedang H, Taberman K (1988) Uncoupling of somatic and otolith growth rates in Arctic char (*Salvelinus alpinus*) as an effect of differences in temperature response. Can J Fish Aquat Sci 45:1514–1524

Oxenford HA, Hunte W, Deane R, Campana SE (1994) Otolith age validation and growth-rate variation in flyingfish (*Hirundichthys affinis*) from the eastern Caribbean. Mar Biol 118:585–592

Paddack MJ, Sponaugle S (in press) Recruitment and habitat selection of newly settled Sparisoma viride to low coral-cover reefs. Mar Ecol Prog Ser

Pannella G (1971) Fish otoliths: daily growth layers and periodical patterns. Science 173:1124–1127

Paris CB, Cowen RK (2004) Direct evidence of a biophysical retention mechanism for coral reef fish larvae. Limnol Oceanogr 49:1964–1979

Parsons GR, KM Peters (1989) Age determination in larval and juvenile sheepshead, *Archosargus probatocephalus*. Fish Bull 87:985–988

Pepin P, Shears TH (1997) Variability and capture efficiency of bongo and Tucker trawl samplers in the collection of ichthyoplankton and other macrozooplankton. Can J Fish Aquat Sci 54:765–773

Peters KM, McMichael RH (1987) Early life-history of the red drum, *Sciaenops ocellatus* (Pisces, Sciaenidae), in Tampa Bay, Florida. Estuaries 10:92–107

Pitcher CR (1988) Validation of a technique for reconstructing daily patterns in the recruitment of a coral reef damselfish. Coral Reefs 7:105–112

Powell AB, Laban EH, Holt SA, Holt JG (2000) Validation of age estimates from otoliths of larval and juvenile spotted seatrout, *Cynoscion nebulosus*. Fish Bull 98:650–654

Prince E, Lee D, Brothers E, Zweifel J (1991) Estimating age and growth of young Atlantic blue marlin *Makaira nigricans* from otolith microstructure. Fish Bull 89:441–459

Radtke RL (1984a) Formation and structural composition of larval striped mullet otoliths. Trans Am Fish Soc 113:186–191

Radtke RL (1984b) Otolith formation and increment deposition in laboratory-reared skipjack tuna, *Euthynnus pelamis*, larvae. Proceedings of the international workshop on age determination in oceanic pelagic fishes: Tunas, billfishes and sharks. NOAA Tech Rep 8:99–103

Radtke R (1984c) Scanning electron microscope evidence for yearly growth zones in giant bluefin tuna, *Thunnus thynnus*, otoliths from daily increments. Fish Bull 82:434–440

Radtke R, Dean J (1981) Morphological features of the otoliths of the sailfish, *Istiophorus platypterus*, useful in age determination. Fish Bull 79:360–367

Ralston S, Miyamoto GT (1983) Analyzing the width of daily otolith increments to age the Hawaiian snapper, *Pristipomoides filamentosus*. Fish Bull 81:523–535

Raventos N, MacPherson E (2001) Planktonic larval duration and settlement marks on the otoliths of Mediterranean littoral fishes. Mar Biol 138:1115–1120

Ré P (1984) Evidence of daily and hourly growth in pilchard larvae based on otolith growth increments, *Sardina pilchardus* (Walbaum, 1792). Cybium 8:33–38

Reyns N, Sponaugle S (1999) Patterns and processes of brachyuran crab settlement to Caribbean coral reefs. Mar Ecol Prog Ser 185:155–170

Reznick D, Lindbeck E, Bryga H (1989) Slower growth results in larger otoliths: an experimental test with guppies (*Poecilia reticulate*). Can J Fish Aquat Sci 46:108–112

Riginos C, Victor BC (2001) Larval spatial distributions and other early life-history characteristics predict genetic differentiation in eastern Pacific blennioid fishes. Proc R Soc Lond 268:1931–1936

Robertson DR, Green DG, Victor BC (1988) Temporal coupling of production and recruitment of larvae of a Caribbean reef fish. Ecology 69:370–381

Rogers JS, Hare JA, Lindquist DG (2001) Otolith record of age, growth, and ontogeny in larval and pelagic juvenile *Stephanolepis hispidus* (Pisces: Monacanthidae). Mar Biol 138:945–953

Rosa HC, Ré P (1985) Influence of exogenous factors on the formation of daily microgrowth increments in otoliths of *Tilapia mariae* juveniles. Cybium 9:341–357

Ross SW (2003) The relative value of different estuarine nursery areas in North Carolina for transient juvenile marine fishes. Fish Bull 101:384–404

Sale PF (1980) The ecology of fishes on coral reefs. Ocean Mar Biol Ann Rev 18:367–421

Sale PF, Douglas WA, Doherty PJ (1984) Choice of microhabiatat by coral reef fishes at settlement. Coral Reefs 3:91–99

Schmitt PD (1984) Marking growth increments in otoliths of larval and juvenile fish by immersion in tetracycline to examine the rate of increment formation. Fish Bull 82:237–242

Searcy S, Sponaugle S (2000) Variable larval growth in a coral reef fish. Mar Ecol Prog Ser 206:213–226

Searcy S, Sponaugle S (2001) Selective mortality during the larval-juvenile transition in two coral reef fishes. Ecology 82:2452–2470

Secor DH, Dean JM (1992) Comparison of otolith-based back-calculation methods to determine individual growth histories of larval striped bass, *Morone saxatilis*. Can J Fish Aquat Sci 49:1439–1454

Secor DH, Dean JM, Laban EH (1991) Manual for otolith removal and preparation for microstructural examination. Belle W. Baruch Institute for Marine Biology and Coastal Research, 85pp. Online: http://cbl.umces.edu/~secor/otolith–manual.html

Secor DH, Dean JM, Laban EH (1992) Otolith removal and preparation for microstructural examination. Can Spec Publ Fish Aquat Sci 117:19–57

Serafy JE, Cowen RK, Paris CB, Capo TR, Luthy SA (2003) Evidence of blue marlin, *Makaira nigricans*, spawning in the vicinity of Exuma Sound, Bahamas. Mar Freshwater Res 54:299–306

Shenker JM, Maddox ED, Wishinski R, Pearl A, Thorrold SR, Smith N (1993). Onshore transport of settlement-stage Nassau grouper *Epinephelus striatus* and other fishes in Exuma Sound, Bahamas. Mar Ecol Prog Ser 98:31–43

Shulman MJ, Bermingham E (1995) Early life histories, ocean currents, and the population genetics of Caribbean reef fishes. Evolution 49:897–910

Sogard SM (1991) Interpretation of otolith microstructure in juvenile winter flounder (*Pseudopleuronectes americanus*): ontogenetic development, daily increment validation, and somatic growth relationships. Can J Fish Aquat Sci 48:1862–1871

Sogard SM (1997) Size-selective mortality in the juvenile stage of teleost fishes: a review. Bull Mar Sci 60:1129–1157

Sponaugle S, Cowen RK (1994) Larval durations and recruitment patterns of two Caribbean gobies (Gobiidae): contrasting early life histories in demersal spawners. Mar Biol 120:133–143

Sponaugle S, Cowen RK (1996a) Nearshore patterns of larval supply to Barbados, West Indies. Mar Ecol Prog Ser 133:13–28

Sponaugle S, Cowen RK (1996b) Larval supply and patterns of recruitment for two Caribbean fishes, *Stegastes partitus* and *Acanthurus bahianus*. Mar Freshwater Res 47:344–347

Sponaugle S, Cowen RK (1997) Early life history traits and recruitment patterns of Caribbean wrasses (Labridae). Ecol Monogr 67:177–202

Sponaugle S, Cowen RK, Shanks A, Morgan SG, Leis JM, Pineda J, Boehlert GW, Kingsford MJ, Lindeman K, Grimes C, Munro JL (2002) Predicting self–recruitment in marine populations: biophysical correlates and mechanisms. Bull Mar Sci 70S:341–375

Sponaugle S, Grorud-Colvert K (2006) Environmental variability, early life history traits, and survival of new recruits of a coral reef fish. Integr Comp Biol 46:623–633

Sponaugle S, Grorud-Colvert K, Pinkard D (2006) Temperature-mediated variation in early life history traits and recruitment success of the coral reef fish *Thalassoma bifasciatum* in the Florida Keys. Mar Ecol Prog Ser 308:1–15

Sponaugle S, Pinkard DR (2004a) Impact of variable pelagic environments on natural larval growth and recruitment of the reef fish *Thalassoma bifasciatum*. J Fish Biol 64:34–54

Sponaugle S, Pinkard DR (2004b) Lunar cyclic population replenishment of a coral reef fish: shifting patterns following oceanic events. Mar Ecol Prog Ser 267:267–280

Struhsaker P, Uchiyama JH (1976) Age and growth of the nehu, *Stolephorus purpureus* (Pisces: Engraulidae), from the Hawaiian Islands as indicated by daily growth increments of sagittae. Fish Bull 74:9–17

Sugeha Hagi Y, Shinoda A, Marui M, Arai T, Tsukamoto K (2001) Validation of otolith daily increments in the tropical eel *Anguilla marmorata*. Mar Ecol Prog Ser 220:291–294

Taylor MS, Hellberg ME (2003) Genetic evidence for local retention of pelagic larvae in a Caribbean reef fish. Science 299:107–109

Thorrold SR (1989) Estimating some early life history parameters in a tropical clupeid *Herklotsichthys castelnaui* from daily growth increments in otoliths. Fish Bull 87:73–84

Thorrold SR, Hare JA (2002) Otolith applications in reef fish ecology. In: Sale PF (Ed) Coral reef fishes: dynamics and diversity in a complex ecosystem. San Diego, CA, pp 243–264

Thorrold SR, Milicich MJ (1990) Comparison of larval duration and presettlement and postsettlement growth in two species of damselfish, *Chromis atripectoralis* and *Pomacentrus coelestis* (Pisces, Pomacentridae), from the Great Barrier Reef. Mar Biol 105:375–384

Thresher RE, Colin PL, Bell LJ (1989) Planktonic duration, distribution and population structure of western and central Pacific damselfishes (Pomacentridae). Copeia 1989:420–434

Tolimieri N (1995) Effects of microhabitat characteristics on the settlement and recruitment of a coral reef fish at two spatial scales. Oecologia 102:52–63

Troadec H, Benzinou A (2002) Computer-assisted age determination. In: Panfili J, de Pontual H, Troadec H, Wright PJ (Eds) Manual of fish sclerochronology. Brest, France, pp 199–241

Tsuji S, Aoyama T (1982) Daily growth increments observed in otoliths of the larvae of Japanese Red Sea Bream (*Pagrus major*). Bull Jpn Soc Sci Fish 48:1559–1562

Tsukamoto K (1988) Otolith tagging of ayu embryo with flourescent substances. Bull Jpn Soc Sci Fish 54:1289–1295

Tsukamoto K (1989) Otolith daily growth increments in the Japanese eel. Bull Jpn Soc Sci Fish 55:1017–1021

Tsukamoto Y, Okiyama M (1993) Growth during the early life history of the Pacific tarpon, *Megalops cyprinoides*. Jpn J Ichthyol 39:379–386

Tucker JW, Warlen SM (1986) Ageing of common snook *Centropomus undecimalis* larvae using sagittal daily growth rings. NE Gulf Sci 8:173–176

Tzeng WN, Yu SY (1989) Validation of daily growth increments in otoliths of milkfish larvae by oxytetracycline labeling. Trans Am Fish Soc 118:168–174

Uchiyama JH, Strusaker P (1981) Age and growth of skipjack tuna, *Katsuwonus pelamis*, and yellowfin tuna, *Thunnus albacares*, as indicated by daily growth increments of sagittae. Fish Bull 79:151–162

Valles H, Sponaugle S, Oxenford H (2001) Larval supply to a marine reserve and adjacent fished area in the Soufriere Marine Management Area, St. Lucia, West Indies. J Fish Biol 59:152–177

Victor BC (1982) Daily otolith increments and recruitment in two coral-reef wrasses, *Thalassoma bifasciatum* and *Halichoeres bivittatus*. Mar Biol 71:203–208

Victor BC (1983) Settlement and larval metamorphosis produce distinct marks on the otoliths of the slippery dick *Halichoeres bivittatus*. In: Reaka ML (Ed) The ecology of deep and shallow coral reefs. National Oceanic and Atmospheric Administration (Symposium Series on Underwater Research Vol 1), Rockville, pp 47–51

Victor BC (1986) Larval settlement and juvenile mortality in a recruitment-limited coral-reef fish population. Ecol Monogr 56:145–160

Victor BC, Wellington GM (2000) Endemism and the pelagic larval duration of reef fishes in the eastern Pacific Ocean. Mar Ecol Prog Ser 205:241–248

Vigliola L, Meekan MG (2002) Size at hatching and planktonic growth determine post-settlement survivorship of a coral reef fish. Oecologia 131:89–93

Villanueva R, Moli B (1997) Validation of the otolith increment deposition ratio using alizarin marks in juveniles of the sparid fishes, *Diplodus vulgaris* and *D. puntazzo*. Fish Res 30:257–260

Walker SM, McCormick MI (2004) Otolith check formation and accelerated growth associated with sex change in an annual protogynous tropical fish. Mar Ecol 266:201–212

Wellington GM, Victor BC (1989) Planktonic larval duration of one hundred species of Pacific and Atlantic damselfishes (Pomacentridae). Mar Biol 101:557–567

Wilson CA (1995) Clinic: chemical marking in otoliths. In: Secor DH, Dean JM, Campana SE (Eds) Recent developments in fish otolith research. University of South Carolina Press, Columbia, p 719

Wilson DT, McCormick MI (1997) Spatial and temporal validation of settlement marks in the otoliths of tropical reef fishes. Mar Ecol Prog Ser 153:259–271

Wilson DT, McCormick MI (1999) Microstructure of settlement marks in the otoliths of tropical reef fishes. Mar Biol 134:29–41

Wilson DT, Meekan MG (2002) Growth-related advantages for survival to the point of replenishment in the coral reef fish *Stegastes partitus* (Pomacentridae). Mar Ecol Prog Ser 231:247–260

Wright PJ, Panfili J, Folkvord A, Mosegaard H, Meunier FJ (2002b) Validation and verification methods A. Direct validation In: Panfili J, de Pontual H, Troadec H, Wright PJ (Eds) Manual of fish sclerochronology. Brest, France, pp 114–127

Wright PJ, Panfili J, Morales-Nin B, Geffen AJ (2002a) Types of calcified structures. A. Otoliths In: Panfili J, de Pontual H, Troadec H, Wright PJ (Eds) Manual of fish sclerochronology. Brest, France, pp 31–57

Xie S, Watanabe Y, Saruwatari T, Masuda R, Yamashita Y, Sassa C, Konishi Y (2005) Growth and morphological development of sagittal otoliths of larval and early juvenile *Trachurus japonicus*. J Fish Biol 66:1704–1719

Zapata FA, Herron PA (2002) Pelagic larval duration and geographic distribution of tropical eastern Pacific snappers (Pisces: Lutjanidae). Mar Ecol Prog Ser 230:295–300

Zar JH (1984) Biostatistical analysis. Prentice-Hall, Englewood Cliffs, NJ

Zerbi A, Aliaume C, Joyeux J-C (2001) Growth of juvenile tarpon in Puerto Rican estuaries. ICES J Mar Sci 58:87–95

5. ALTERNATIVES TO SECTIONED OTOLITHS: THE USE OF OTHER STRUCTURES AND CHEMICAL TECHNIQUES TO ESTIMATE AGE AND GROWTH FOR MARINE VERTEBRATES AND INVERTEBRATES

NATALIE MOLTSCHANIWSKYJ[1] AND MIKE CAPPO[2]

[1] *National Centre for Marine Conservation & Resource Sustainability, University of Tasmania, Locked Bag 1370, Launceston, Tasmania 7250, Australia*

[2] *Australian Institute of Marine Science, PMB 3, Townsville MC, Queensland 4810, Australia*

1 Introduction

Knowledge of age-related processes and growth rates of individuals is critical to understanding how tropical fish and invertebrate populations grow and recover from additional sources of mortality from fisheries and habitat disturbance. Understanding the ages of maturity and longevities of these organisms can help predict risks, refine management strategies, and understand the roles of these organisms in maintaining and modifying the ecosystems they inhabit. The choice of method used for estimating age is guided by a suite of factors: the anatomy of the study animal; the ease of obtaining, preparing and reading hard parts; and the accuracy and precision of the ageing techniques. Detailed knowledge of how these structures grow, and how the observed increments are formed, is also an integral part of developing an accurate protocol for age estimation (Francis 1995, Gauldie 1988, Durholtz et al. 1999).

The history of development of ageing protocols for tropical marine organisms is relatively short, partly because of the belief that a lack of seasonal temperature variation in lower latitudes would preclude the formation of visible increments on calcified hard parts and partly due to prevalence of small-scale, multi-species fisheries (Longhurst & Pauly 1987). Protocols using sectioned otoliths to age tropical teleosts are now well established (Choat & Robertson 2002) but the universality of otoliths as a tool for age determination in marine species is limited. Size and microstructure render otoliths unreadable for some fish families, elasmobranchs do not have otoliths, and there is no structure homologous to the otolith than can be used for age determination in tropical invertebrates, except for statoliths in some molluscs (Jackson 1994, Richardson 2001).

Fortunately, biologists have found a variety of other hard body structures that they use to obtain age estimates for invertebrates, fishes, sharks, and rays. These structures include scales, vertebrae, bones and fin spines for the vertebrates and, for the invertebrates, hard external shells, spines, ossicles, opercula, statoliths, pens, cuttlebones, and beaks. More recently, novel chemical techniques, such as the measurement of natural and anthropogenic isotopes in calcified structures and lipofuscin pigments in tissues, have been developed and used to validate and verify age estimates of tropical animals.

B.S. Green et al. (eds.), *Tropical Fish Otoliths: Information for Assessment, Management and Ecology*, Reviews: Methods and Technologies in Fish Biology and Fisheries 11,
DOI 10.1007/978-1-4020-5775-5_5, © Springer Science+Business Media B.V. 2009

Invertebrate fisheries focus on the crustaceans (lobsters, prawns and shrimps), bivalves (oysters), gastropods (abalone), and cephalopods (squid, cuttlefish and octopus). Successful management of any harvest species relies on information about life history traits, most notably recruitment rates, growth, and age at maturity. Most invertebrates have a typical asymptotic growth pattern, with rapid growth rates during the juvenile period followed by a slowing of growth rates once individuals reach sexual maturity. The cephalopods are an exception, with many squid species in particular having non-asymptotic growth patterns (Moltschaniwskyj 2004). Size frequency distributions sometimes can be used for cohort analysis and the identification of year classes, especially for species with rapid growth and discrete spawning events, but use of length frequency analysis can lead to misleading age and growth estimates (e.g., squid, Jackson et al. 2000). Cohort analysis also has been problematic for species whose cohorts are not easily identifiable on the basis of size, which makes age determination and growth estimates problematic. Many invertebrate species have hard structures that contain annual or daily records of growth that are being used to estimate age to avoid such problems.

In this chapter we describe structures, other than otoliths, and methods that are being used to estimate ages of fishes, sharks, molluscs, cubozoans, and echinoderms. Vertebrates and invertebrates are introduced and treated in separate sections and in each section we describe the structure, its preparation, interpretation and validation as a recorder of age. Our focus is on tropical, marine species but major gaps in the application of the various ageing techniques to this fauna means that we have described potential techniques from some temperate and freshwater studies. The utility of alternative structures to otoliths is illustrated and compared for the lutjanid sea perches, which predominate in published studies of age and growth of tropical fishes. Finally, we review the use of radioisotopes as a novel means of validating the age estimates of tropical animals and describe the technique of lipofuscin quantification to derive age estimates of invertebrates that lack hard body parts which can be used to estimate age.

The definition and standardisation of terms continue to be a priority in reporting studies of age and growth (see Cailliet et al. 2006 for review). Throughout this chapter the term "annulus" refers to a ring-like configuration, zone, check, part, structure or marking (Casselman 1983, Cailliet & Goldman 2004, Cailliet et al. 2006) that forms each year. The terms "verification" and "validation" describe the process of evaluating the timing and periodicity of growth zone deposition in hard structures. Verification was defined by Cailliet and Goldman (2004) as "confirming an age estimate by comparison with other indeterminate methods", such as modal progression in length frequency analysis of catches, whilst validation involves "proving the accuracy of age estimates by comparison with a determinate method", such as the rearing of known-age fish.

2 The vertebrates – use of scales, vertebrae, bones and fin spines

A wide variety of hard body parts aside from otoliths have been used in studies of age and growth of fishes, including scales, vertebral centra, urohyal bones, cleithra, opercular bones, eye lenses and fin spines (Figure 1). Validation has seldom been attempted in tropical studies using these structures and marginal increment analysis has been the prevailing means of verification. Sharks and rays have no otoliths and counts of bands on vertebral centra are used routinely for age determination. Some teleost fish, such as the marlins, sailfish, and broadbill swordfish, have otoliths that are very small and difficult to obtain, prepare and interpret, so fin spines and vertebral centra are used.

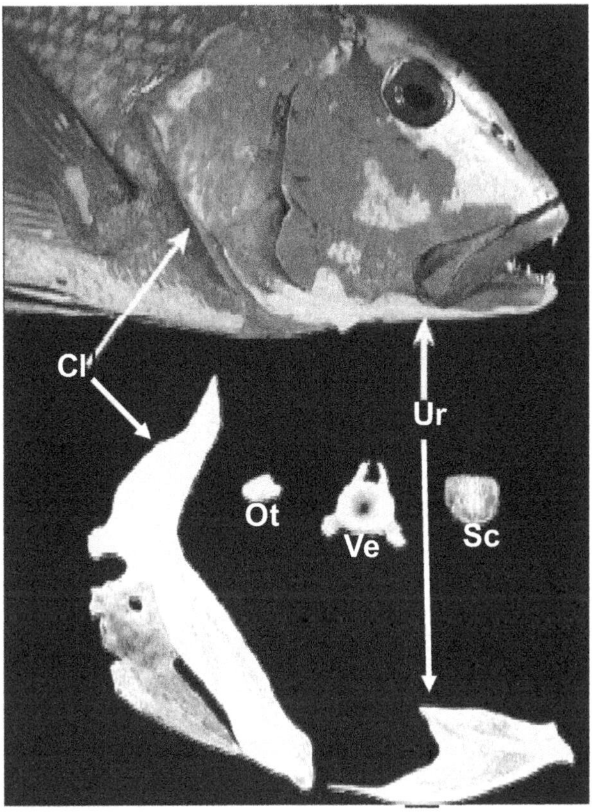

Figure 1. The location and relative size of five structures that have been used to estimate age and growth parameters for tropical snappers of the genus *Lutjanus*. Head of a red bass *Lutjanus bohar* (total length = 688 mm) showing the cleithra (Cl) and urohyal bones (Ur), whole otolith (Ot), scale (Sc), and vertebral centrum (Ve) removed from the same fish.

Studies of age and growth of tropical fishes, sharks, and rays are relatively recent compared to the long history of study of temperate families. The early studies of age and growth of tropical marine fishes mostly concerned economically important lutjanids, serranids and lethrinids and used the techniques with the simplest preparations, such as reading of scales and whole otoliths, developed for temperate species (Manooch 1987). Their results have mostly been superceded by much higher longevities identified in recent development of verified or validated ageing procedures based on sectioned otoliths (Table 1).

Table 1. Estimates of von Bertalanffy growth parameters (L_∞ and K) and longevity for four *Lutjanus* species (red snappers) by calcified structure used for age estimation and geographic region. *Legend: Structure* – TS = transverse section, W = whole, Bomb = Bomb radiocarbon validation, ^{210}Pb:^{226}Ra = ^{210}Pb:^{226}Ra radiometry; *Region* – AS = Arafura Sea, GBR = Great Barrier Reef, GoA = Gulf of Aden, GoC = Gulf of Carpentaria, GoM = Gulf of Mexico, NWA = North West Australia, WF = West Florida.

Species	L_∞ (mm)	K	Longevity	Structure	Region	Source
L. campechanus	–	–	6	Scales	GoM	Moseley (1966)
L. campechanus	941	0.17	13	Scales	GoM	Manooch (1987)
L. campechanus	941	0.17	11	Scales	WF	Nelson and Manooch (1982)
L. campechanus	969	0.192	34	Otoliths (TS)	GoM	Patterson et al. (2001)
L. campechanus	941	0.18	53	Otoliths (TS)	GoM	Wilson and Nieland (2001), Wilson et al. (2001)
L. campechanus	–	–	55	Otoliths (TS) Bomb	GoM	Baker and Wilson (2001)
L. erythropterus	721	0.213	7	Vertebrae	AS	Ju et al. (1989)
L. erythropterus	726	0.209	7	Vertebrae	NWA	Ju et al. (1989)
L. erythropterus (M)	600	0.41	7	Otoliths (W)	GBR	McPherson and Squire (1992)
L. erythropterus (F)	600	0.44	7	Otoliths (W)	GBR	McPherson and Squire (1992)
L. erythropterus	565	0.3	6	Otoliths (W); ^{210}Pb:^{226}Ra	GoC	Milton et al. (1995)
L. erythropterus	584	0.392	32	Otoliths (TS)	GBR	Newman et al. (2000a)
L. malabaricus	964	0.119	8	Vertebrae	AS	Lai and Liu (1979)
L. malabaricus	937	0.126	8	Vertebrae	NWA	Lai and Liu (1979)
L. malabaricus	707	0.168	10	Vertebrae/Scales	NWA	Edwards (1985)
L. malabaricus (M)	987	0.18	7	Otoliths (W)	GBR	McPherson and Squire (1992)
L. malabaricus (F)	838	0.23	7	Otoliths (W)	GBR	McPherson and Squire (1992)
L. malabaricus	592	0.22	9	Otoliths (W); ^{210}Pb:^{226}Ra	GoC	Milton et al. (1995)
L. malabaricus	727	0.286	20	Otoliths (TS)	GBR	Newman et al. (2000a)
L. malabaricus	623	0.225	31	Otoliths (TS)	NWA	Newman (2002)
L. sebae	660	0.16	11	Scales	GoA	Druzhinin and Filatova (1980)
L. sebae	678	0.13	10	Vertebrae	NWA	Yeh et al. (1986)
L. sebae	841	0.166	11	Vertebrae	AS	Liu and Yeh (1991)
L. sebae	1,483	0.06	9	Otoliths (W); ^{210}Pb:^{226}Ra	GoC	Milton et al. (1995)
L. sebae (M)	1,025	0.15	8	Otoliths (W)	GBR	McPherson and Squire (1992)
L. sebae (F)	887	0.18	8	Otoliths (W)	GBR	McPherson and Squire (1992)
L. sebae	792	0.139	22	Otoliths (TS)	GBR	Newman et al. (2000a)
L. sebae	524	0.233	34	Otoliths (TS)	NWA	Newman and Dunk (2002)

2.1 SCALES

Tropical centropomids, lethrinids, lutjanids, nemipterids, serranids, and sciaenids have been aged using scales (Bortone & Hollingworth 1980, Davis & Kirkwood 1984, Manooch 1987, Matlock et al. 1993, Rao & Rao 1986, Talbot 1960, Wassef & Bawazeer 1990a,b, Ihde & Chittenden 2002). Scales are usually taken from the area immediately behind the pectoral fin where they are relatively large and may be less prone to loss or damage. They are rinsed free of mucus, dried, and stored in paper envelopes. Thin scales are read at low magnification using transmitted light through microprojectors. For thicker scales, an imprint of the scale is taken by pressing it upon cellulose acetate film and the imprinted film is read at low magnification (Matlock et al. 1987, 1993).

Scales with an interpretable pattern of growth increments and unbroken radial rays joined at or near the focus are selected for estimating age (Figure 2). Annuli on the outer surface of scales are identified as narrow, more or less continuous growth increments of different optical density, with circuli that cut over each other in the lateral field of the scale. Scale radii and distances from the focus to successive annuli are measured along a diagonal line to the right antero-lateral scale corner (Figure 2c, d). Longhurst and Pauly (1987) illustrated an increase in the ability to discern such annuli on nemipterid scales with increasing latitude and suggested that seasonal temperature fluctuations of 4–5°C are needed to induce the formation of annuli.

A major disadvantage with the technique is that while scales which are lost through mechanical abrasion or disease are replaced by the fish these replacement scales lack a regular pattern of circuli. Tropical demersal fishes inhabiting rugose topography, such as coral reefs and mangrove swamps, may be expected to have very high rates of scale loss and replacement caused by contact with their habitat. For example, less than half of the scales collected at recapture of tagged red drum *Sciaenops ocellatus* were readable and not regenerated (Matlock et al. 1987) and only 5% of the sample of 40 scales collected from the reef-dwelling *Lutjanus bohar* in Figure 1 were identified as original scales suitable for reading.

There have been no attempts that we know of to verify or validate the timing and periodicity of ciruli on scales using calciphilic fluorochromes, although oxy-tetracycline hydrochloride (OTC) stains them very well (Marriott & Cappo 2000). Marginal increment analyses have been used, however, to verify the nature of growth increments in the spacing of circuli on tropical marine fish scales (Seshappa 1969, 1999).

2.2 VERTEBRAE, FIN SPINES, AND FIN RAYS

Whole or sectioned vertebral centra are routinely used to age tropical elasmobranchs. Extensive reviews by Goldman (2005) and Cailliet and Goldman (2004) precisely describe and generously illustrate the methods used to estimate, verify, and validate age in sharks, rays, and skates – covering age and growth information for over 110 species, including numerous tropical representatives of the genera *Pristis, Dasyatis, Himantura, Rhinobatos, Ginglymostoma, Carcharhinus, Galeocerdo, Rhizoprionodon*, and *Sphyrna*. Vertebral centra also have been used in early studies of large tropical lutjanids, while fin spines are used to age ictalurids, balistids, monacanthids and billfishes because of the very small size of their otoliths or their lack of cycloid scales (Manooch & Drennon 1987, Grove-Jones & Burnell 1991, Hill et al. 1989, Drew et al. 2006). There has been a recent resurgence in use of sections of dorsal fin rays and spines of large serranids and labrids requiring non-lethal ageing protocols in vulnerable populations (NMFS 2006), marine parks (Metcalf & Swearer 2005) or spawning aggregations (Debicella 2005). The following is intended only as a brief overview of the techniques using vertebral centra and fin spines, and we direct our readers to reviews by Cailliet et al. (2006) and Prince and Pulos (1983) for deeper synthesis of information.

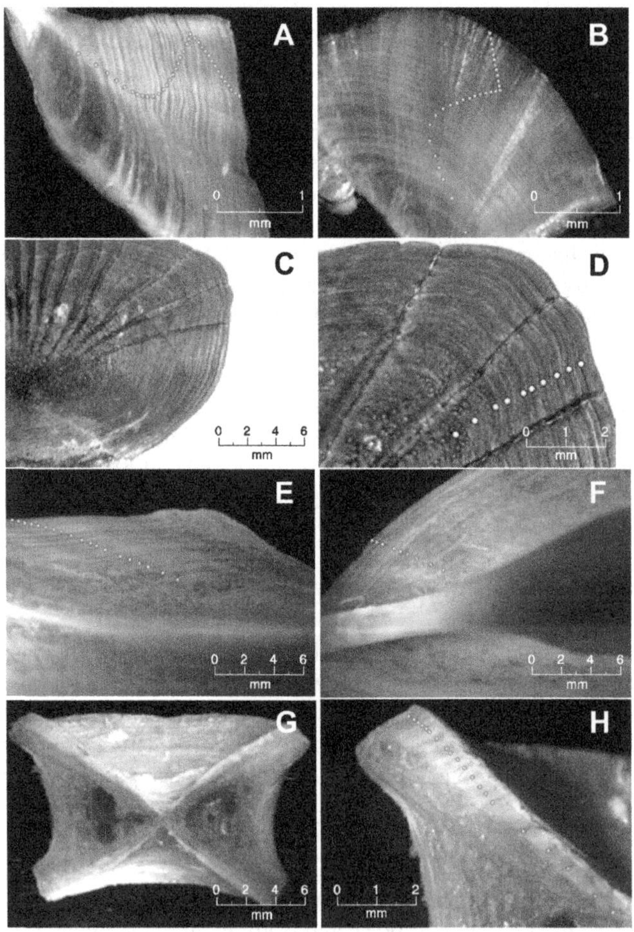

Figure 2. Photomicrographs of five different structures from a red bass *Lutjanus bohar* (TL= 688 mm) from the central Great Barrier Reef, with growth increments marked by white circles. Dorsal and ventral portion of sectioned sagitta showing 36 (**A**) and 30 (**B**) annuli, scale (**C**) with 10 of 17 increments marked (**D**), dorsal region of urohyal bone showing 19 increments (**E**), ventral region of cleithrum showing 9 increments (**F**), and sectioned 5th vertebra (**G**) showing 20 increments (**H**).

2.2.1 Vertebrae

The cartilaginous vertebral centra of elasmobranchs contain calcium phosphate laid down in distinct concentric walls of mineralised cartilage, enclosing seams of unmineralised hyaline cartilage. These consist of concentrically deposited fibres of collagen (Moulton et al. 1992). The location in the vertebral column from which samples are collected can have a statistically significant effect on counts of annuli (Officer et al. 1996). The larger, more anterior (thoracic) centra are mainly used for estimating age of both teleosts and elasmobranchs. The importance of standardising the vertebral sampling region for all ageing studies has been emphasised by Cailliet et al. (2006) and Goldman (2005) to allow for precise, valid comparisons among individuals within a population and for more accurate comparisons between populations. Annuli are read on the face of whole centra or on longitudinal sections made though the core (see Figure 2g, h). Annuli are more tightly grouped at the outer edge of vertebrae as the animals become older and somatic growth declines. These annuli may be inadvertently grouped and counted together if transverse sections or whole centra are used, producing underestimates of age. Longitudinally (sagittally) sectioned vertebrae should be used for ageing elasmobranchs unless it can be unequivocally demonstrated that identical counts of annuli can be obtained repeatedly by reading whole centra (Cailliet & Goldman 2004, MacNeil & Campana 2002).

Vertebral centra are prepared after removal by soaking in sodium hypochlorite (2.5% solution) or potassium hydroxide (5% solution) to allow removal of fascial material. The centra are stored in 70–95% ethyl alcohol or 95% isopropyl alcohol but Cailliet and Goldman (2004) recommend that a sub-sample of centra is stored frozen for staining and because long-term exposure to alcohol may reduce resolution of the banding pattern.

Vertebral sectioning is typically done with a low-speed wafering saw with diamond blades. Sections of very small centra can be cut with scalpels or by grinding away half of the centrum with fine sandpaper. Very large shark vertebrae may be secured in a vice and cut with a handsaw or small circular saw attachment on an electric drill. The face of the section can be polished with wet fine-grit sandpaper and annuli can be read using microscopy with reflected light. This technique has been used to estimate the age of large tropical lutjanids (Lai & Liu 1979). Most commonly for elasmobranchs, another cut is made on the other side of the core to obtain a thin section. This procedure is described and illustrated in detail by Goldman (2005). Cailliet and Goldman (2004) recommend that thin sections (0.3–0.5 mm in thickness) from elasmobranch centra are mounted onto slides, sanded and polished in a series of grit sizes (grades 320, 400 and 600), and finally air-dried. Microscopy with transmitted light generally is used for identification of annuli and image analysis of these thin sections.

Calciphilic compounds, or stains that have an affinity for protein, have been applied in a range of treatments and preparations to enhance the visibility of annuli in elasmobranch vertebral centra. Examples listed by Cailliet and Goldman (2004) include alcohol immersion, xylene impregnation, histology, X-radiography, X-ray spectrometry, cedar wood oil, alizarin red, silver nitrate, crystal violet, graphite microtopography, and the use of copper-, lead-, and iron-based salts. These enhancement techniques vary in their efficiency, cost, persistence, and ease of use, but most have not been evaluated thoroughly. Goldman (2005) suggests that the cobalt nitrate and ammonium sulfide stain used on the vertebral centra of the tropical lemon shark, *Negaprion brevisrostris,* could be particularly useful for other species but that simpler and cheaper staining techniques (such as crystal violet) should be tested first.

Consistent ageing protocols to increase precision in counts of annuli on vertebral centra, and to minimise ageing error, have been described only for the elasmobranchs (Goldman 2005). Sectioned elasmobranch centra have an "X" or hourglass appearance, with the arms representing the corpus calcareum separated by the intermedialia. Microscopic examination reveals "opaque" wide bands separated by distinct "translucent" narrow bands. Each pair of wide/narrow bands extends across one arm of the corpus calcareum, across the intermedialia, and across the opposing corpus calcareum arm. The sections from some species have notches visible in the outer surface of the corpus calcareum. These pairs of bands are considered to represent an annual growth cycle, and the translucent narrow bands on the corpus calcareum are counted as annuli (Goldman 2005, Cailliet & Goldman 2004). Staining has shown that opaque (calcified) bands form during summer, and less calcified and translucent (protein rich) bands form during winter or spring (Davenport & Stevens 1988, Moulton et al. 1992).

Lai and Liu (1979) identified annuli as continuous "translucent" zones, continuous or broken growth ridges, or growth ridges bordering continuous translucent zones on the centrum of whole and sectioned lutjanid vertebrae. Growth rings visible as ridges on the posterior centrum of the largest, fifth thoracic vertebra (Figure 2g) were counted under reflected light to estimate age. Age was estimated by counting only darker, denser increments on the surface of longitudinal sections (Figure 2h). These correspond to the "translucent" zones shown in Lai and Liu (1979), which must also be visibly associated with the growth ridges recognised on the concave face of the centrum to classify them as annuli in their ageing protocol. Sectioning of lutjanid vertebrae into halves through the core reveals darker zones at the margins of the centra and allows additional annuli to be distinguished within the discontinuous series of outer growth ridges on the centrum face (Marriott & Cappo 2000). Vertebral growth is likely to slow or cease altogether when teleosts and elasmobranchs approach asymptotic length. Sectioning of the vertebrae may reveal some of the growth increments crowded at the margin of teleost vertebrae but others may not be discernible, rendering the technique unsuitable for very old individuals, especially for those species with "flat-topped" growth curves (Choat & Robertson 2002).

A variety of validation techniques using known-aged individuals, tag and recapture and chemical marking have been applied to elasmobranch ageing protocols but not to the use of vertebral centra in tropical fish. The marking, release and recapture of wild elasmobranchs with calciphilic fluorochromes (Oxytetracycline, OTC) has enabled identification of sharp OTC marks on whole and sectioned vertebral centra – for some individuals at liberty for as long as 8–20 years (Brown & Gruber 1988, Smith et al. 2003, McAuley et al. 2006).

Relative marginal increment analysis has been the most common verification technique applied to ageing protocols using vertebral centra in both tropical fish and elasmobranchs (Cailliet & Goldman 2004). Such validation has proven useful for some species, especially when supported by other validation methods (Cailliet et al. 2006) but Lessa et al. (2006) reported three main sources of bias confounding the results of this technique for tropical sharks. These were the extended seasons of parturition of tropical sharks, the extended time of collection within seasons and the limited number of individuals available within particular months and size classes. These same sources of bias apply to many tropical teleosts.

2.2.2 Fin spines and fin rays

Sections of dorsal and anal fin spines have been used to estimate ages of tropical species lacking otoliths with readily interpretable internal structure, such as scombrids (Franks et al. 2000), billfishes (Hill et al. 1989, Tserpes & Tsimenides 1995, Sun et al. 2002, Drew et al. 2006), balistids and monacanthids (Manooch & Drennon 1987, Grove-Jones & Burnell 1991) and some chondrichthyans (Tanaka 1990, Clarke & Irvine 2006). Standard procedures for selection of billfish spines and the location of sections have been developed by examining the spine diameter, the degree to which the core is vascularised and the ease of interpretation of presumed annuli and other growth increments (Hill et al. 1989). Annuli in the core of the spine are replaced by vascularised tissue as billfish age (Figure 3). Billfish age estimation protocols therefore require an understanding of the vascularisation rate (Drew et al. 2006) so that these spine cores can be replaced in statistical analyses by the number of "missing" increments predicted from the radius of the spine and radius of its core (see Hill et al. 1989, Speare 2003). Oxytetracycline hydrochloride marks are easily visible on fin spines and have been used to directly validate the frequency of growth increments (Speare 1992) but marginal increment analysis has been the most common verification method (Tserpes & Tsimenides 1995, Sun et al. 2002).

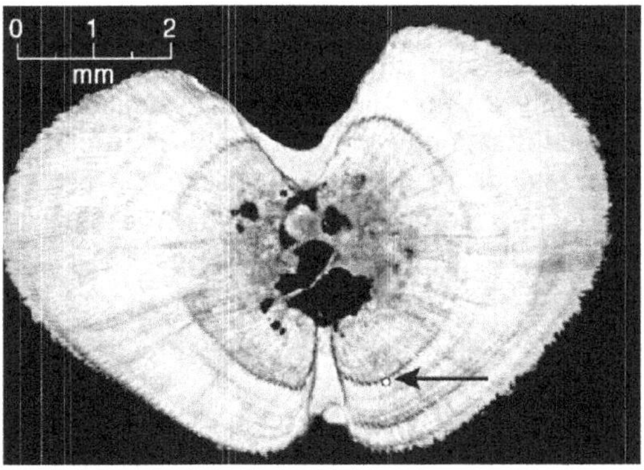

Figure 3. Section of the third anal spine of a juvenile black marlin *Makaira indica* of body length = 1,147 mm and total weight = 8.5 kg, showing one annulus and resorption and vascularisation of the core (see Speare 1992, 2003).

Several species in the warm-temperate Gulf of Mexico have been aged using fin rays or fin spines. A 90% agreement between age estimates from sectioned otoliths and fin rays was reported by Murie and Parkyn (1999) for white grunt (*Haemulon plumieri*) up to 10 years old. Fish beyond that age accumulated increments close together on the outer edge of the fin rays, leading to under-estimates of true longevity. Sections from the large, first dorsal spine of the gray triggerfish (*Balistes capriscus*) revealed up to 13 distinct, dark bands interpreted as annuli by Johnson and Saloman (1984) with agreement between readers of 98%.

Following these promising results, and to meet a demand for non-lethal ageing techniques, Debicella (2005) assessed the fin ray method in detail using the serranid *Mycteroperca microlepis* (gag) as a model species. Transverse sections of the otoliths, fin spines and fin rays of gag up to 20 years old were compared to assess average percentage error between techniques, within and amongst readers. Frayed, damaged or fused rays and spines were avoided by sampling only the mid-central region of the gag dorsal fin. Dorsal fin rays 3–7 and spines II–V were clipped off at the base of the pterygiophores and boiled for 1–10 min to allow removal of the skin with forceps or a soft-bristled brush. Cleaned and dried fin rays were coated with epoxy adhesive before 0.8 mm thick cross-sections were cut with a diamond wafering saw. Mounted sections were viewed at magnification × 4–10 with light transmitted through a green filter (wavelength 550 nm) to enhance contrast between alternating translucent and opaque zones. Annuli were enumerated from counts of the narrow translucent zones, following the procedure of Chilton and Beamish (1982). All annuli were present in sections made at the base of the dorsal fin and those from just above the pterygiophores but sections made from positions upward along the fin ray above the level of the dorsum had obscure first increments. The sinus of the internal blood vessel was an irregular shape in those distal regions of the rays, making the first increment difficult to identify. This problem also has been reported for a sciaenid and a percid (Ihde & Chittenden 2002, Isermann et al. 2003).

Within-reader and between-reader measures of precision, or reliability, of age estimates were lower using fin rays compared to sectioned otoliths but there was 87–95% agreement within ±1 year. Qualitative marginal increment analysis was used to conclude that deposition of an opaque zone, during relatively fast somatic growth, was completed at approximately the same time each spring (May–August) in both structures. Biases in age estimates using fin ray sections were postulated by Debicella (2005) to arise from problems in identification of the first annulus, differentiation of annuli from "false checks" and interpretation of the edge of sections – particularly for older fish where outer annuli were compacted into a narrow rim. Procedures to overcome some of these problems were presented by Murie (2003) and Penha et al. (2004).

2.3 UROHYAL BONES AND CLEITHRA

The largest element of the shoulder girdle, the cleithrum, is fused to the other bones of the pectoral girdle in more recently evolved teleost fish families (Figure 1). This is not the case in the esocid pikes and muskellunges, however, for which the cleithra forms the basis of age determination in temperate freshwater fisheries (Casselman 1990, Casselman et al. 1999, Harrison & Hadley 1979). In a novel esocid (a genus of pikes) monitoring program, anglers pull away and clean the cleithrum, which lies alone behind the operculum just below the skin, and send it to researchers. Termed "The Cleithrum Project", this cooperative venture began over three decades ago and continues to provide critical demographic information on *Esox masquinongy* (Casselman &

Crossman 1986, Casselman 2007). This cooperative approach has been extended to the temperate freshwater fishery for walleye *Stizostedion vitreum* (Isermann et al. 2003). The growth of annuli in esocid cleithra are readily interpreted, unlike the structure shown for *Lutjanus bohar* in Figure 2f, but we could find no application of this method for tropical fishes.

The urohyal bone is an element of the hyoidean arch (Figure 1) and is rarely used to estimate age in tropical species (Davis & West 1992), despite having interpretable growth increments (Figure 2e). Though easily accessible, this bone is not always available in tropical fisheries because fishers destroy the urohyal when making cuts to "bleed" the fish or when "gutting and gilling" fish during processing. Counts of annuli in urohyal bones, validated with marginal increment analysis, supported an estimated longevity of eight years for small *Lutjanus vittus* (Davis & West 1992) but Newman et al. (2000b) estimated longevity at 12 years using sectioned otoliths supported by mark and recapture with oxy-tetracycline (Cappo et al. 2000). No comparison of these structures within the same individuals was available to comment on the sources of this difference.

The use of bones in studies of age and growth may generally be problematic because of potential resorption by osteoclasia, where the bone surface is broken down by osteoclasts, or by osteolysis, where osteocytes are responsible for internal resorption and remodelling. Mineral resorption and redeposition can also occur. In the large swordfish *Xiphias gladius,* the bone mineral hydroxyapatite is redeposited in the centre of the vertebrae, increasing its density, and presumably its strength, in support of the swimming muscles (Kalish 2003). This might not affect the appearance of some growth bands used for ageing, and may actually play a role in the formation of the check rings in billfish fin spines, but it can violate the circumstances required for radiometric ageing and radiocarbon dating [see Section 7.1 and Section 7.2].

2.4 COMPARISONS OF DIFFERENT STRUCTURES IN ESTIMATING TROPICAL FISH AGES

Numerous studies have shown that perfect agreement over a broad range of ages is unlikely when age estimates from different structures are compared. Sometimes these comparisons are used to verify techniques and there has been a tendency to assume that the structure that provided the oldest assessment was the most reliable (Casselman 1983). Comparisons of age estimates from sectioned otoliths, whole otoliths, and scales have been most common and generally show discrepancies that increase with age and length of fish in the population. There is often a marked divergence at a threshold age, where annuli counted on otolith sections are not visible on the other structures (Ihde & Chittenden 2002). This divergence in age reached a magnitude of two to threefold for a sciaenid, a lutjanid, and a temperate percichthyid (Lowerre-Barbieri et al. 1994, Rocha-Olivares 1998, Secor et al. 1995).

The usefulness of different structures to age a tropical lutjanid was assessed by Marriott and Cappo (2000) by comparing the accuracy and precision of whole and sectioned otoliths, whole and sectioned vertebral centra and scales from the spotted-scale sea perch *Lutjanus johnii*, including recaptured individuals marked with OTC. Large differences were evident among structures, even though very old fish (25+ year) known to occur in the population were not sampled. The absence of an increment between the OTC marks and the edges of whole and sectioned vertebrae indicated an inherent inaccuracy in the use of vertebrae to age *L. johnii*. This inaccuracy was probably due to both a lag time in appearance of an identifiable increment and to errors in interpretation of the annuli at the margins of the centra.

We followed published preparation protocols (Lai & Liu 1979, Davis & West 1992, Newman et al. 1996, Matlock et al. 1993) and counted as many increments as possible that might be interpreted as annuli in our comparison here of hard parts to estimate age of *Lutjanus bohar* (Figure 2). The number of presumed annuli derived from these structures were: 36 (dorsal) and 30 (ventral) along the sulcus of sectioned otoliths; 17 on scales; 20 on sectioned vertebrae; 19 on the urohyal bone; and only 9 on the cleithrum. This individual was in the upper size range for the species, which attains a longevity of at least 55 years (Marriott et al. 2007). Whatever the true age of this fish, it is obvious from the images that sectioned otoliths are superior preparations from which to count, measure, and interpret annuli.

Otoliths from older fish are not always available, however, when the fish populations to be monitored are of special conservation significance or subject to strong catch-and-release ethics. The use of scales (or fin spines and rays) might offer a non-destructive means to determine approximate age of fish in catch-and-release programs in such cases. For example, Rocha-Olivares (1998) found that scales were reliable until *Lutjanus peru* reached 500 mm FL or an age of 5+ years and whole otoliths could be used for fish up to 800 mm FL and 16+ years but otoliths must be sectioned to observe annuli for fish 17–32+ years. For the striped bass *Morone saxatilis*, age estimates derived from scales were, on average, 9 years less than ages estimated from sectioned otoliths in fish older than 20 years but age determination using scales was adequate up to 12 years (Secor et al. 1995). Welch et al. (1993) compared scales, spines, and fin rays for the same species and recommended for fish >900 mm long (and >10 years old) the choice of a structure for age determinations should depend on whether the improved accuracy and precision expected from otoliths was worth sacrificing the fish.

The non-lethal use of fin ray sections is now recognised as an especially promising method for ageing threatened serranids, such as the goliath grouper *Epinephelus itajara*, since it would allow for monitoring the recovery of the age structure of their spawning stocks without sacrificing a large number of already scarce fish (Debicella 2005, NMFS 2006).

It is notable for the tropical lutjanids that the von Bertalanffy growth parameters K and L_∞ do not vary much among ageing methods, despite major differences in maximum longevity estimated when using the different structures for the same species (Table 1). These parameters do not capture some major life history features important for tropical fishes, however, such as the reproductive lifetime and the timing of onset of sex change (see Choat & Robertson 2002 for review), nor do they show evidence of the major changes in fisheries yield implied by different longevities. Systematic under-ageing does cause erroneous estimates of natural mortality (up to a threefold increase) that might lead to recruitment overfishing (Newman et al. 2000a).

3 The cubozoans – statoliths

The cubozoans, or box-jellyfish, have four statoliths located in the bell margin that are made of gypsum ($CaSO_4 2H_2O$) (Chapman 1985) and are used for orientation. These statoliths, like fish otoliths, have regular internal structures that indirect evidence suggests may well be growth increments (Ueno et al. 1995, Gordon et al. 2004). Statoliths from two species have been investigated to date: *Carybdea rastoni* (Ueno et al. 1995) and *Chiropsalmus* sp. (Gordon et al. 2004). The number of apparent growth increments was positively correlated with statolith size (Ueno et al. 1995), putative age estimates based on the number of increments were supported by the ecology of the animals (Ueno et al. 1995), and there was a strong relationship between increment numbers and theoretical age based on size-frequency distributions (Gordon et al. 2004).

Statoliths from both species were prepared for reading by mounting the cleaned and dried statoliths in epoxy resin and grinding the statolith using either fine emery paper or polishing paper with toothpaste or alumina paste (Ueno et al. 1995, Gordon et al. 2004). The only reported difficulty was in the resolution of increments in the centre and on the edge of the statoliths (Ueno et al. 1995).

Validation that the observed increments are deposited daily still needs to be done. Staining using tetracycline was unsuccessful as it caused death of the individuals (Gordon et al. 2004). It is also unclear exactly when the statoliths are formed, which may be as early as during the polyp phase (Ueno et al. 1995).

4 The echinoderms – ossicles and spines

The hard outer structure, or test, of echinoderms is rigid and formed by a series of radiating, plate-like ossicles or plates. The ossicles have numerous tiny crystals of calcite deposited forming a sponge-like microstructure called stereom in which a banding pattern of alternating translucent and opaque lines is observed. The banding pattern is thought to be due to the density of calcite crystals deposited in the stereom during growth, with coarse-textured stereom deposited during fast growth and fine-textured stereom deposited during slow growth (Pearse & Pearse 1975). Additional organic material and pigmentation can also be associated with growth bands (Gage 1990). Ossicles from different regions of the test can be used to estimate age in echinoderms, including ossicles from the aboral, ambital, interambulacrum, genital, and oral regions and rotula from Aristotle's lantern.

No single method of preparation appears to work on all species and a suitable method usually is determined by ossicle thickness (Jensen 1969). One popular method is to char ossicles over an alcohol flame until they become dark brown. Alternating dark and light zones (growth increments) become visible when the cooked ossicles are transferred to xylene (Jensen 1969). Alternatively, ossicles can be sanded lightly and wetted with xylene (Lumingas & Guillou 1994, Russell & Meredith 2000). Thicker ossicles from older animals may require sanding on both sides to produce a 0.5–1 mm section to allow increments to be viewed using transmitted light when the section is wetted with xylene (Lumingas & Guillou 1994). A combination of both sanding and charring is required in some cases for increments to become visible under a low-power stereo-microscope (Sano et al. 2001).

Russell and Meredith (2000) reviewed the validation of these apparent growth increments in ossciles as measures of chronological age and found that in five of eleven species the growth increments were annual. Of the remaining six species, five showed no evidence of age-related increment deposition and one was inconclusive. Problems using the increments to estimate age differed among the species, but included inter-plate variation in the number of increments, increment formation associated with feeding activity, and asymmetry in ossicle growth (Russell & Meredith 2000). The use of validated annuli to estimate age and determine growth rates for those species that have been validated, however, is proving useful and feasible (e.g. *Sphaerechinus granularis* Jordana et al. 1997).

Vertebral ossicles in many of the ophiuroids (brittle stars) have "a fine labyrinthic stereom with a ring-like pattern" (Dahm & Brey 1998) which is visible using scanning electron microscopy (SEM) of vertebral ossicles sputter-coated with gold (Gage 1990). Growth increments have been found in a range of brittle-star species (*Ophiothrix fragilis, Amphiura filiformis, A. chiajei, Ophiura ophiura*, and *Ophiura albida*) and are assumed to be formed annually, based on indirect evidence of marginal increment analysis and cohort analysis (Gage 1990). The increments in the ossicles of the brittle-stars become wider as individuals get larger (Gage 1990), unlike fish otoliths, in which increments become narrower as fish grow larger and older. All increments are visible only in juvenile brittle-stars, however, due to the process by which increments are deposited. In bigger and older animals, more recently deposited increments overlie older increments and, as a result, the older increments are not visible and the first visible increment is very wide (Dahm & Brey 1998). It is assumed when ageing older individuals, therefore, that the "hidden" increment widths in the bigger and older individuals are the same as the visible increments in younger individuals, to avoid underestimating the age of the older individuals (Dahm 1993, Dahm & Brey 1998). In this way, a number of "hidden" rings within a very wide increment in the older individual can be estimated by dividing the width of the wide increments by the width of the narrower increments in younger individuals.

The spines of the tropical crown-of-thorns starfish *Acanthaster planci* have alternating regions of dark and light pigmentation along the shaft. It is hypothesised that these regions potentially record variations in growth but contradictory conclusions about the usefulness of these spines for ageing have been reached in two studies (Stump & Lucas 1990, Souter et al. 1997). Stump and Lucas (1990) validated the frequency of the alternating pigmented regions by injecting animals with doxycycline hydrochloride and later recapturing the chemically tagged individuals. They concluded that each set of light and dark bands represented annual growth. Souter et al. (1997), however, found little concordance in the number and pattern of the bands on different spines from the same individual. Starfish are vulnerable to damage from predators and will lose and regrow spines, meaning that the bands on the spines are not a reflection of an individual starfish growth but the growth of the spine (Souter et al. 1997). Consequently, it is important that spines are not selected randomly but rather that the longest spines are selected from the proximal area of arms that show no obvious damage (Stump & Lucas 1999).

5 The molluscs – shells, opercula, statoliths, and beaks

A record of growth in the hard structures of mollusc species has been recognised for at least five decades now. The earliest estimates of age used the ring structure on the surface of bivalve shells. A recent review by Richardson (2001) examines how the periodicity and nature of ring formation and deposition can provide information about the environmental conditions experienced by a range of bivalve, gastropod, and cephalopod species. Most marine invertebrates for which age determination has been attempted have hard external structures (shells) that are formed during the larval phase and so have considerable potential as tools to determine the age of individuals. The cephalopods, however, have a number of internal hard structures that potentially can be used for age determination. Statoliths (analogous to otoliths) are most commonly used to age squid species but the statoliths of the octopus and many cuttlefish species are not suitable for age determination. The increment structure visible in beaks, the cuttlebone of *Sepia* spp. and the vestigial shell of *Octopus* spp. have been explored with some success for ageing octopus and squid species.

5.1 SHELLS

5.1.1 Gastropods

The extensive spiralling in the shell of gastropods limits the viewing of a continuous pattern of growth increments in the mesogastropods (e.g., top shells) and neogastropods (e.g., whelks and cone shells). Only in the archeogastropods (e.g., abalone, *Haliotis* spp. and limpets, *Patella* spp.), which have a flattened shell with limited spiral structure, are there structural features that can be used readily to estimate age.

Shepherd et al. (1995a) provide a description of the microstructure of the *Haliotis* shell with a view to developing a consistent use of terms. The shell of *Haliotis* is composed of three layers. On the outside is the periostracum, which is an organic protective layer. Beneath this lies a prismatic layer composed of calcite or aragonite. The innermost layer is nacre, composed of aragonite, within which increments are visible (Figure 4). The shell increases in width longitudinally at its anterior margin and in thickness through deposits of aragonite alternating with prismatic material, with one band of each deposited annually and each pair forming an annulus (Shepherd et al. 2000).

The tallest point on the shell is the spire, which is present at the time the juvenile settles to adopt an adult lifestyle. As result, the shell beneath the spire provides a record of adult age in some species (e.g., *H. fulgens* Shepherd et al. 1995a, *H. kamtschatkana* Shepard et al. 2000, *H. corrugate* Gluyas-Millan & Talavera-Maya 2003). Horizontal sections of *Haliotis* shell are prepared by grinding the spire of the shell with a disk grinder, followed by fine sanding and polishing of the ground surface. The shell is then etched with dilute hydrochloric acid (Shepherd et al. 1995a,b, 2000). It is also possible to obtain vertical sections of shell by cutting across the mid-point on the spire using a diamond saw, followed by polishing and etching as for the horizontal section. The number of annuli is comparable between the vertical and horizontal sections but annuli are easier to identify in vertical sections, although horizontal sections are easier to prepare (Shepherd et al. 2000).

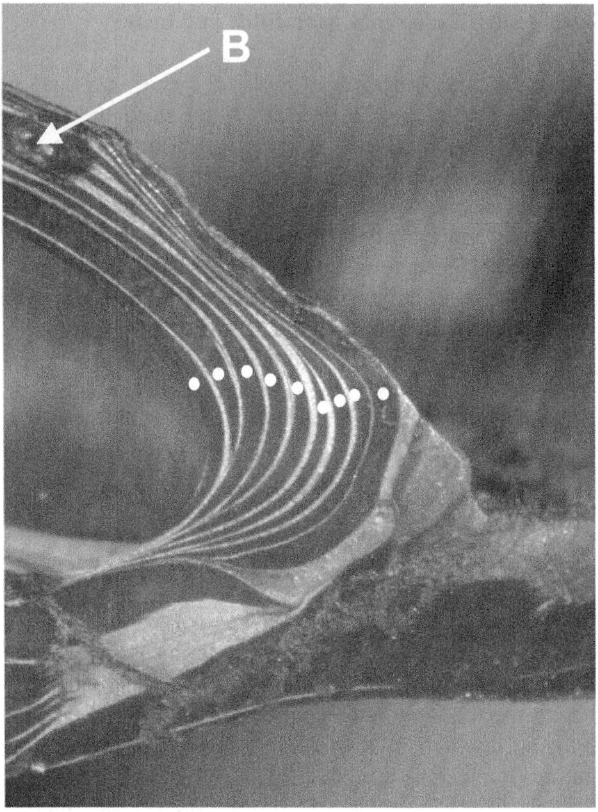

Figure 4. Section of an abalone shell (*Haliotis rubra*). This shell was sectioned through the spire and stained with a carbonate stain. There are nine presumed annuli in this shell, indicted by the *white dots*. Damage to the shell has occurred by the action of borers (B). Photos provided by the Tasmanian Aquaculture and Fisheries Institute.

Validation of the increments in the inner layer has been attempted using a number of techniques. One way is to physically mark the shell by forming a notch in the shell (Murray 1986), providing a mark produced at a known time. This method runs the risk of stressing the animal and interrupting growth (Day et al. 1995). An alternative way to mark the shell is by immersion of animals in fluorochrome stain, providing a mark at the growing edge but not on the spire where increment counts are typically obtained (Pirker & Schiel 1993, Day et al. 1995). Day et al. (1995) compared the performance of fluorochrome stains and found calcein at high concentrations to be the most successful, with bright readable marks consistently produced. Although other fluorochrome stains were successful, foaming when aeration was present (tetracycline and oxytetracycline), insolubility (alizarin), and long wavelengths (alizarin and xylenol) made these alternatives less suitable than calcein (Day et al. 1995). Tag and recapture was used to validate annual growth increments in the shells of *H. mariae* over a range of sizes (Shepherd et al. 1995b).

Some gastropod species have an operculum, a corneous or calcareous structure, which lies at the aperture of the shell. This structure does have striae and increments that can be counted and used to estimate age of temperate and tropical gastropods (e.g., Sire & Bonnet 1984, Chen & Soong 2002, Ilano et al. 2004, Richardson et al. 2005a). In gastropod species that have a thin corneous operculum, the striae on the inner side, the side attached to the foot, are readily visible using a stereomicroscope with no preparation (e.g., *Coralliophila violacea* Chen & Soog 2002, *Buccinum isaotakii* Ilano et al. 2004). Species of gastropod that have a robust calcareous operculum have an internal increment structure that is evident by preparing ground cross-sections (e.g., *Turbo setosus* Sire & Bonnet 1984) or using acetate peels of unetched polished cross-sections (e.g., *Nepyunea antiqua* Richardson et al. 2005a). The increments in the opercula appear to be deposited annually (Chen & Soong 2002, Ilano et al. 2004, Richardson et al. 2005) or daily (Sire & Bonnet 1984). Validation of deposition rate has been achieved using growth data from mark-capture studies (Chen & Soong 2002), marginal increment growth analysis (Ilano et al. 2004), staining with alizarin and fluorescein (Sire & Bonnet 1984) and correlation with age estimates using other hard structures (Richardson et al. 2005a). Like any external structure used to estimate age, the structure is exposed to damage associated with predator attacks and damage when in contact with the substratum (Richardson et al. 2005a).

5.1.2 Bivalves

Cerrato (2000) in a recent review makes comparisons between growth increments in bivalve shells and fish otoliths. Richardson (2001) provides an extensive review of the use of bivalve shells to estimate age. Bivalve shells are calcified structures made up of calcium carbonate crystals and organic material. The external surface of the shell has clearly recognisable growth increments that are assumed to be annual (e.g., scallop *Placopecten magellanicus,* Merrill et al. 1965). More recently, sections of shell have been prepared allowing greater resolution of shell microstructure. Most studies prepare shells for age determination using one of two methods. A common method uses acetate peels, in which a polished cross-section of the shell is etched with hydrochloric acid and a sheet of acetate applied providing an imprint from which the increments are read (Cerrato 2000). This method is less labour-intensive than the second method, which involves preparing a thin (150–200 µm) section allowing variations in transparency and micro-increments to be seen (Cerrato 2000). It is difficult to produce sections of uniform thickness from larger shells, however (Cerrato 2000), and preparations of shell sections from only about 5% of individuals is possible (Sejr et al. 2002). A more recent and less used method uses X-ray photographs, which allow bands of higher density to be visualised (e.g., the scallop *Adamussium colbecki* Heilmayer et al. 2003), although accurate resolution of the bands close to the shell edge is not possible.

Increments in the shell also are deposited daily (e.g., scallops Parsons et al. 1993) and the micro-increments are deposited sub-daily, correlating with tidal rhythms (e.g., Richardson 1989, Schöne et al. 2002, 2003). A prominent dark micro-increment and a faint micro-increment are produced during each tidal cycle, effectively forming a pair of micro-increments deposited each full tidal cycle. For example, in *Phacosoma japonicum* a single line and increment is produced with each tidal cycle, resulting in the appearance of two micro-growth lines and micro-growth increments during each lunar day. The micro-growth increments produced each lunar day are present within the annual increments and are resistant to the etching process. The etching process therefore helps differentiate micro-growth increments from annual increments (Schöne et al. 2003). The width of the micro-increment correlates with water temperature in a species of *Chione*, with slowest growth during the warmest and coolest months of the year (Schöne et al. 2002). Growth of the shell is controlled by temperature and may cease at cooler temperatures (14.2–16.8°C) (Schöne et al. 2003). Growth lines are also present on the resilium of the ligament of scallops and oysters and in the umbone (see review by Richardson 2001).

Validating the periodicity of micro-increments has been achieved using a range of methods. These include notching of the shell using an electrical grinder or file (Merrill et al. 1965, Sejr et al. 2002), measuring oxygen isotope levels (Keller et al. 2002, Heilmayer et al. 2003) and fluorochrome calcein markers (Kaehler & McQuiad 1999). Marking the animals through physical damage to the shell affects the condition and growth of individuals and compromises evidence of annual increment formation (Sejr et al. 2002). It is possible to validate dense and less dense regions of the increments with seasons if variations in the levels of oxygen isotope in the shell can be correlated with seasonal changes in the isotopic composition of seawater. Changes in oxygen isotope in the shell are consistent with seasonal changes in temperature, suggesting that the increments are annual, in the bivalves *Chamela gallina* (Keller et al. 2002), *Phacosoma japonica* (Schöne et al. 2003) and *Adamussium colbecki* (Heilmayer et al. 2003). The fluorochrome stain calcein can be administered either by injection into the mantle cavity or by immersion of the animal in a solution. Calcein is incorporated into the growing edge of the shell and to a lesser extent the nacreous layer of the umbonal region (Kaehler & McQuaid 1999). In the mussel, *Perna perna*, the injection produces a far more distinct fluorescent mark than immersion (Kaehler & McQuaid 1999).

Shell erosion and shell damage in both gastropods and bivalves potentially cause problems in preparing sections of shell in which all the increments can be read clearly (Figure 4). Many predators of abalone (e.g., crabs and whelks) damage the shell when trying to access the tissue (Shepherd & Breen 1992) and this damage disrupts the pattern of increments in the shell. Fouling and sand scour also potentially modify and interrupt the deposition process as well as reducing or eliminating the deposited prismatic material. Growth checks, anomalies, and minor increments are present in the nacre layer, and the identification of minor, major, and compound increments (Shepherd et al. 2000) is a problem when trying to identify annual increments. Interruptions to the depositional process due to stress events (e.g., dredging and storms) can obscure or interfere with the annual periodicity of the growth process required for the shell to be used in age determination (Merrill et al. 1965). There is no evidence that bivalve mollusc shells are resorbed during nutritional stress (e.g. Lewis & Cerrato 1997), meaning that the microstructure of the shell does provide a permanent record of growth.

The internal shell surface of some bivalves, namely fan mussels *Pinna* spp., has posterior adductor muscle scars which are produced as the posterior adductor muscle migrates with shell growth. The scars can be viewed, with no preparation, using oblique lighting across the scar. Each scar is assumed to represent seasonal growth (Butler 1987, Richardson et al. 1999) and this assumption is supported by oxygen isotope analysis of shells (Richardson et al. 1999) and "in situ" estimates of growth (Butler 1987). It appears, however, that either a muscle-scar ring is not formed or is not easily seen during the first year of growth (Richardson et al. 1999, Garcia-March & Márquez-Aliaga 2007) and so the first observed muscle scar represents two years of growth.

5.1.3 Stable oxygen isotopes

The concentration of stable oxygen isotopes has been used to age the shells of both bivalves and gastropods based on the relationship between temperature and the relative abundance of stable oxygen isotopes in the calcium carbonate of the shell (Epstein et al. 1951). In other words, levels of oxygen isotopes in the shell should oscillate with the same frequency as changes in seawater temperature and the growth in the shell between two points of high concentrations of oxygen isotope represents growth from one summer to the next, hence reflecting approximately a year of shell growth. The relative abundance of stable oxygen isotopes in the shell has been used both to validate growth increments in the shell of the scallop *Adamussium colbecki* (Heilmayer et al. 2003), the operculum of the gastropod *Buccinum undatum* (Santarelli & Gros 1985) and to determine the age of abalone *Haliotis rubra* (Gurney et al. 2005) and *Haliotis iris* (Naylor et al. 2007).

5.2 STATOLITHS

5.2.1 Gastropods

Growth increments are present in the calcareous statoliths of larval prosobranch and opistobranch larvae (e.g. Bell 1982) and juvenile and adults of some neogastropods and mesogastropods (Richardson 2001). The statoliths of larvae can be read while still in the animal (Richardson 2001). The statoliths of juveniles and adults can either be dissected out, once the pedal ganglion is located (Barroso et al. 2005), or the tissue can be extracted from the shell and digested in 1M NaOH leaving behind the statoliths (Richardson et al. 2005a,b). The statoliths can then be rinsed in tap water and alcohol and air dried before preparation for reading. Statoliths can be read whole once mounted in DPX (Richardson et al. 2005a,b), resin epoxy (Barroso et al. 2005), or CrystalBond™ 509 resin (Chatzinikolaou & Richardson 2007). Alternatively, statoliths can be resin-embedded and ground using wet abrasive paper until the nucleus is exposed and then polished with diamond paste, effectively preparing a half-section of the statolith (Richardson et al. 2005a).

Increment structure is clear and each increment is made up of a light and dark zone, apparently representing one year of growth, in three gastropod species studied recently (*Neptunea antique*, Richardson et al. 2005a, *Polinices pulchellus*, Richardson et al. 2005b, *Nassarius reticulates*, Chatzinikolaou & Richardson 2007). Validation of the one increment – one year hypothesis has been achieved using laboratory reared larvae and juveniles (Chatzinikolaou & Richardson 2007), length-frequency data analysis (Richardson et al. 2005b), or correlation with age estimates derived from other hard structures (Richardson et al. 2005a).

In addition, the presence of checks or marks in increment structure associated with hatching and the transition from a larval to adult way of life (Barroso et al. 2005,

Richardson et al. 2005a,b) provides a powerful tool to derive information about growth rates and life history strategies of gastropod molluscs. Considerable care needs to be taken to fully develop this tool, however, as for fish otoliths, with particular attention to independent validation of the assumption of annual increment formation over the entire lifetime and how preparation of statoliths affects their readability (Richardson et al. 2005b).

5.2.2 Cephalopods

The statolith in squid is now commonly and extensively used as a tool to obtain age estimates of individuals, allowing valuable life history and biological information to be derived for these unusual molluscs. A review of the techniques used to prepare statoliths and problems with using statoliths to estimate squid age is published in workshop proceedings (Jereb et al. 1991). Jackson (1994) has since reviewed the application and potential of statoliths in determining squid age. We provided here a brief overview of the use of statoliths for age determination.

Cephalopods have a single pair of statoliths encased in a cartilaginous case located posteriorly to the brain (Figure 5). The adult statolith is a tooth-shaped structure with specific regions: dorsal dome, lateral dome, wing, and rostrum (Figure 6a).

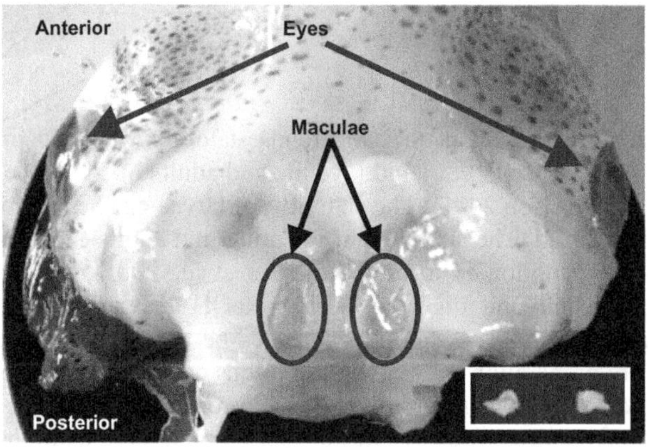

Figure 5. The location of the statoliths in the head of the loliginid squid *Sepioteuthis australis* (mantle length = 290 mm). This ventral view of the head severed from the body shows the translucent cephalic cartilage within which the statoliths are contained (*circled*). The white statoliths (inset) often can be seen sitting in the maculae.

The structure and function of the statolith is very similar to that of the otolith and, like fish otoliths, statoliths are composed of crystalline calcium carbonate in aragonite form (Radtke 1983). The periodic growth increments in statoliths are bipartite structures (Figure 6b), consisting of a discontinuous zone and an incremental zone composed of aragonite ($CaCO_3$) crystals (Lipinski 1986). Methods using statoliths to derive age estimates of squid were first developed in the late 1970s and early 1980s (Spratt 1978, Kristensen 1980) and are now commonly used on a range of squid species (see review by Jackson 1994). Many of the teuthoid squid, including the giant squid

Architeuthis kirki (Gauldie & West 1994), some cuttlefish (*Sepia officinalis,* Bettencourt & Guerra 2000 and *S. hierredda*, Raya et al. 1994), and *Idiosepius* spp (Jackson 1989, Tracey et al. 2003) have periodic structures in the statoliths that are used to estimate age. Increment structure can be seen in any of the regions, including the rostrum (*Loligo vulgaris*, Bettencourt & Guerra 2000), the wing (*Loligo vulgaris reynaudii,* Durholtz et al. 2002) and superficial to the medial inclusion of the wing (Lipinski et al. 1998). Most studies prepare the statolith through the lateral dome, however, so that the focus (Figure 6b) is evident and increments can be viewed from the nucleus through to the edge of the statolith.

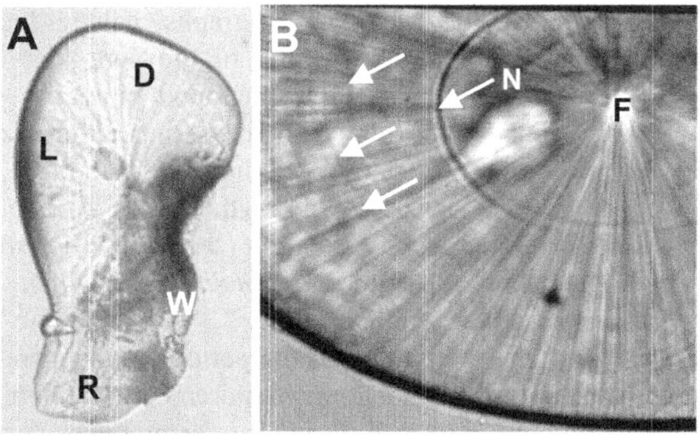

Figure 6. (**A**) A whole statolith from the tropical squid, *Photololigo* sp., showing the dorsal dome (D), lateral dome (L), wing (W), and rostrum (R). Terms defined by Clarke (1978). (**B**) The microstructure of an adult *Photololigo* sp., showing the growth increments (*arrows*), focus (F), and natal (*hatching*) check (N).

The microstructure of statoliths taken from juveniles is often clear enough without needing to prepare a section but a polish of the posterior or anterior surface using 0.9 μm alumina powder often aids resolution of the increments. The statoliths of adults of some species can be read directly without any grinding or polishing (e.g., *Loliolus noctiluca*, Jackson 1990, *Lolliguncula brevis*, Jackson et al. 1997).

Preparation of statoliths for reading, when necessary, is typically done by preparing a sagittal section. The statolith is ground or polished on the posterior (convex) or the anterior (concave) surface or both. Alternatively, a cross-section of the statolith is prepared by grinding both the ventral and dorsal ends of the statolith until a thin section containing the nucleus is obtained. Sufficient grinding is needed to ensure that increments are clearly visible (González et al. 2000) but it is also important that over-grinding does not result in removal of the focus. The ground surface of the statolith section is gently sanded with wet-and-dry polishing paper (e.g., 1,200 μm grit) or lapping film (e.g., 9 μm) and finally polished using 0.9 μm alumina powder on a polishing cloth. Statoliths are small and fragile relative to fish otoliths, so they are usually attached to a glass slide using a hard, clear resin (e.g., Eukitt, CrystalBond™, orthodontic acrylic resin, or Polarbed 812 resin). The statoliths are then viewed under a compound microscope, for which the use of a polarised light source, immersion oil, glycerol, Canada balsam, or DPX can help in the resolution of the daily growth increments.

Unfortunately, the crystalline structure of the statoliths in octopus species, the sepiolids (bobtail and dumpling squids) and many cuttlefish species either lacks visible regular increments (e.g., octopus) or has an irregular increment structure (e.g., cuttlefish). It is thought that the crystalline structure of the statolith is a function of the biochemical composition of the endolymph surrounding the statolith. Hence, the relatively lower organic content in *Sepia officinalis* statoliths may explain the poor visibility of increments (Bettencourt & Guerra 2001).

The growth increments in the statolith generally are deposited daily (Jackson 1994) and as most cephalopod species live for a year or less (3–12 months), annual growth increments are not seen in the statolith. Reading and counting of several hundred increments is aided by the use of photographs, camera-lucida and, more recently, computer imaging technology. Validation of the periodicity of the growth increments continues to be a requirement for cephalopod biologists wishing to use statoliths for age determination. Reassuringly, an increasing number of studies have evidence to support the presumption of daily periodicity in increment formation (e.g. Jackson 1994, Durholtz et al. 2002, Jackson & Forsythe 2002). A range of methods is used to validate the frequency of increment formation (see review Jackson 1994) but for only one species, the tropical broadfin squid, *Sepioteuthis lessoniana* (Jackson et al. 1993) has validation encompassed the whole lifetime.

Unfortunately, no study has yet validated the periodicity of increment formation across a range of temperatures. It appears that this requirement is becoming increasingly important, however, because slower growth during winter produces narrow increments that cannot always be resolved using light microscopy but are evident using scanning electron microscopy (e.g., *Loligo vulgaris*, Villanueva 2000 and *Lolliguncula brevis*, Durholtz & Lipinski 2000). Failure to resolve the narrow increments using light microscopy will result in underestimating age and overestimating growth rates of individuals. Very narrow increments will be especially problematic when ageing individuals that have experienced predominantly winter temperatures during their lifetime. It is possible to validate statolith increment formation by rearing or holding animals in captivity for a time but captive animals tend to have less distinct increment formation than do wild animals, especially close to the statolith margin (Jackson et al. 1993). These issues suggest that both light and scanning microscopy needs to be used to view increments if using statoliths for age determination (Lipinski & Durholtz 1994).

As for any hard structure, factors such as the angle and amount of grinding affect the view of increments in the peripheral zone (Rodhouse & Hatfield 1990, Takagi & Kitahara 2002). Furthermore, there is considerable inter-specific variation in the amount of preparation needed before the increments are seen and counted. For example, the statoliths of *Loligo vulgaris reynaudii* (Durholtz et al. 2002) and *Sepioteuthis australis* (Pecl 2000) must be sectioned, while, in contrast, *Lolliguncula brevis* statoliths require no preparation (Jackson et al. 1997). However, counts of increments are repeatable both within and between observers, suggesting that the increments are easily viewed and are not ambiguous (Jackson & Moltschaniwskyj 1999, Durholtz et al. 2002).

5.3 CUTTLEBONE, PEN, AND VESTIGIAL SHELL

Both cuttlefish and squid have a hard structure that lies in the dorsal musculature of the mantle and along the long axis of the animal. The calcium carbonate "cuttlebone" in cuttlefish is large and soft, while the "pen" or "gladius" in the squids is a less robust chitinous structure. The structure is all but lost in the octopus and is found as a vestigial shell or stylet (Figure 7) in the mantle muscle close to the attachment point of the gills (Reis & Fernandes 2002). No remnant hard structure is present in the mantle tissue of the temperate sepiolid *Euprymna tasmanica* (dumpling squid) (Moltschaniwskyj, pers. obs.) and there is nothing in the literature to suggest this is not so for all the sepiolids.

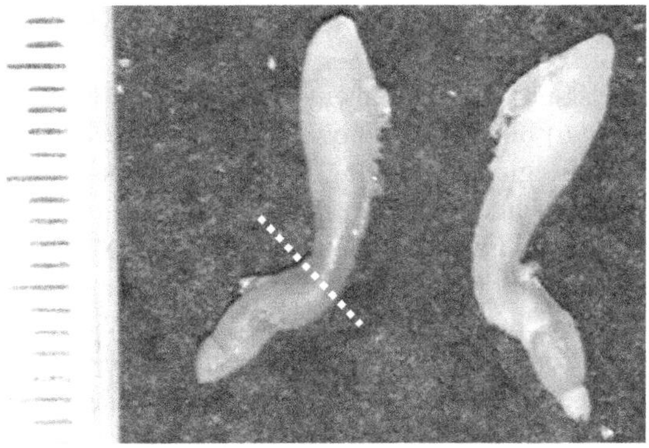

Figure 7. Paired vestigial shells of *Octopus pallidus* found at base of gills. Scale bar is in millimetres. The *dashed line* indicates the position along which the transverse section is prepared. Photo provided by J. Semmens, Tasmanian Aquaculture and Fisheries Institute.

The cuttlebone consists of an outer dorsal layer of hard, thin calcium carbonate (a hypostracum or shield) beneath which is the calcareous phragmocone. The phragmocone is made up of layers of calcareous lamellae deposited on top of one another during growth, increasing the length, depth, and width of the cuttlebone. Each layer forms a chamber, with the most recent filled with liquid and the rest filled with gas or gas and liquid. The chambers in the cuttlebone allow the animal to control its buoyancy (Denton & Gilpin-Brown 1961). Each layer of calcareous lamellae is attached to the underside of the hypostracum and it is possible therefore to remove the lamellae material along the length of cuttlebone until the hypostracum is exposed and the number of lamellae deposited along the hypostracum can be counted. The frequency with which the lamellae are deposited depends upon the biotic and abiotic conditions experienced by the animal. Periodicity of lamellae formation in *Sepia officinalis* depends on temperature (Goff et al. 1998, Bettencourt & Guerra 2001), ranging from 3.75 lamellae per month at 13°C to 18.75 lamellae per month at 25°C (Goff et al. 1998). As a result, the age for *Sepia officinalis* based on the number of lamellae in the cuttlebone can be estimated only if the thermal history of the individual is known, though this phenomenon is not consistent among all cuttlefish. Early research on three cuttlefish species (*Sepia esculenta, S. subaculeata*, and *Sepiella maindroni*) suggested that if animals were provided with sufficient food at 19–13°C then the lamellae were deposited daily (Choe 1963). More recent research on *Sepia esculenta* confirms that the periodicity of lamellae formation is not temperature dependent and therefore may indeed provide a record of age (Natsukari et al. 1991).

The squid gladius, or pen, is composed of three layers of chiton and it increases in length as the animal grows, thereby providing a record of growth (Spratt 1978, Bizikov 1991). Bizikov (1991) provides full details of the preparation of the gladius to allow increments to be seen using light microscopy. Growth increments can be seen in whole gladii in some squid or in longitudinal sections or cross-sections of the posterior tip of the gladius. There is a correlation between the number of growth increments in the statolith and the gladius in the oegopsid squids, *Sthenoteuthis oulaniensis* and *Berrytheuthis magiste* (Arkhipkin & Bizikov 1991). Other species, however, do not display such close correlations (e.g., *Sepioteuthis sepioidea*, LaRoe 1971, *Moroteuthis ingens*, Arkhipkin & Bizikov 1991, *Sepioteuthis lessoniana*, Jackson et al. 1993). It is possible that in juveniles sub-daily increments are deposited in the gladius (Jackson et al. 1993, Perez et al. 1996) but little is known about the processes involved in growth increment formation in the gladius, making it difficult to assess the validity of using increments as a measure of age. In contrast, in the oegopsid squid *Illex illecebrosus* growth increments in the gladius approximates daily growth in dorsal mantle length (Perez et al. 1996) and statolith (González et al. 2000) and increment analysis has been used to examine life history characteristics (Perez & O'Dor 2000).

The vestigial shell or stylet in octopus is a cartilage-like rod, which in *Octopus pallidus* is approximately 2 mm long and 55 μm wide (Doubleday et al. 2006). Growth increments in the stylet are present in *Octopus vulgaris* (Reis & Fernandes 2002) and *O. pallidus* (Doubleday et al. 2006). Transverse sections of the stylet are prepared either by using a microtome and mounting the section in "Aqua-mounting" under a cover slip (Reis & Fernandes 2002) or by grinding the stylet using carborundum sandpaper and lapping film (Doubleday et al. 2006). Validation using laboratory-reared animals confirms that increment formation is daily, providing a very useful tool for age determination of a group of commercially important cephalopods. Poor increment visualisation and difficulties in preparation, however, resulted in approximately only 54% of readable stylets being used in the Doubleday et al. (2006) study, suggesting that further work on preparation is needed.

5.4 BEAK

The presence of concentric increments in cephalopod beaks (Figure 8) have been noted since the 1960s (Clarke 1965). There is little evidence in the literature that the increments can be used for age determination, although they have been examined in *Octopus vulgaris* more recently (Raya & Hernández-González 1998). The presumption of one-ring – one-day could be supported in only 48.1% of the beaks of paralarval *Octopus vulgaris* of known ages (3–26 days) (Hernández-López et al. 2001). Both the upper and lower beaks are prepared by sagitally sectioning along the mid-section (Figure 8a). The increments are viewed on the inner (lateral) wall of the beak (Figure 8b) using a stereo-microscope with no further preparation (Hernández-López et al. 2001). That study found that increments at the tip of the beak, the region of the beak used for biting and tearing prey, are incomplete because of erosion and age is considered to be underestimated for adults.

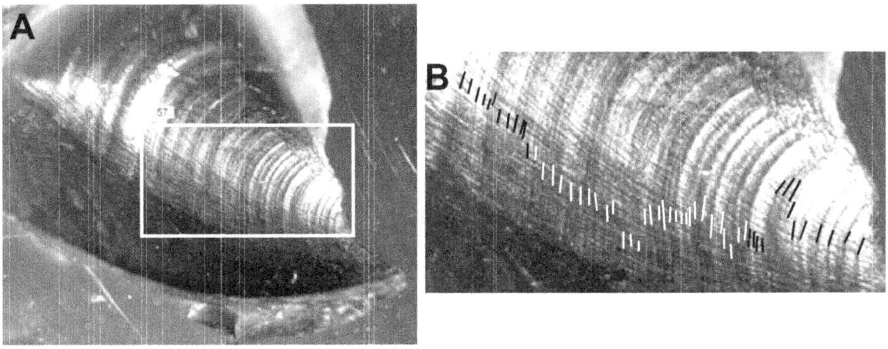

Figure 8. (**A**) A longitudinally sectioned beak from a laboratory-reared *Octopus pallidus,* which was 257 days old and weighed 97.1 g. (**B**) The increment structure of this beak. The *black and white lines* indicate the positions of increments. Photos provided by J. Semmens, Tasmanian Aquaculture and Fisheries Institute.

6 Lipofuscin – a tissue-based technique to age invertebrate species

Some invertebrates, most notably the crustaceans, do not have permanent hard structures that contain a record of somatic growth. Crustaceans shed the outer hard shell to allow somatic growth to occur. This process of moulting occurs repeatedly during the lifetime of the animals. It is possible to age post-larval blue shrimp (*Litopenaeus stylirostris*) using the number of dorsal rostral spines, although this is limited to the first 21 days of the postlarval stage (Aragón-Noriega 2005)

It has been discovered recently that in a number of crustacean and molluscan species deposits of autofluorescent neurolipofuscin accumulate in post-mitotic tissue (Figure 9). Lipofuscin is produced in crustaceans as a result of peroxidation reactions in the eyestalk ganglion and the concentration of lipofuscin is correlated to the physiological age of the individual (Sheehy 1990, Sheehy et al. 1999). The neurolipofuscin pigment can be extracted from the tissue using solvents and the intensity of florescence is used as a measure of neurolipofuscin concentration (Ju et al. 2001). Alternatively, tissue sections prepared histologically can be viewed under a specialised microscope and the density of fluorescing granules estimated (Sheehy et al. 1999). The rates of deposition are modified by environmental factors such as temperature, however, as deposition rates of neurolipofuscin are related to the physiological, not chronological, age of individuals (Sheehy & Bannister 2002). It is still possible, however, to estimate growth rates and determine the year-class strength in populations for the purpose of fisheries management.

Lipofuscin deposits are also present in the connective tissue of bivalve species (e.g. *Eurhomalea exalbida*, Lomovasky et al. 2002). This opens the possibility of using lipofuscin for age determination in mollusc species when using shell structure for age determination proves problematic.

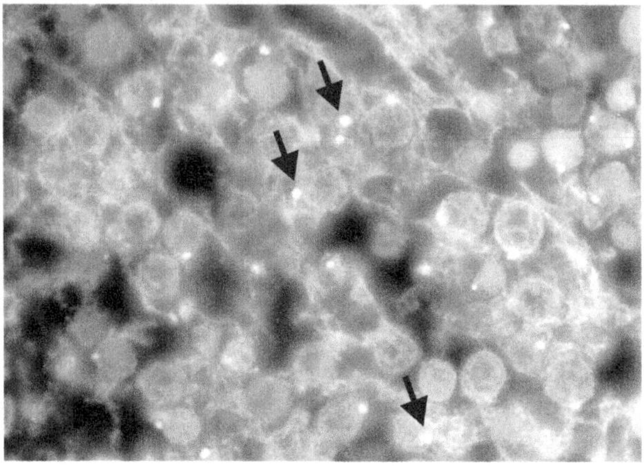

Figure 9. Photomicrograph of fluorescing lipofuscin (indicated by *arrows*) in the neurons of brain tissue taken from *Octopus maorum*. Photo provided by J. Semmens, Tasmanian Aquaculture and Fisheries Institute.

7 Ageing techniques using natural and anthropogenic radioisotopes

Direct confirmation of the timing and periodicity of growth increments in any calcified structure is difficult, apart from the use of "known age" individuals hatched and reared in captivity or tagged and released into the wild. Mark-release-recapture of tagged individuals injected with calciphilic fluorochromes or strontium was considered to be the most accurate way of confirming the frequency of formation of presumed annuli on calcified structures prior to the development of the bomb radiocarbon chronometer (see Campana 2001 for review). The number and position of presumed annuli deposited outside the chemical mark on calcified structures of recaptured individuals was compared with the known time at liberty since the chemical mark was adminstered. This approach has been used in the tropics for fish otoliths, fin spines and shark vertebral centra but not for marine fish scales. These studies are not a validation of all presumed annuli in the structure, however, and can confirm only the temporal nature of presumed annuli deposited since tagging, not necessarily absolute age. Success of mark-recapture programs relies on tagging large numbers of animals and high recovery rates several years later. There is typically little control over the size, age, and sex of animals released or recaptured and recovery rates are usually very low for animals at liberty for more than 2–3 years.

Analysis of the seasonal progression of marginal increments (MIA) has been a common verification method for sectioned otoliths, scales, fin rays and spines and vertebral centra. The technique relies on the notion that if increments are formed on an annual basis, the cycle of widths of increments on the outer edge of a structure should be sinusoidal when plotted against calendar month or season of fish collection. The resulting curves should attain values close to a maximum when the increment is completed, and values close to zero when a new increment begins to form. A temporal correlation between these cycles and events in the life history and environment should be regular and predictable (Lessa et al. 2006). Verification through MIA is often inconclusive, however, despite knowledge of such relationships. It is not appropriate for slow-growing tropical fishes because annnuli are closely grouped on the edges of calcified structures in old individuals, making them very difficult to measure in relation to the edge or margin. Secondly, the accuracy of MIA depends on large, regular, comprehensive samples from all age classes, including the rarer, oldest classes. Marginal increment analysis also suffers from the lack of an independent means of evaluation (see Campana 2001 for review). Many coral reef fish (such as large serranids) and elasmobranchs have special conservation status and often are rare on a local scale, making it difficult to sample sufficient animals for mark-recapture and marginal increment validation studies (Choat & Robertson 2002). This gap in our ability to validate or verify the age interpretations of long-lived tropical species has been partially filled by the recent application of age determination studies using natural and anthropogenic radioisotopes.

7.1 RADIOMETRIC AGEING TECHNIQUES

The chemical and microstructural features of the otolith, once deposited, remain unchanged throughout the life of a fish. This provides a basis for radiometric age determination because natural radioactive isotopes of the uranium and thorium decay series are incorporated into the accreting aragonite structure and retained thereafter (Fenton & Short 1992). These incorporated radioisotopes decay into radioactive daughter products that are retained within the otolith. The physical principles governing this radioactive decay are used in radiochemical dating. The half-lives of the parent and daughter isotopes are known (and fixed) and the "activity ratio" between them is an index of time elapsed since incorporation of the parent isotope into the otolith. An equilibrium is approached as the rate of decay of the daughter comes to equal the rate of decay of the parent.

Radiochemical dating has been attempted for fish otoliths, elasmobranch vertebral centra, nautilus and clam shells and corals (see Fenton & Short 1992, Campana 1999 for reviews). Natural radioisotopes have extremely low concentrations in both the environment and calcified tissues, so measurements of only three parent–daughter pairs have been used to infer age directly at the scale of years or verify results of traditional ageing methods. The ^{210}Lead:^{226}Radium, ^{228}Thorium:^{228}Radium, and ^{210}Polonium:^{210}Lead pairs have verified age determinations for red snappers and red drum (Milton et al. 1995, Baker & Wilson 2001, Baker et al. 2001) and tarpon (Andrews et al. 2001).

Radiochemical dating of extracted otolith cores appears to be both objective and accurate in the estimation of fish age. The cores are usually "sculpted" from the otoliths with fine, high-speed drills or mechanical grinders. Discriminatory power was considered by Campana (1999) to be in the order of five years for ^{210}Pb:^{226}Ra over the age range of 0–50 years and 1–2 years for ^{228}Th:^{228}Ra over the age range of 0–8 years. Dating the whole otolith is a useful way of discriminating among widely divergent interpretations of fish age, such as in the studies of deep-water *Sebastes* and *Hoplostethus* (Campana 1999).

The key assumptions in this discrimination have been listed by Fenton and Short (1992) and Campana (1999) and have received vigorous debate in the scientific literature (Campana & Jones 1998, West & Gauldie 1994). The greatest concerns have come from modification of the radioisotope decay equations to accommodate variable otolith growth. Circular logic has sometimes been applied, by using counts of annuli to calculate the growth rate required by the decay equation, which is in turn applied to verify the counts of annuli (see Francis 1995).

7.2 THE "BOMB RADIOCARBON CHRONOMETER"

The recent finding that nuclear testing leaves a dated mark in the calcified structures of bivalves, fishes and elasmobranchs provides a significant breakthrough in the ability to determine accurate, absolute ages for long-lived individuals (see Kalish 1993, Campana 1999 for reviews). The widespread atmospheric detonation of thermo-nuclear bombs in the 1950s and 1960s resulted in a doubling of background atmospheric radiocarbon (^{14}C), which was quickly incorporated into the surface of the oceans. Analysis of annual growth bands in longitudinal cores through *Pavona* and *Porites* coral colonies demonstrates that "bomb radiocarbon" was incorporated into the accreting coralline skeleton proportional to concentrations present in the water column at the same time (Druffel 1989). The time series of bomb radiocarbon recorded in the coral cores increased by about 20% to a plateau between 1950 and 1970, providing a benchmark or "radiocarbon bomb chronometer" against which to compare the radiocarbon levels in other calcified structures.

Using accelerator mass spectrometry (AMS) on sculpted otolith cores, Kalish (1993) was the first to demonstrate that otoliths of a marine fish (the New Zealand sparid *Pagrus auratus*) also incorporated ^{14}C. The time series of radiocarbon reconstructed from the otolith increments is similar to that present in bands from hermatypic coral cores collected in Fiji and the southern Great Barrier Reef (GBR). Kalish (1993) was therefore able to infer directly that the otolith increments had been interpreted appropriately as annuli and the fish aged correctly from 2 to 54 years because systematic underestimation or overestimation of age would have resulted in a "phase shift" between the otolith ^{14}C and the coral ^{14}C time series.

The uptake of ^{14}C in fish otoliths is synchronous with that of both corals and bivalves (Campana 1999), implying that the ^{14}C time series reconstructed from the otolith cores can be compared to one of the other radiocarbon time series in the same oceanic region and habitat type. Marine waters with ^{14}C values greater than zero parts per thousand ($^{0}/_{00}$) did not generally exist prior to the late 1950s, so otolith cores from coastal fish with zero values must have formed before the late 1950s. Even contamination with material of more recent origin could only increase, not decrease, the ^{14}C values. Thus, the ^{14}C value sets a minimum age to the sample and the years 1958–1965 become the most sensitive years for ageing based on ^{14}C levels (Campana 1999).

Subsequent work has confirmed the value of the bomb radiocarbon technique for solving problems of age validation in tropical fish and elasmobranchs, including tropical lutjanids, sciaenids, lethrinids, and serranids (Campana & Jones 1998, Baker & Wilson 2001, Baker et al. 2001, Kalish 2003), and temperate sharks (Campana et al. 2002, Kalish 2003). One of the key requirements in use of the technique, however, is that the period of increase in the ^{14}C levels in calcified structures of the study species must be synchronous with that of reference chronologies, or any lags must be understood and accounted for.

There can be a lack of synchrony for fish from both deep-sea and estuarine environments due to different rates of mixing of ^{14}C between the atmosphere, freshwaters, coastal waters and the deep ocean (Campana & Jones 1998, Kalish 2003). The ^{14}C chronology in well-mixed estuarine environments, subject to riverine input, is much closer to an atmospheric ^{14}C chronology, which started to increase in the mid-1950s and is shifted 2–4 years earlier. The opposite is true of deep-sea waters, with much slower mixing and a delayed ^{14}C chronology. Campana et al. (2002) found the delta ^{14}C of individual stable organic portions of vertebral bands from porbeagle sharks lags behind a reference carbonate chronology. This was thought to be due to the depleted ^{14}C incorporated in the non-calcified cartilaginous tissue when older porbeagles switched their diet to older prey in deeper water containing less ^{14}C.

Comparison of recent studies of the ^{14}C time series in otoliths of the tropical lutjanids *Lutjanus campechanus* and *L. erythropterus* shows the possible influence of variable salinity regimes on ^{14}C chronologies (Figure 10). The time series for *L. campechanus* in the northern Gulf of Mexico (Figure 10a) closely follows those for coral cores from Bermuda and Florida and confirms that age estimates from sectioned otoliths up to 55 years are accurate, on average, to within 1–3 years. In contrast, the ^{14}C levels for *L. erythropterus* otolith cores from the central GBR lie between a coral series from the southern GBR and atmospheric radiocarbon levels from Wellington, New Zealand, which serve as a proxy for radiocarbon in freshwater (Figure 10b). This verified the longevity of at least 32 years for this species (reported by Newman et al. 2000a) in the central GBR but also outlines the need for a better understanding of radiocarbon chronologies in coastal tropical waters subject to regular or episodic freshwater input. Ontogenetic shifts in habitat and diet must also be recognised.

The benefits of the bomb radiocarbon method have been discussed extensively and it is considered to be the most advanced and accurate method of fish age validation that is currently available (Kalish 1993, 1995, 2003, Campana 2001). As little as 0.2 mg of carbon is required for assays by AMS. The only constraints to this procedure are the relatively high cost of AMS and the requirement for samples of fish hatched before or during the 1958–65 period to take advantage of the unique ^{14}C levels during that period. It is feasible to use archival collections of otoliths held in research laboratories, if available, to apply the method to species where the fish are not long-lived (<30 years), though archival collections may be rare for tropical species, given the general scarcity of long-time series studies on tropical species (Longhurst & Pauly 1987).

7.3 TROPICAL EXAMPLES OF RADIOCHEMICAL SUPPORT FOR DISTINGUISHING LONGEVITIES

The radiochemical dating of otolith cores provides conclusive evidence that the longevity of the tropical Atlantic tarpon is >30 years for males and >50 years for females (Andrews et al. 2001). Baker et al. (2001) reported some difficulties in applying radioassays of $^{210}Po:^{226}Pb$ for *Lutjanus campechanus* and red drum *Sciaenops ocellatus* but the radiometric age estimates closely approximate ages derived from otolith sections. In contrast, the radioassays of $^{210}Pb:^{226}Ra$ in cores and whole otoliths of three red snapper species by Milton et al. (1995) produced much younger age estimates than those obtained from sectioned otoliths (Table 1). These differences have produced uncertainty about the nature of potential development of stocks shared by Indonesian and northern Australian fisheries and raised important questions regarding intraspecific, latitudinal variation in otolith interpretation and demographic parameters.

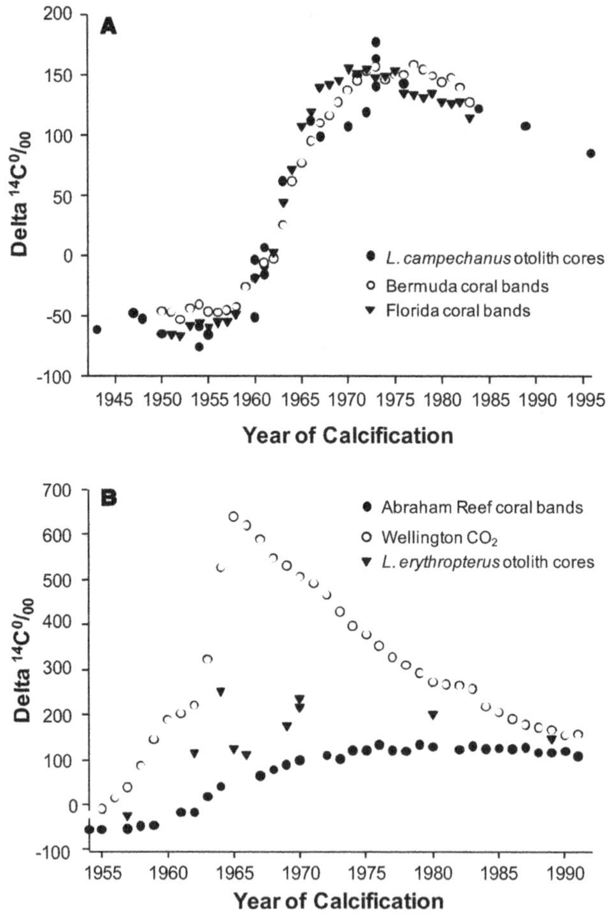

Figure 10. Radiocarbon time series by year of calcification from centres of tropical *Lutjanus* otoliths and within bands in longitudinal cores from corals. (**A**) Coral from Bermuda and Florida and otoliths from *Lutjanus campechanus*. (**B**) Coral from the southern Great Barrier Reef, CO_2 from New Zealand and otoliths from *Lutjanus erythropterus*. These data were redrawn from average values presented by Druffel (1989), Druffel and Griffin (1995), Manning and Melhuish (1994), Baker and Wilson (2001), and Kalish (2003).

Radiocarbon dating by Kalish (2003) supports the use of counts from sectioned otoliths to age Indo-Pacific lutjanids, lethrinids and serranids exploited in commercial and recreational fisheries. Longevities of at least 46 years for *Epinephelus octofasciatus*, 33 years for *Etelis carbunculus*, 27 years for *Lethrinus nebulosus* and 32 years for *Lutjanus erythropterus* have been verified this way. These verifications have greatly advanced the emerging paradigm that tropical species can be aged using sectioned otoliths and that warmer waters do not necessarily impose faster, continuous growth rates, higher natural mortality, continuous spawning seasons or reduced longevities (Longhurst & Pauly 1987).

8 Conclusion

Age-based studies of tropical fishes, elasmobranchs and squids have shown a decoupling between size and age and differences in demographic parameters between and within lineages. As a result, it will be difficult to generalise about responses to environmental changes or human impacts without a substantial database on life history characteristics (Choat & Robertson 2002, Smith et al. 1998). Consequently, there is a definitive need to develop and refine protocols to obtain accurate and precise age estimates for tropical marine vertebrates and invertebrates.

The utility of sectioned otoliths for fish, sectioned vertebrae for elasmobranchs and statoliths in squid is well established in tropical studies and these methods are improving with better understanding of how these calcified structures grow. Confidence in the age estimates provided by these protocols is growing with application of independent validation techniques. Similar procedures have been successfully applied to some marine invertebrates that have hard structures such as tests and shells. Incorrect interpretation of the growth increments in different hard structures, however, can result in serious and systematic ageing error and underestimation of age can lead to overestimates of the resilience to disturbance and harvesting of tropical species.

The comparison of methods and structures presented in this chapter compels the use of sections of teleost otoliths to estimate age if increments are interpretable. There might also be opportunities to develop reliable, cost-effective ageing methods using other methods for selected research questions once comprehensive determinations of longevity and growth rate are obtained using sectioned otoliths. Such an approach might suit the routine monitoring of heavily fished populations truncated to relatively few younger year classes, recognised quickly by reading validated annuli on whole otoliths or scales, and few larger older fish that can be aged using sectioned otoliths. The non-lethal removal of scales and fin rays offers the only chance of estimating age in studies of rare or threatened teleosts or those in marine reserves and catch-and-release sport-fisheries. The choice of a structure for age determinations in these cases should depend on whether the improved accuracy and precision expected from sectioned otoliths is worth sacrificing the fish.

Vertebral centra will continue to form the basis of elasmobranch age estimation, with recent reviews providing refined protocols, but accuracy of counts made for the older age classes of some long-lived species remain problematic. The otoliths of the tropical billfishes pose special problems in preparation and interpretation. Better understanding of bone remodelling processes and more validation and verification of the methods to interpret fin spines are needed to determine their ages using these alternative structures. The development of the bomb radiocarbon chronometer has provided a powerful tool to verify and validate ageing protocols using some of these calcified structures if material is available from older individuals, born or hatched prior to the 1970s.

Options of hard structures for use in age determination for invertebrates are limited and still being fully explored. There is evidence for some species (most notably squids and bivalves), however, that hard structures can provide a record of growth and be used for age determination. Unfortunately, for many invertebrate groups the increments in the structures are not deposited with temporal predictability or they are damaged, obscured, or obliterated in older animals, thereby preventing their use for reliable age prediction. Exploration of the structures for age determination will need to progress on a species by species basis, particularly with respect to validation and method of preparation. This is going to be very important where daily increment structures are being counted and interpreted. Using statoliths from squid to estimate age has been a powerful and useful technique and has identified critical elements about their life history that are being used by fisheries managers. In particular, the short lifespan and the shape of the growth curve were only recognised once this method had been developed and used. The density of lipofuscin deposits is potentially a useful technique to obtain an estimate the age of some species and for the crustaceans is currently the only technique to estimate age. Its greatest potential, however, is most probably only for relatively sedentary species whose thermal history is known.

The choice of structures used for estimating age in both vertebrates and invertebrates of the tropics has been guided in the past by the anatomy of the study animal, the ease of obtaining, preparing and reading the hard parts and, more recently, by verification and validation of the accuracy and precision of the ageing techniques. There is increasing recognition of the need to match these applications with appropriate knowledge of how these structures grow, and how visible increments are formed.

Acknowledgements

We thank Tarra Anderson and two anonymous reviewers for constructive comments on drafts of the chapter and Craig Mundy and Jason Seimmens for some of the photographs we used.

References

Andrews AH, Burton EJ, Coale KH, Cailliet GM, Crabtree RE (2001) Radiometric age validation of Atlantic tarpon, *Megalops atlanticus*. Fish Bull 99:389–398

Aragón-Noriega EA (2005) Dorsal rostral spines as a hard structure to determine age of blue shrimp, *Litopenaeus stylirostris* postlarvae (Decapoda, Penaeidae). Crustaceana 77:193–1202

Arkhipkin AI, Bizikov VA (1991) Comparative analysis of age and growth rates estimation using statoliths and gladius in squid. In: Jereb P, Ragonese S, Von Boletzky S (Eds) Squid age determination using statoliths. Proceedings of the International Workshop held in the Istituto di Tecnologia della Pesca e del Pescato, Vol Spec Publ No 1. Note techniche e Reprints dell'Istituto di Technologia della Pesca e del Pescato, Via Luigi Vaccara, 61-91026-Mazara del Vallo (TP), Italy, pp 19–34

Baker MJ, Wilson CA (2001) Use of bomb radiocarbon to validate otolith section ages of red snapper *Lutjanus campechanus* from the northern Gulf of Mexico. Limnol Oceanogr 46:1819–1824

Baker MJ, Wilson CA, Van Gent DL (2001) Age validation of red snapper, *Lutjanus campechanus*, and red drum, *Sciaenops ocellatus*, from the northern Gulf of Mexico using $^{210}Po/^{226}Ra$ disequilibria in otoliths. Proc Gulf Caribb Fish Inst 52:63–73

Barroso CM, Nunes M, Richardson CA, Moreira MH (2005) The gastropod statolith: a tool for determining the age of *Nassarius reticulates*. Mar Biol 146:1139–1144

Bell JL (1982) Daily increments in the statoliths of gastropod larvae: their use in age determination. Am Zool 22:861

Bettencourt V, Guerra A (2000) Growth increments and biomineralization process in cephalopod statoliths. J Exp Mar Biol Ecol 248:191–205

Bettencourt V, Guerra A (2001) Age studies based on daily growth increments in statoliths and growth lamellae in cuttlebone of cultured *Sepia officinalis*. Mar Biol 139:327–334

Bizikov VA (1991) A new method of squid age determination using the gladius. In: Jereb P, Ragonese S, Von Boletzky S (Eds) Squid age determination using statoliths. Proceedings of the International Workshop held in the Istituto di Tecnologia della Pesca e del Pescato, Vol Spec Publ No 1. Note techniche e Reprints dell'Istituto di Technologia della Pesca e del Pescato, Via Luigi Vaccara, 61-91026-Mazara del Vallo (TP), Italy, pp 39–52

Bortone SA, Hollingsworth CL (1980) Ageing red snapper, *Lutjanus campechanus*, with otoliths, scales, and vertebrae. Northeast Gulf Sci 4:60–63

Brown CA, Gruber SH (1988) Age assessment of the lemon shark, *Negaprion brevirostris*, using tetracycline validated vertebral centra. Copeia 3:747–753

Butler AJ (1987) Ecology of *Pinna bicolor* Gmelin (Mollusca: Bivalvia) in Gulf St. Vincent, South Australia: density, reproductive cycle, recruitment, growth and mortality at three sites. Aust J Mar Freshwater Res 38:743–769

Cailliet GM, Goldman KJ (2004) Age determination and validation in chondrichthyan fishes. In: Carrier J, Musick JA, Heithaus M (Eds) The biology of sharks and their relatives. CRC Press, Boca Raton, FL, pp 399–447

Cailliet GM, Smith WD, Mollet HF, Goldman KJ (2006) Age and growth studies of chondrichthyan fishes: the need for consistency in terminology, verification, validation, and growth function fitting. Environ Biol Fishes 77:211–228

Campana SE (1999) Chemistry and composition of fish otoliths: pathways, mechanisms and applications. Mar Ecol Prog Ser 188:263–297

Campana SE (2001) Accuracy, precision and quality control in age determination, including a review of the use and abuse of age validation methods. J Fish Biol 59:197–242

Campana SE, Jones CM (1998) Radiocarbon from nuclear testing applied to age validation of black drum, *Pogonias cromis*. Fish Bull 96:185–192

Campana SE, Natanson LJ, Myklevoll S (2002) Bomb dating and age determination of large pelagic sharks. Can J Fish Aquat Sci 59:450–455

Cappo M, Eden P, Newman SJ, Robertson S (2000) A new approach to validation of periodicity and timing of opaque zone formation in the otoliths of eleven species of *Lutjanus* from the central Great Barrier Reef. Fish Bull 98:474–488

Carlander D (1987) A history of scale age and growth studies of North American freshwater fish. In: Summerfelt RC, Hall GE (Eds) Age and growth of fish. Papers presented at the International Symposium on Age and Growth of fish, Des Moines, Iowa, June 9–12, 1985, Iowa State University Press, Ames, Iowa, pp 3–14

Casselman JM (1983) Age and growth assessment of fish from their calcified structures – techniques and tools. In Prince ED, Pulos LM (Eds) Proceedings of the International workshop on age determination of oceanic pelagic fishes: tunas, billfishes and sharks. US Department of Commerce, NOAA Technical Report NMFS-8, NOAA, Washington, DC, pp 1–17

Casselman JM (1990) Growth and relative size of calcified structures of fish. Trans Am Fish Soc 119:673–688

Casselman JM (2007) Dr. E.J. (Ed) Crossman's scientific contributions on muskellunge: celebrating a lasting legacy. Env Biol Fish 79:5–10

Casselman JM, Crossman EJ (1986) Size, age, and growth of trophy muskellunge and muskellunge-northern pike hybrids – the Cleithrum Project, 1979–1983. In: Hall GE (Ed) Managing muskies. American Fisheries Society Special Publication 15, pp 93–110

Casselman JM, Robinson CJ, Crossman EJ (1999) Growth and ultimate length of muskellunge from Ontario water bodies. North Am J Fish Man 19:271–290

Cerrato RM (2000) What fish biologists should know about bivalve shells. Fish Res 46:39–49

Chapman DM (1985) X-ray microanalysis of selected coelenterate statoliths. J Mar Biol Assoc UK 65:617–627

Chatzinikolaou E, Richardson CA (2007) Evaluating growth and age of netted whelk *Nassarius reticulatus* (Gastropoda: Nassariidae) using statolith growth increments. Mar Ecol Prog Ser 342:163–176

Chen M-H, Soong K (2002) Estimation of age in the sex-changing, coral-inhabiting snail *Coralliophila violacea* from the growth striae on opercula and a mark-recapture experiment. Mar Biol 140:337–342

Chilton DE, Beamish RJ (1982) Age determination methods of fishes studied by the Pacific Biological Station. Can Spec Publ Fish Aquat Sci 60:102pp

Choat JH, Robertson JR (2002) Age-based studies. In: PF Sale (Ed) Coral reef fishes. Dynamics and diversity in a complex ecosystem. Academic Press, San Diego, pp 57–80

Choe S (1963) Daily age markings on the shell of cuttlefish. Nature 197:306–307

Clarke M (1965) "Growth rings" in the beaks of the squid *Moroteuthis ingens* (Oegopsida: Onychoteuthidae). Malacologia 3:287–307

Clarke MR (1978) The cephalopod statolith – an introduction to its form. J Mar Biol Assoc UK 58:701–712

Clarke MW, Irvine SB (2006) Terminology for the ageing of chondrichthyan fish using dorsal-fin spines. Environ Biol Fish 77:273–277

Dahm C (1993) Growth, production and ecological significance of *Ophiura albida* and *O. ophiura* (Echinodermata: Ophiuroidea) in the German Bight. Mar Biol 116:431–437

Dahm C, Brey T (1998) Determination of growth and age of slow growing brittle stars (Echinodermata: Ophiuroidea) from natural growth bands. J Mar Biol Assoc UK 78:941–951

Davenport S, Stevens JD (1988) Age and growth of two commercially important sharks (*Carcharhinus tilstoni* and *C. sorrah*) from northern Australia. Aust J Mar Freshwater Res 39:417–433

Davis TLO, Kirkwood GP (1984) Age and growth studies on barramundi, *Lates calcarifer* (Bloch), in northern Australia. Aust J Mar Freshwater Res 35:673–689

Davis TLO, West GJ (1992) Growth and mortality of *Lutjanus vittus* (Quoy and Gaimard) from the North West Shelf of Australia. Fish Bull 90:395–404

Day R, Williams M, Hawkes G (1995) A comparison of fluorochromes for marking abalone shells. Mar Freshwater Res 46:599–605

Debicella JM (2005) Accuracy and precision of fin-ray aging for gag (*Mycteroperca microlepis*). MSc Thesis, University of Florida, 65pp

Denton EJ, Gilpin-Brown JB (1961) The buoyancy of the cuttlefish, *Sepia officinalis* (L.). J Mar Biol Assoc UK 41:319–342

Doubleday Z, Semmens JM, Pecl G, Jackson G (2006) Assessing the validity of stylets as ageing tools in *Octopus pallidus*. J Exp Mar Biol Ecol 338:35–42

Drew K, Die DJ, Arocha F (2006) Understanding vascularization in fin spines of white marlin (*Tetrapturus albidus*). Bull Mar Sci 79:847–852

Druffel ERM (1989) Decade time scale variability of ventilation in the North Atlantic: high-precision measurements of bomb radiocarbon in banded corals. J Geophys Res 94:3271–3285

Druffel ERM, Griffin S (1995) Regional variability of surface ocean radiocarbon from southern Great Barrier Reef corals. Radiocarbon 37:1–9

Druzhinin AD, Filatova NA (1980) Some data on Lutjanidae from the Gulf of Aden area. J Ichthyol 20:8–14

Durholtz MD, Lipinski MR (2000) Influence of temperature on the microstructure of statoliths of the thumbstall squid *Lolliguncula brevis*. Mar Biol 136:1029–1037

Durholtz MD, Kretsinger RH, Lipinski MR (1999) Unique proteins from the statoliths of Lolliguncula brevis (Cephalopoda: Loliginidae). Comparative Biochemistry and Physiology B-Biochemistry & Molecular Biology 123:381–388

Durholtz MD, Lipinski MR, Field JG (2002) Laboratory validation of periodicity of incrementation in statoliths of the South African chokka squid *Loligo vulgaris reynaudii* (d'Orbigny, 1845): a reevaluation. J Exp Mar Biol Ecol 279:41–59

Edwards RCC (1985) Growth rates of Lutjanidae (snappers) in tropical Australian waters. J Fish Biol 26:1–4

Epstein S, Buchsbaum R, Lowenstam H, Urey HC (1951) Carbonate-water isotopic temperature scale. Geol Soc Am Bull 62:417–426

Fenton GE, Short SA (1992) Fish age validation by radiometric analysis of otoliths. Aust J Mar Freshwater Res 43:913–922

Francis RICC (1995) The problem of specifying otolith-mass growth parameters in the radiometric estimation of fish age using whole otoliths. Mar Biol 124:169–176

Franks JS, Brown-Peterson N, Griggs MS, Garber NM, Warren JR, Larsen KM (2000) Potential of the first dorsal spine for estimating the age of wahoo *Acanthocybium solandri* from the northern Gulf of Mexico with comments on specimens from Bimini Bahamas. Proc Gulf Carib Fish Inst 51:428–440

Gage J (1990) Skeletal growth bands in brittle stars: microstructure and significance as age markers. J Mar Biol Assoc UK 70:209–224

Garcia-March JR, Márquez-Aliaga A (2007). *Pinna noblis* L. 1758 age determination by shell internal register. Mar Biol 151:1077–1085

Gauldie RW (1988) Function, form and time-keeping properties of fish otoliths. Comp Biochem Physiol 91A:395–402

Gauldie RW, West IF (1994) Statocyst, statolith, and age estimation of the giant squid, *Architeuthis kirki*. Veliger 37:93–109

Gluyas-Millan MG, Talavera-Maya J (2003) Size and age composition of the populations of abalone *Haliotis fulgens* and *H. corrugata* of Bahia Tortugas, Baja California Sur, Mexico. Ceinc Mar 29:89–101

Goff R, Gauvrit E, Pinczon Du Sel G, Daguzan J (1998) Age group determination by analysis of the cuttlebone of the cuttlefish *Sepia officinalis* L. in reproduction in the Bay of Biscay. J Molluscan Stud 64:183–193

Goldman KJ (2005) Age and growth of elasmobranch fishes, Chapter 5. In Musick JA, Bonfil R (Eds) Elasmobranch fisheries management techniques. FAO Fisheries Technical Paper 474, Food and Agriculture Organization of The United Nations, Rome, pp 76–102

González AF, Dawe EG, Beck PC, Perez JAA (2000) Bias associated with statolith-based methodologies for ageing squid: a comparative study on *Illex illecebrosus* (Cephalopoda: Ommastrephidae). J Exp Mar Biol Ecol 244:161–180

Gordon M, Hatcher C, Seymour J (2004) Growth and age determination of the tropical Australian cubozoan *Chiropsalmus* sp. Hydrobiologia 530/531:339–345

Grove-Jones RP, Burnell AF (1991) Fisheries biology of the ocean jacket (Monacanthidae: *Nelusetta ayraudi*) in the eastern waters of the Great Australian Bight, South Australia. Unpubl Final Report FRDC DFS01Z. S Aust Dept Fish, 107pp

Gurney LJ, Mundy C, Porteus MC (2005) Determining age and growth of abalone using stable oxygen isotopes: a tool for fisheries management. Fish Res 72:353–360

Harrison EJ, Hadley WF (1979) A comparison of the use of cleithra to the use of scales for age and growth studies. Trans Am Fish Soc 108:452–456

Heilmayer O, Brey T, Chiantore M, Cattaneo-Vietti R, Arntz WE (2003) Age and productivity of the Antarctic scallop, *Adamussium colbecki*, in Terra Nova Bay (Ross Sea, Antarctica). J Exp Mar Biol Ecol 288:239–256

Hernández-López JL, Castro-Hernendez JJ, Hernandez-Garcia V (2001) Age determined from the daily deposition of concentric rings on common octopus (*Octopus vulgaris*) beaks. Fish Bull 99:679–684

Hill KT, Cailliet GM, Radtke RL (1989) A comparative analysis of growth zones in four calcified structures of Pacific blue marlin, *Makaira nigricans*. Fish Bull 87:829–843

Ihde TF, Chittenden ME (2002) Comparison of calcified structures for aging spotted seatrout. Trans Am Fish Soc 131:634–642

Ilano AS, Ito A, Fujinaga K, Nakao S (2004) Age determination of *Buccinum isatakii* (Gastropod: Buccinidae) from the growth striae on the operculum and growth under laboratory conditions. Aquaculture 242:181–195

Isermann DA, Meerbeek JR, Scholten GD, Willis DW (2003) Evaluation of three different structures used for walleye age estimation with emphasis on removal and processing times. Nth Am J Fish Mgmt 23:625–631

Jackson GD (1989) The use of statolith microstructures to analyse life-history events in the small tropical cephalopod *Idiosepius pygmaeus*. Fish Bull 87:265–272

Jackson GD (1990) The use of tetracycline staining techniques to determine statolith growth ring periodicity in the tropical loliginid squids *Loliolus noctiluca* and *Loligo chinensis*. Veliger 33:389–393

Jackson GD (1994) Application and future potential of statolith increment analysis in squids and sepioids. Can J Fish Aquat Sci 51:2312–2625

Jackson GD, Arkhipkin AI, Bizikov VA, Hanlon RT (1993) Laboratory and field corroboration of age and growth from statoliths and gladii of the loliginid squid *Sepioteuthis lessoniana* (Mollusc: Cephalopoda). In: Okutani T, O'Dor RK, Kubodera T (Eds) Recent advances in cephalopod fisheries biology. Tokai University Press, Tokyo, pp 189–199

Jackson GD, Alford RA, Choat JH (2000) Can length frequency analysis be used to determine squid growth? – An assessment of ELEFAN. ICES J Mar Sci 57:948–954

Jackson GD, Forsythe JW (2002) Statolith age validation and growth of *Loligo plei* (Cephalopoda: Loliginidae) in the north-west Gulf of Mexico during spring/summer. J Mar Biol Assoc UK 82:677–678

Jackson GD, Forsythe JW, Hixon RF, Hanlon RT (1997) Age, growth, and maturation of *Lolliguncula brevis* (Cephalopoda: Loliginidae) in the northwestern Gulf of Mexico with a comparison of length–frequency versus statolith age analysis. Can J Fish Aquat Sci 54:2907–2919

Jackson GD, Moltschaniwskyj NA (1999) Analysis of precision in statolith derived age estimates of the tropical squid *Photololigo* (Cephalopoda: Loliginidae). ICES J Mar Sci 56:221–227

Jensen M (1969) Age determination of echinoids. Sarsia 37:41–44

Jereb P, Ragonese S, Von Boletzky S (Eds) (1991) Squid age determination using statoliths. Proceedings of the International Workshop held in the Istituto di Tecnologia della Pesca e del Pescato (ITPP-CNR), Mazara del Vallo, Italy, 9–14 October 1989. NTR-ITPP Spec Publ No 1, 128pp

Johnson AG, Saloman CH (1984) Age, growth, and mortality of gray triggerfish from the Northeastern Gulf of Mexico. Fish Bull 82:485–492

Jordana E, Guillou M, Lumingas LJL (1997) Age and growth of the sea urchin *Sphaerechinus granularis* in southern Brittany. J Mar Biol Assoc UK 77:1199–1212

Ju D, Yeh S, Liu H (1989) Age and growth studies on *Lutjanus altifrontalis* from the Arafura Sea region. Acta Oceanogr Taiw 22:68–82

Ju S-J, Secor DH, Harvey HR (2001) Growth rate variability and lipofuscin accumulation rates in the blue crab *Callinectes sapidus*. Mar Ecol Prog Ser 244:197–205

Kaehler S, McQuaid CD (1999) Use of the fluorochrome calcein as an 'in situ' growth marker in the brown mussel *Perna perna*. Mar Biol 133:455–460

Kalish JM (1993) Pre- and post-bomb radiocarbon in fish otoliths. Earth Plan Sci Lett 114:549–554

Kalish JM (1995) Radiocarbon and fish biology. In: Secor DH, Dean JM, Campana SE (Eds) Recent developments in fish otolith research. University South Carolina Press, pp 637–654

Kalish JM (2003) Use of the bomb radiocarbon chronometer to validate fish age. FRDC Report 93/109. ANU Canberra, 384pp

Keller N, Del Piero D, Longinelli A (2002) Isotopic composition, growth rates and biological behaviour of *Chamelea gallina* and *Callista chione* from the Gulf of Trieste (Italy). Mar Biol 140:9–15

Kristensen TK (1980) Periodic growth rings in cephalopod statoliths. Dana 1:39–51

Lai H, Liu H (1979) Age and growth of *Lutjanus sanguineus* in the Arafura Sea and north-west shelf. Acta Oceanogr Taiw 10:160–171

LaRoe ET (1971) The culture and maintenance of the loliginid squids *Sepioteuthis sepioidea* and *Doryteuthis plei*. Mar Biol 9:9–25

Lessa R, Santana FM, Duarte-Neto P (2006) A critical appraisal of marginal increment analysis for assessing temporal periodicity in band formation among tropical sharks. Environ Biol Fish 77:309–315

Lewis D, Cerrato RM (1997) Growth uncoupling and the relationship between shell growth and metabolism in the soft shell clam *Mya arenaria*. Mar Ecol Prog Ser 158:177–189

Lipinski M (1986) Methods for the validation of squid age from statoliths. J Mar Biol Assoc UK 66:505–526

Lipinski MR, Durholtz MD (1994) Problems associated with ageing squid from their statoliths: towards a more structured approach. Ant Sci 6:215–222

Lipinski MR, Durholtz MD, Underhill LG (1998) Field validation of age readings from the statoliths of chokka squid (*Loligo vulgaris reynaudii* d'Orbigny, 1845) and an assessment of associated errors. ICES J Mar Sci 55:240–257

Liu H, Yeh S (1991) Age determination and growth of red emperor snapper (*Lutjanus sebae*) in the Arafura Sea off north Australia. Acta Oceanogr Taiw 26:36–52

Lomovasky B, Morriconi E, Brey T, Calvo J (2002) Individual age and connective tissue lipofuscin in the hard clam *Eurhomalea exalbida*. J Exp Mar Biol Ecol 276:83–94

Longhurst AR, Pauly D (1987) Ecology of tropical oceans. Academic Press Inc., San Diego, California, 257pp

Lowerre-Barbieri SK, Chittenden ME, Jones CM (1994) A comparison of a validated otolith method to age weakfish, *Cynoscion regalis*, with the traditional scale method. Fish Bull 92:555–568

Lumingas LJL, Guillou M (1994) Growth zones and back-calculation for the sea urchin, *Sphaerechinus granularis*, from the Bay of Brest, France. J Mar Biol Assoc UK 74:671–686

MacNeil MA, Campana SE (2002) Comparison of whole and sectioned vertebrae for determining the age of young blue shark (*Prionace glauca*). J NW Atl Fish Sci 30:77–82

Manning MR, Melhuish WH (1994) Delta 14C carbon dioxide record from Wellington. In Boden TA, Kaiser DP, Sepanski RJ, Stoss FW (Eds) Trends 93 – A compendium of data on global change: and online updates (Online Trends). ORNL/CDIAC–65. Carbon Dioxide Information Analysis Centre. Oak Ridge National Laboratory, Oak Ridge, TN, pp 173 202

Manooch CI (1987) Age and growth of snappers and groupers. In: Polovina JJ, Ralston S (Eds) Tropical snappers and groupers: biology and fisheries management. Westview Press, Boulder, Colorado, pp 329–373

Manooch CI, Drennon CL (1987) Age and growth of yellowtail snapper and queen triggerfish collected from the U.S. Virgin Islands and Puerto Rico. Fish Res 6:53–68

Marriott R, Cappo M (2000) Comparative precision and bias of five different ageing methods for the large tropical snapper *Lutjanus johnii*. Asian Fish Sci 13:149–160

Marriott RJ, Mapstone BD, Begg GA (2007) Age-specific demographic parameters, and their implications for management of the red bass, *Lutjanus bohar* (Forsskal 1775): a large, long-lived reef fish. Fish Res 83:204–215

Matlock GC, Colura RL, Maciorowski AF, McEachron LW (1987) Use of on-going tagging programs to validate scale readings. In: Summerfelt RC, Hall GE (Eds) Age and growth of fish. Papers presented at the international symposium on age and growth of fish, Des Moines, Iowa, June 9–12, 1985, Iowa State University Press, Ames, Iowa, pp 279–286

Matlock GC, Colura RL, McEachron LW (1993) Direct validation of black drum (*Pogonias cromis*) ages determined from scales. Fish Bull 91:558–563

McAuley RB, Simpfendorfer CA, Hyndes GA, Allison RR, Chidlow JA, Newman SJ (2006) Validated age and growth of the sandbar shark, *Carcharhinus plumbeus* (Nardo 1827) in the waters off Western Australia. Environ Biol Fish 77:385–400

McPherson GR, Squire L (1992) Age and growth of three dominant *Lutjanus* species of the Great Barrier Reef inter-reef fishery. Asian Fish Sci 5:25–36

Merrill A, Posgay J, Nichy F (1965) Annual marks on shell and ligament of sea scallop (*Plactopecten magellanicus*). Fish Bull 65:299–311

Metcalf SJ, Swearer SE (2005) Non-destructive ageing in *Notolabrus tetricus* using dorsal spines with an emphasis on the benefits for protected, endangered and fished species. J Fish Biol 66:1740–1747

Milton DA, Short SA, O'Neill MF, Blaber SJM (1995) Ageing of three species of tropical snapper (Lutjanidae) from the Gulf of Carpentaria, Australia, using radiometry and otolith ring counts. Fish Bull 93:103–115

Moltschaniwskyj NA (2004) Understanding the processes of growth in cephalopods. Mar Freshwater Res 55:379–386

Moseley FN (1966) Biology of the red snapper, *Lutjanus aya* Bloch, of the northwestern Gulf of Mexico. Publ Inst Mar Sci Univ Texas 11:90–101

Moulton PL, Walker TI, Saddlier SR (1992) Age and growth studies of Gummy Shark, *Mustelus antarcticus* Gunther, and School Shark, *Galeorhinus galeus* (Linnaeus), from southern Australian waters. Aus Mar Freshwater Res 43:1241–1267

Murie DJ (2003) Section 3.6.3.2. Sectioning fin rays. In: VanderKooy S, Guindon-Tisdel K (Eds) A practical handbook for determining the ages of Gulf of Mexico fishes. Gulf States Marine Fisheries Commission, Ocean Springs, MS, Publication No. 111, 114pp

Murie DJ, Parkyn DC (1999) Age, growth, and sexual maturity of white grunt. Prepared for Florida Marine Research Institute, St. Petersburg, USA. Final Report Part II. 57pp

Murray T (1986) Paua shell and its growth. NZ Ministry of Agriculture and Fisheries Shellfish Newsletter 30:10

Natsukari Y, Hirata S, Washizaki M (1991) Growth and seasonal changes in cuttlebone characteristics of *Sepia esculenta*. In: Boucaud-Camou E (Ed) First international symposium on the cuttlefish *Sepia*. Université de Caen, Caen, pp 49–67

Naylor JR, Manighetti BM, Neil HL, Kim SW (2007) Validated estimation of growth and age in the New Zealand abalone *Haliotis iris* using stable oxygen isotopes. Mar Freshwater Res 58:354–362

Nelson RS, Manooch CS (1982) Growth and Mortality of Red Snappers in the West-Central Atlantic-Ocean and Northern Gulf of Mexico. Transactions of the American Fisheries Society 111:465–475

Newman SJ (2002) Growth rate, age determination, natural mortality and production potential of the scarlet seaperch, *Lutjanus malabaricus* Schneider 1801, off the Pilbara coast of north-western Australia. Fish Res 58:215–225

Newman SJ, Cappo M, Williams DM (2000a) Age, growth, mortality rates and corresponding yield estimates using otoliths of the tropical red snappers, *Lutjanus erythropterus*, *L. malabaricus* and *L. sebae*, from the central Great Barrier Reef. Fish Res 1:1–14

Newman SJ, Cappo M, Williams DM (2000b) Age, growth and mortality of the stripey, *Lutjanus carponotatus* (Richardson) and the brown-stripe snapper, *L. vitta* (Quoy and Gaimard) from the central Great Barrier Reef, Australia. Fish Res 3:263–275

Newman SJ, Dunk IJ (2002) Growth, age validation, mortality, and other population characteristics of the red emperor snapper, *Lutjanus sebae* (Cuvier 1828) off the Kimberley coast of north-western Australia. Estuar Coast Shelf Sci 55:67–80

Newman SJ, Williams DM, Russ GR (1996) Age validation, growth and mortality rates of the tropical snappers (Pisces: Lutjanidae) *Lutjanus adetii* (Castelnau, 1873) and *L. quinquelineatus* (Bloch 1790) from the central Great Barrier Reef, Australia. Mar Freshwater Res 47:575–584

NMFS (2006) Status report on the continental United States distinct population segment of the goliath grouper (*Epinephelus itajara*). NOAA/NMFS, January 12, 2006, 49pp

Officer RA, Gason AS, Walker TI, Clement JG (1996) Sources of variation in counts of growth increments in vertebrae from gummy shark, *Mustelus antarcticus*, and school shark, *Galeorhinus galeus*: implications for age determination. Can J Fish Aquat Sci 53:1765–1777

Parsons G, Robinson S, Roff J, Daswell M (1993) Daily growth rates as indicated by valve ridges in postlarval giant scallop *Plactopecten magellanicus* (Bivalvia: Pectinidae). Can J Fish Aquat Sci 50:456–464

Patterson WI, Cowan JJ, Wilson CA, Shipp RL (2001) Age and growth of red snapper, *Lutjanus campechanus*, from an artificial reef area off Alabama in the northern Gulf of Mexico. Fish Bull 99:617–627

Pearse J, Pearse V (1975) Growth zones in the echinoid skeleton. Amer Zool 15:731–753

Pecl GT (2000) Comparative life-history of tropical and temperate *Sepioteuthis* squids in Australian waters. PhD Thesis, James Cook University, Australia

Penha JMF, Mateus LAF, Petrere MP (2004) A procedure to improve confidence in identification of the first annulus in fin-spines of fishes. Fish Mgmnt Ecol 11:135–137

Perez JAA, O'Dor RK (2000) Critical transitions in early life histories of short–finned squid, *Illex illecebrosus* as reconstructed from gladius growth. J Mar Biol Assoc UK 80:509–514

Perez JAA, O'Dor RK, Beck PC, Dawe EG (1996) Evaluation of gladius dorsal surface structure for age and growth studies of the short-finned squid *Illex illecebrosus* (Teuthoidea: Ommastrephidae). Can J Fish Aquat Sci 53:2837–2846

Pirker J, Schiel D (1993) Tetracycline as a fluorescent shell-marker in the abalone *Haliotis iris*. Mar Biol 116:81–86

Prince ED, Pulos LM (1983) Proceedings of the International Workshop on Age determination of oceanic pelagic fishes: tunas, billfishes, and sharks held at Miami, Florida on February 15–18, 1982. NOAA-TR-NMFS-8

Radtke RL (1983) Chemical and structural characteristics of statoliths from the short-finned squid *Illex illecebrosus*. Mar Biol 76:47–54

Rao DM, Rao KS (1986) Studies on the age determination and growth of *Nemipterus japonicus* (Bloch) off Visakhapatnam. Indian J Fish 33:426–439

Raya C, Hernández-González C (1998) Growth lines within the beak microstructure of the *Octopus vulgaris* Cuvier, 1797. S Afr J Mar Sci 20:135–142

Raya CP, Fernandez-Nunez M, Balguerias E, Hernandez-Gonzalez CL (1994) Progress towards ageing cuttlefish *Sepia hierredda* from the northwestern African coast using statoliths. Mar Ecol Prog Ser 114:139–147

Reis CS, Fernandes R (2002) Growth observations on *Octopus vulgaris* Cuvier, 1797 from the Portuguese waters: growth lines in the vestigial shell as possible tools for age determination. Bull Mar Sci 71:1099–1103

Richardson CA (1989) An analysis of the growth bands in the shell of the common mussel *Mytilus edulis*. J Mar Biol Assoc UK 69:477–491

Richardson CA (2001) Molluscs as archives of environmental change. Oceanogr Mar Biol 39:103–164

Richardson CA, Kennedy H, Duarte CM, Kennedy DP, Proud SV (1999) Age and growth of the fan mussel *Pinna nobilis* from south-east Spanish Mediterranean seagrass (*Posidonia oceanica*) meadows. Mar Biol 133:205–212

Richardson CA, Kingsley PR, Seed R, Chatzinikolaou E (2005b) Age and growth of the naticid gastropod *Polinices pulchellus* (Gastropoda: Nacticidae) based on length frequency analysis and statolith growth rings. Mar Biol 148:319–326

Richardson CA, Saurel C, Barroso CM, Thain J (2005a) Evaluation of the age of the welk *Neptunea antique* using statoliths, opercula and element ratios in the shell. J Exp Mar Biol Ecol 325:55–64

Rocha-Olivares A (1998) Age, growth, mortality, and population characteristics of the Pacific red snapper, *Lutjanus peru*, off the southeast coast of Baja California, Mexico. Fish Bull 96:562–574

Rodhouse PG, Hatfield EMC (1990) Age determination in squid using statolith growth increments. Fish Res 8:323–334

Russell MP, Meredith RW (2000) Natural growth lines in echinoid ossicles are not reliable indicators of age: a test using *Stronglyocentrotus droebachiensis*. Invertebr Biol 119:410–420

Sano M, Omori M, Taniguchi K, Seki T (2001) Age distribution of the sea urchin *Strongylocentrotus nudus* (A. Agassiz) in relation to algal zonation in a rocky coastal area on Oshika Peninsula, northern Japan. Fish Sci 67:628–639

Santarelli L, Gros P (1985) Age and growth of the whelk *Buccinum undatum* L. (Gastropod: Prosobranchia) using stable isotope of the shell and operculum striae. Oceanol Acta 8:221–229

Schöne BR, Goodwin DH, Flessa KW, Dettman DL, Roopnarine PD (2002) Sclerochronology and growth of the bivalve mollusks *Chione* (*Chionista*) *fluctifraga* and *C.* (*Chionista*) *cortezi* in the northern Gulf of California, Mexico. Veliger 45:45–54

Schöne BR, Tanabe K, Dettman DL, Sato S (2003) Environmental controls on shell growth rates and delta O-18 of the shallow-marine bivalve mollusk *Phacosoma japonicum* in Japan. Mar Biol 142:473–485

Secor DH, Trice TM, Hornick HT (1995) Validation of otolith-based ageing and a comparison of otolith and scale-based ageing in mark-recaptured Chesapeake Bay striped bass, *Morone saxatilis*. Fish Bull 93:186–190

Sejr MK, Jensen KT, Rysgaard S (2002) Annual growth bands in the bivalve *Hiatella arctica* validated by a mark-recapture study in NE Greenland. Polar Biol 25:794–796

Seshappa G (1969) The problem of age determination in the Indian mackerel, *Rastrelliger kanagurta*, by means of scales and otoliths. Indian J Fish 16:14–28

Seshappa G (1999) Recent studies on age determination of Indian fishes using scales, otoliths and other hard parts. Indian J Fish 46:1–11

Sheehy MRJ (1990) The potential of morphological lipofuscin age – pigment as an index of crustacean age. Mar Biol 107:439–442

Sheehy MRJ, Bannister RCA (2002) Year-class detection reveals climatic modulation of settlement strength in the European lobster, *Homarus gammarus*. Can J Fish Aquat Sci 59:1132–1143

Sheehy MRJ, Bannister RCA, Wickins JF, Shelton PMJ (1999) New perspectives on the growth and longevity of the European lobster, *Homarus gammarus*. Can J Fish Aquat Sci 56:1904–1915

Shepherd SA, Al-Wahaibi D, Al-Azri A (1995b) Shell growth checks and growth of the Omani abalone *Haliotis mariae*. Mar Freshwater Res 46:575–582

Shepherd SA, Avalos-Borja M, Quintanilla M (1995a) Toward a chronology of *Haliotis fulgens*, with a review of abalone shell microstructure. Mar Freshwater Res 46:607–615

Shepherd SA, Breen PA (1992) Mortality in abalone: its estimation, variability, and causes. In: Shepherd SA, Tegner M, Guzman del Proo S (Eds) Abalone of the world: biology, fisheries and culture. Blackwell Scientific, Oxford, pp 276–303

Shepherd SA, Woodby D, Rumble JM, Avalos-Borja M (2000) Microstructure, chronology and growth of the pinto abalone, *Haliotis kamtschatkana*, in Alaska. J Shellfish Res 19:219–228

Sire J-Y, Bonnet P (1984) Croissance et structure de l'opercule calcifié du gastéropode *Turbo setosus* (Prosobranchia: Turbinidae): détermination de l'âge individuel. Mar Biol 79:75–87

Smith SE, Au DW, Show C (1998) Intrinsic rebound potential of 26 species of Pacific sharks. Mar Freshwater Res 48:663–678

Smith SE, Mitchell RA, Fuller D (2003) Age-validation of a leopard shark (*Triakis semifasciata*) recaptured after 20 years. Fish Bull 101:194–198

Souter D, Cameron A, Endean R (1997) Implications of sublethal predation, autotomy and regeneration: pigment bands on their spines can not be used to determine the ages of adult specimens of the corallivore *Acanthaster planci*. Mar Freshwater Res 48:321–328

Speare P (1992) A technique for tetracycline injecting and tagging billfish. Bull Mar Sci 51:197–203

Speare P (2003) Age and growth of black marlin, *Makaira indica*, in east coast Australian waters. Mar Freshwater Res 54:1–8

Spratt JD (1978) Age and growth of the market squid, *Loligo opalescens* Berry, in Monterey Bay. Calif Fish Game 169:35–44

Stump R, Lucas J (1990) Linear growth in spines from *Acanthaster planci* (L.) involving growth lines and periodic pigment bands. Coral Reefs 9:149–154

Stump R, Lucas J (1999) Age estimation and patterns of growth in *Acanthaster planci*: a reply to Souter et al. (1997). Mar Freshwater Res 50:71–72

Sun CL, Wang SP, Yeh SZ (2002) Age and growth of the swordfish (*Xiphias gladius* L.) in the waters around Taiwan determined from anal-fin rays. Fish Bull 100:822–835

Takagi K, Kitahara T (2002) Bias in age estimation of Japanese common squid due to grinding degree of statoliths. Nippon Suisan Gakkaishi 68:351–355

Talbot FH (1960) Notes on the biology of the Lutjanidae (Pisces) of the east African coast, with specific reference to *L. bohar* (Forskal). Ann S Afr Mus 65:549–573

Tanaka S (1990) The structure of the dorsal spine of the deep sea squaloid shark *Centrophorus acus* and its utility for age determination. Bull Jap Soc Sci Fish 56:903–909

Tracey S, Steer M, Pecl G (2003) Life history traits of the temperate mini-maximalist *Idiospeius notoides* (Cephalopoda: Sepioidea). J Mar Biol Assoc UK 83:1297–1300

Tserpes G, Tsimenides N (1995) Determination of age and growth of swordfish, *Xiphias gladius* L., 1758, in the eastern Mediterranean using anal-fin spines. Fish Bull 93:594–602

Ueno S, Imai C, Mitsutani A (1995) Fine growth rings found in statolith of a cubomedusa *Carybdea rastoni*. J Plankton Res 17:1381–1384

Villanueva R (2000) Effect of temperature on statolith growth of the European squid *Loligo vulgaris* during early life. Mar Biol 136:449–460

Wassef EA, Bawazeer F (1990a) Comparative growth studies on *Lethrinus lentjan*, Lacepede 1802 and *L. mahsena*, Forsskal. 1775 (Pisces, Lethrinidae) in the Red Sea. Bull Nat Inst Oceanogr Fish Cairo 16:85–101

Wassef EA, Bawazeer F (1990b) The biology of *Lethrinus elongatus*, Val. 1830 (Teleostii: Lethrinidae) in the Red Sea. Bull Nat Inst Oceanogr Fish Cairo 16:103–124

Welch TJ, Avyle MD, Betsill RK, Driebe EM (1993) Precision and relative accuracy of striped bass age estimates from otoliths, scales, and anal fin rays and spines. N Am J Fish Man 13:616–620

West IF, Gauldie RW (1994) Determination of fish age using ^{210}Pb:^{226}Ra disequilibrium methods. Can J Fish Aquat Sci 51:2333–2340

Wilson CA, Nieland DL (2001) Age and growth of red snapper, *Lutjanus campechanus*, from the northern Gulf of Mexico off Louisiana. Fish Bull 99:653–664

Wilson CA, Stanley AL, Nieland DL (2001) Age estimates from annuli in otoliths of red snapper, *Lutjanus campechanus*, from the northern Gulf of Mexico. Proc Gulf Carib Fish Inst 52:48–62

Yeh S, Chen C, Liu H (1986) Age and growth of *Lutjanus sebae* in the waters off northwestern Australia. Acta Oceanogr Taiw 16:90–102

6. THE BACK-CALCULATION OF FISH GROWTH FROM OTOLITHS

LAURENT VIGLIOLA[1] AND MARK G. MEEKAN[2]

[1] *Institut de Recherche pour le Développement, BP 70, 29280 Plouzané, France, e-mail: laurent.vigliola@ird.fr*

[2] *Australian Institute of Marine Science, PO Box 40197, Casuarina MC 0811, NT Australia*

1 Introduction

There is nothing more practical than a good theory.
Leonid Ilich Brezhnev, *quoted in Nature, 1977, 270, pp 470–1*

The earliest example of using earbones or otoliths to provide estimates of fish ages dates back to at least 1899 (Reibisch, cited in Jones 1992). Back-calculation to reconstruct growth patterns from hard parts of fish (otoliths, bones and scales) followed soon after (Lea 1910). The approach involves using measurements made on these bony structures to infer, or back-calculate, body length at ages prior to capture. Back-calculation has been used to generate individual growth histories of fishes for almost a century (Francis 1990) and has proved to be an invaluable tool for fisheries scientists and fish ecologists.

Otoliths can show annual, and for younger fish, daily patterns of growth (Pannella 1971, 1974). The analysis of daily increments within otoliths has been very popular over the last 30 years due to the ability of the technique to provide growth data during the larval and juvenile phases of the life history, when growth rates are critical to year-class success and, ultimately, population size (Stevenson & Campana 1992). Typically, growth data can be generated from otoliths in 2 ways. When sample sizes are large and the fish collected by a study encompass a wide range of age and length, then ageing fish may suffice to estimate growth curves for a population. In many cases, however, the use of growth back-calculation techniques to estimate fish growth is not a choice, it is a necessity. Sample sizes are often small, due to rarity in multi-species assemblages (a particular issue in species-rich tropical habitats such as coral reefs) or because of the diversity of catches in some fisheries. The difficulty of collecting particular life history stages of interest to researchers, such as pelagic larvae or juveniles, may also result in small sample sizes. Back-calculation may be required where cohorts or populations must be sampled sequentially, as in studies of size selective mortality (Hovenkamp 1992, Meekan & Fortier 1996). Only back-calculation of histories held within otoliths will allow the reconstruction of growth trajectories in such situations, both increasing sample sizes and filling gaps in life history information at earlier ages.

Back-calculation of daily or annual growth from otoliths requires firstly that the rate of deposition of increments in otoliths does not vary. This usually can be verified experimentally (Geffen 1992). Secondly, it assumes that these increments can be read

B.S. Green et al. (eds.), *Tropical Fish Otoliths: Information for Assessment, Management and Ecology*, Reviews: Methods and Technologies in Fish Biology and Fisheries 11,
DOI 10.1007/978-1-4020-5775-5_6,

with accuracy and precision (Campana 1992). Thirdly, all back-calculation models assume that there is a relationship between the growth of the otolith (increment width) and the somatic growth, usually length, of the fish. Evidence for this phenomenon usually is derived from strong correlations between the size of the otolith and body size of fish. There have been relatively few studies that have shown a complete lack of correlation between fish and otolith size and most of those that have done so have been confounded by problems in regression analysis and sample selection (Meekan 1997, Meekan et al. 1998, but see Bang & Gronkjaer 2005). Somatic and otolith growth can be uncoupled at least in the short term, however, (Mosegaard et al. 1988, Secor & Dean 1989, Wright et al. 1990, Fey 2006) and this is thought to result from two causes: (i) a "growth effect" whereby otoliths from slow-growing fish at a given size are larger than those of fast-growing fish of the same size (Templeman & Squires 1956); and (ii) an "age effect" where some constant or proportional amount of calcification occurs onto the otolith despite daily fluctuations in somatic growth rate (Secor & Dean 1992). Otolith growth is a conservative daily process (Mugiya 1987, 1990) and it is easiest to describe an age effect in the extreme case where somatic growth ceases but otolith growth continues. Age effects, however, can be important during negative, static, and positive somatic growth phases (Secor & Dean 1992). Because growth varies over time, particularly between ontogenetic stages, growth effects may also vary through time, introducing curvature into individual fish – otolith size trajectories (Campana 1990). Furthermore, the increase in otolith size in non-growing fish is consistent with disproportionately larger otoliths in slower-growing individuals and so the results of growth and age effects may be similar and both can therefore induce a bias into the otolith – fish length relationship upon which back-calculation procedures are based.

Some authors have recently opted to analyse otolith radius at age in order to generate data sets of growth rate, rather than back-calculated fish length at age, to avoid the potential errors associated with back-calculation or in cases where age and growth effects might be a problem (Hare & Cowen 1995). The principal argument for this approach is that there is no reason to bother with back-calculation because otolith radius is a proxy of fish length. Furthermore, it is thought to avoid bias due to the use of complex back-calculation models (Campana 1990, Francis 1990) and the effect of selective mortality on fish-otolith size relationships (Ricker 1969, Gleason & Bengston 1996, Grimes & Isley 1996). In reality, these points are arguable. Figure 1 compares the growth trajectories and fish length (L) – otolith radius (R) relationships of three individuals. In this example, the three fish have the same growth characteristics, but fish 2 has a slightly different L – R relationship (i.e. morphology) than the others. When comparing individuals of very different ages (e.g. 1 vs 2 or 3), then otolith radius will indeed be a good proxy of length and the same conclusion will hold whether individuals are compared on the basis of otolith radius or body length: individual 1 is smaller than 2 and 3, primarily because it is much younger. When comparing fish of about the same age, however, fish with larger otoliths at age might not necessarily be larger in size (e.g. 2 vs 3). In our example, fish 3 always has a larger otolith radius at age than fish 2, despite both having exactly the same body length at age (growth trajectory). This demonstrates that otolith radius is a poor proxy of fish length when comparing size-at-age. The confounding effect of morphology may be relevant not only when comparing otolith radii among individuals but also when comparing among populations or samples with different L – R relationships.

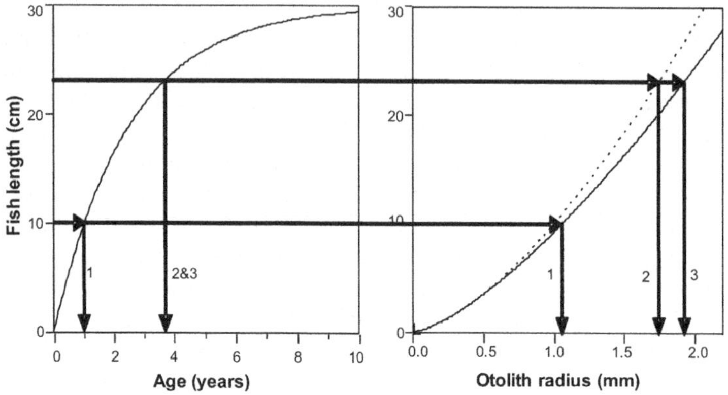

Figure 1. The growth trajectories and fish length (L) – otolith radius (R) relationships of three fish from a simulated data set. The plot of age vs size (*left*) shows that the three individuals have the same growth curve. The plot of otolith vs fish size on the right, however, shows fish 2 has a slightly different L – R relationship (i.e. morphology) than the others. The lines overlaid on the plots show that otolith radius will be a good proxy of length when fish are of different ages but fish with larger otoliths at age might not necessarily be larger in size than other fish of the same age.

It is also important to recognise that the use of otolith radii to generate growth data does not in fact avoid back-calculation. Rather, the approach inherently assumes that the simplest and most primitive back-calculation model applies ($L_i = aR_i$, where "a" is a constant and L_i and R_i are fish length and otolith radius at age i, respectively). Because "a" is the same for all fish, the approach is equivalent to back-calculating fish size by the "simple regression method" (see below), which is not recommended since it ignores individual differences in L – R relationships (Francis 1990). Effectively, a century of research and refinement in back-calculation techniques is discarded. Perhaps the only situation where such an approach could be preferred is in the analysis of fossil otoliths where there may be no extant populations of the study species and so relationships between otolith and fish size cannot be constructed empirically.

So, if back-calculation is inevitable even when we try to avoid it, what then is the best approach? The purpose of this chapter is not to describe all of the various back-calculation models that have been developed over the last century, these have been expertly reviewed by Francis (1990, 1995). Rather, we will update the lists provided by Francis (1990, 1995) and examine the theoretical and experimental evidence for and against the use of the various models published in the last decade. This is the first aim of our chapter. We also propose to take the issue of back-calculation one step further and ask the question: once you have a back-calculated data set of size at age, how do you go about analysing it? Chambers and Miller's review in 1995 recognised the unique nature of these data sets, which are longitudinal, auto-correlated and invariably unbalanced. These features provide some challenges for statistical analysis that recent developments in software have addressed, principally through the use of mixed-effects models (Pinheiro & Bates 2000). A description of the use of these models for analysis of back-calculated data is the second aim of our chapter.

2 Selection of a back-calculation model

2.1 BACK-CALCULATION METHODS

The back-calculation of fish length (L_i) from otolith radius (R_i) at age i requires the development of a back-calculation model, which is a two-step procedure. Firstly, the shape (linear or curvilinear) of the relationship between fish length (L) and otolith radius (R) must be determined. This can be described by two functions, "f" and "g" where R = f(L) and L = g(R) (Francis 1990). It is important to realise that f and g define families of lines (that may be straight or curved) and that any given line is defined by a unique set of function parameters. Because back-calculation for a particular fish will involve just one line from the family for a given function, the second step of model development will involve determining which particular line (i.e., the set of function parameters) should be used for a given individual. This can be done using three different methods. Paradoxically, the first method considers that all fish follow the same L – R line calculated by regression of fish length on otolith radius for the function g. Size-at-age estimated by this "simple regression" method (*sensu* Secor & Dean 1992) is given by the following back-calculation model (Francis 1990):

$$L_i = g(R_i) \tag{1}$$

This "simple regression" method is insensitive to changes in individual growth histories due to its averaging effect (Secor et al. 1989) and will often produce biased size-at-age estimates. It is hard to see any reason for its use (Francis 1990). The second and third methods both consider that each individual has a unique L – R line that passes through otolith and fish size at capture (R_{cpt}, L_{cpt}), but differ in how the parameters of individual lines are estimated. The second method back-calculates size by assuming proportionality between measurements at individual and population levels. Two hypotheses of proportionality exist. Whitney and Carlander (1956) stated these as: "if the otolith of one individual was 10% larger at capture than the mean size of otoliths from a group of fish of the same length, then its otolith had also been 10% larger than the average throughout life"; and "if one individual was 10% larger in length at capture than the average length for fish with the same otolith radius, then this individual had also been 10% larger than the average throughout life". These are known as the "Scale Proportional Hypothesis" (SPH) and "Body Proportional Hypothesis" (BPH) (*sensu* Francis 1990) respectively. Size-at-age estimated by "proportionality" methods are given by the following back-calculation models:

SPH $$f(L_i) = \frac{R_i}{R_{cpt}} f(L_{cpt}) \tag{2}$$

BPH $$L_i = \frac{L_{cpt}}{g(R_{cpt})} g(R_i) \tag{3}$$

"Proportionality" methods can be applied to any function f or g of any mathematical form, providing that the parameters of f are estimated by R-on-L regression and those of g by L-on-R regression.

The third back-calculation method constrains individual L – R lines to go through one or more known points such as otolith size and body length at capture (R_{cpt}, L_{cpt}) or the origin of the growth curve (R_0, L_0, usually at hatching), which can be estimated by regression or fixed at a known biological intercept (BI, Campana 1990). This "Constraint" method thus requires as many known points as there are function parameters, and so can be difficult to apply for complex functions.

2.2 EXISTING BACK-CALCULATION MODELS

The use of "constraint" or "proportionality" methods coupled with back-calculation functions f and g lead to the development of different back-calculation models (BCMs). In his review of the subject, Francis (1990) compiled 6 back-calculation functions, to which two others have subsequently been added by Tremblay and Giguère (1992) and by Morita and Matsuishi (2001). Because most back-calculation functions can be inverted (i.e., $L = g(R) \leftrightarrow R = f(L)$), a total of 8 f and 8 g functions exist that are appropriate for use in back-calculation. A total of 22 back-calculation models have been derived from these functions (Appendix 1).

Differences between "constraint" and "proportionality" models are often subtle (some may say artificial, because models are derived from the same f and g functions), or even non-existent when functions f or g have a single parameter. For example, the Dahl-Lea back-calculation model (BCM 1 "DALE", Appendix 1) can be obtained by constraining the function $L = bR$ to pass through the point at capture (R_{cpt}, L_{cpt}) or by applying a BPH to the function $L = bR$ or a SPH to the function $R = L/b$. When an intercept is added to the linear function (i.e., $L = a + bR$), then the BCMs will differ in the intercept value calculated for each individual. For example, the Fraser-Lee model (BCM 2 "FRALE", Appendix 1) is obtained when individual L – R lines are all constrained to pass through the intercept (R=0, L=a) with "a" estimated by L-on-R regression. Similarly, the linear biological intercept model (BCM 3 "BI", Appendix 1) results in individual L – R lines passing through biologically-determined intercepts ($R=0$, $L = L_{cpt} - R_{cpt}\dfrac{(L_{cpt} - L_{0p})}{(R_{cpt} - R_{0p})}$), although all lines cross at ($R = R_{0p}$, $L = L_{0p}$), whereas the linear BPH (BCM 4 "LBPH", Appendix 1) results in individual L – R lines passing through intercepts ($R=0$, $L = \dfrac{aL_{cpt}}{(a + bR_{cpt})}$) where "a" and "b" are estimated by L-on-R regression. These differences are subtle but important as they can render back-calculation models more or less sensitive to bias induced by growth or age effects.

Campana (1990) showed that growth effects will always result in linear regressions between L and R that overestimate slopes and underestimate intercepts, so that regression-based Fraser-Lee (BCM 2) and linear BPH or SPH (BCM 4 & 6) will inevitably produce biased size-at-age estimates. The linear biological intercept model (BCM 3) developed by Campana (1990) constrained a linear function of back-calculation to pass through biologically-determined intercepts of otolith and fish size to reduce the influence of variable growth rates in the population (i.e., the growth effect). The model was still sensitive to non-linear effects, however, and in particular those induced by growth rate variations through time (i.e., a time-varying growth effect). This can introduce curvature into individual fish – otolith size trajectories (Campana 1990). The Time-Varying Growth model (BCM 5 "TVG", Sirois et al. 1998, Appendix 1) was developed to address non-linear fish – otolith relationships formed because of time-varying growth rates by incorporating a growth effect into the structure of the linear biological intercept model. This was done by weighting the contribution of individual increments in the length back-calculation. The TVG model still assumes the underlying relationship between fish and otolith size to be linear, however, and so may not be appropriate when the relationship is non-linear for reasons other than growth effects.

Vigliola et al. (2000) took a different approach to non-linearity. They constrained an allometric L – R function to go through a biological intercept so their modified Fry model (BCM 14 "MF", Appendix 1) assumed non-linearity and was robust to growth effects. Only the Age Effect (BCM 7 "AE", Appendix 1) model of Morita and Matsuishi (2001), however, was designed to remove bias due to age effects. That model was developed on the assumption that otolith radius was a linear combination of both fish length and age. A SPH was then applied to the 3 dimensional back-calculation function that linked otolith radius, fish length and age to obtain the AE model.

2.3 SELECTION OF A ROUTINE BACK-CALCULATION MODEL

The plethora of models now available for back-calculation creates a dilemma for any researcher intending to back-calculate fish size from otoliths: which of these is the most appropriate to use? The first step for model selection consists of determining which function(s) should be used to describe the relationship between fish length and otolith radius. Back-calculation functions f and g technically could be of any mathematical form (Francis 1990, 1995) but back-calculation usually assumes that there is a proportional relationship between the growth of the otolith and the somatic growth of the fish (Campana 1990, Hare & Cowen 1995, Sirois et al. 1998, Vigliola et al. 2000, Morita & Matsuishi 2001). Thus, we define a function f or g as appropriate for use in back-calculation if it complies with the assumption of proportionality between otolith and somatic growth. Of the 16 functions listed in Appendix 1, this criterion allowed the immediate elimination of functions (f3), (g3), (f6), (g6), (f7), (g7), (f8) and (g8). The remaining functions were all derived from the assumption of proportionality between the relative growth rates of the fish and the otolith and could be written as:

$$\frac{d(L - a)}{(L - a)dt} = c\frac{dR}{Rdt} \tag{4}$$

where "a" is the body length of an individual at otolith formation (Fraser 1916, Lee 1920) and "c" is the proportionality coefficient.

Solution of this differential equation resulted in $L = a + bR^c$ (g5), a generalisation of the well-known function for an allometric relationship between two body parts of an organism (Ricker 1975, 1979). When isometry existed (i.e., c = 1), function (g5) became identical to L = bR (g1) if a = 0 and to L = a + bR (g2) if a ≠ 0. Furthermore, function $L = bR^c$ (g4) was a specific case of (g5) if a = 0. It seems likely that "a" will be greater than 0 in most cases (Francis 1990) since the otoliths of many species form in the period just prior to hatching (Geffen 1992). Thus, only functions (g2) and (g5) (and their inverse f2 and f5) remained appropriate to model the relationship between the length of a fish and the radius of its otolith.

After some back-calculation functions have been selected on the criterion of a proportionality relationship between otolith and somatic growth (Equation 4), it must be determined which of the "constraint" or BPH or SPH "proportionality" methods (Equations 2 & 3) are most appropriate for model development. Models derived by BPH or SPH are sensitive to growth effects because they cannot be constrained to go through a biological intercept. Practically, this means that BPH and SPH proportionality methods can generate unrealistic estimates of fish size, in particular when back-calculating fish size outside the age range of the sample from which the parameters of the back-calculation formulae were estimated by regression (see below, Vigliola et al. 2000). Consequently, a criterion of accuracy in back-calculated size-at-age (i.e., the BCM output) eliminated all models based on proportionality hypotheses SPH and BPH as well as other models that did not contain a biological intercept.

Application of our first (proportionality of otolith – fish growth) and second (to generate realistic size-at-age data) selection criteria to the list of 22 existing back-calculation models thus eliminated all but the linear Biological Intercept model ("BI", BCM 3), the Time-Varying Growth model ("TVG", BCM 5) and the Modified Fry model ("MF", BCM 14). The applicability of these models can be determined by examining the evidence from the few recent studies that have attempted to assess their accuracy and precision as descriptors of fish growth in the field and laboratory.

2.4 RECENT FIELD AND EXPERIMENTAL EVIDENCE OF THE ACCURACY OF BACK-CALCULATION MODELS

Vigliola et al. (2000) compared the outputs of BI, TVG, ABPH (the L – R relationship was allometric in the 3 study species) and MF models to field estimates of size of newly settled individuals of 3 species of sparid (*Diplodus sargus, D. vulgaris* and *D. puntazzo*) from the Mediterranean Sea. Field estimates of size were obtained by underwater visual survey. The error associated with estimating size by this method was determined by comparing visual estimates with the actual sizes of fish after capture and was always <±3.5 mm (Macpherson 1998). Growth curves derived from visual survey were extrapolated using published values for length-at-age of planktonic larvae to include the pre-settlement stage of the life history of the three species. Overall, Vigliola et al. (2000) found that the MF model produced growth curves that were the most similar to those derived from direct observations of juvenile *Diplodus* (Figure 2). As predicted, the ABPH model gave unrealistic estimates of size-at-age of very young fish, as it was not constrained to go through a biological intercept and was sensitive to growth effects. Both the BI and TVG curves consistently overestimated size-at-age in comparison with curves estimated directly by visual survey. These models produced very similar estimates despite the presence of an allometric L – R relationship. This result was consistent with the findings of Sirois et al. (1998) study of rainbow smelt (*Osmerus mordax*) larvae, where the TVG model was originally proposed. Vigliola et al. (2000) concluded that further comparisons were required in situations where the growth rates were highly variable to distinguish between the MF and TVG models.

Wilson et al. (2008) recently completed such comparisons. They collected 2 species of newly settled tropical gobies (*Elacatinus evelynae* and *E. prochilos*) from reefs in the Florida Keys and maintained them in aquaria for 2 months. Fish were marked externally by the subcutaneous injection of coloured paint (Wilson & Osenberg 2002) to identify individuals and measured (standard length, SL) every 2 weeks. Otoliths were tagged at the start of the study and then by regular immersion (every 4 weeks) in Alizarin Red to validate daily increment formation and monitor otolith growth. Growth rates of fishes were manipulated by altering ration so that individuals received high, medium, low or variable (2 weeks high, 2 weeks low) levels of food. Growth curves generated by back-calculation were compared with directly measured longitudinal size-at-tag data by linear mixed effect modelling (LME, Pinheiro & Bates 2000) that included food regime as a fixed factor in the models. Smaller negative log-likelihood in the LME models indicated greater precision in predicted size, whereas slopes of 1 in LME models indicated unbiased predictions (greater accuracy).

Both species examined by Wilson et al. (2008) had linear L – R relationships. Furthermore, analyses showed that age effects, growth effects and time-varying growth effects were present in the dataset. Surprisingly, however, growth curves generated by the MF model gave a better fit to directly measured growth than the LBPH, BI, TVG and AE models. This result was consistent for both species and there was no significant effect of food regime (Table 1). With the exception of the AE model, all back-calculation models yielded lower negative log-likelihoods (i.e., better fits to directly measured data) than the use of otolith radius in the analysis (Table 1), confirming that back-calculation generates a better proxy of fish size than does otolith radius. Most models gave reasonably precise predictions of fish size, although all overestimated size at age, though the AE model was an exception which yielded negative log-likelihoods twice as large as those obtained with other methods (i.e., a poor fit to directly measured data). The biases in these predictions were very small, again with the exception of the AE model. The smallest overestimation (i.e., slope closest to 1) was made by the MF model, with a slope of 0.95 for both species, giving a relatively trivial 5% average overestimation in size (Figure 3). Predictions from the MF model averaged within 0.9 mm of measured size and yielded a maximum error of 4.5 mm for fish ranging in size from 8.5 to 26.6 mm standard length between the start and end of the experiment. Other models gave similar, although slightly larger errors in prediction of up to 5.7 mm greater than measured values. Very large errors in predicted size were often produced by the AE model (36 mm average error and 657 mm maximum error), presumably because of collinearity problems among the variables of fish length, otolith radius and fish age in the multiple linear regression model.

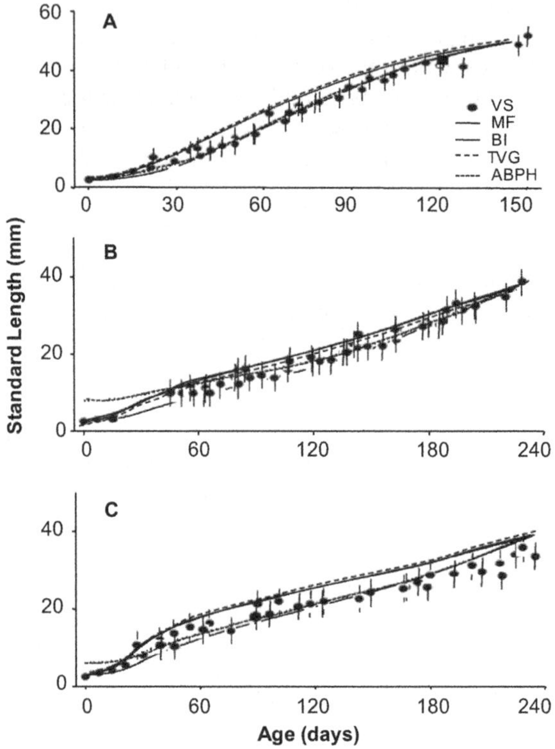

Figure 2. Growth curves back-calculated from otoliths and estimated directly in the field for *Diplodus sargus* (**A**), *D. vulgaris* (**B**) and *D. puntazzo* (**C**). Field size-at-age data are shown as points with error bars of fixed size (±3.5 mm), the maximum error when fish sizes were estimated underwater by divers (Macpherson 1998). Sizes of 30–40 specimens of each species were estimated visually and the specimens then captured and measured using a ruler. All differences between mean estimated and measured sizes were <3.5 mm. Conversion of survey date to age was as follows. First, the average size of all individuals recorded during the last survey was calculated. An age was then assigned to this size by averaging all increment counts of otoliths from fish of this size. Dates of sampling for all surveys were known, so ages could be assigned to average sizes from those surveys. Size-at-age was extrapolated to the planktonic larval stage using growth curves provided by other studies. These points are shown without error bars. Four models of back-calculation were used to generate growth curves from otoliths: the modified Fry (MF), biological intercept (BI), time-varying growth (TVG) and allometric body proportional hypothesis (ABPH) models. After Vigliola et al. (2000).

Table 1. Linear mixed effect (LME) model between longitudinal observations of fish length at tag and otolith radius at tag (R) or between longitudinal observations of fish length at tag and predicted fish length at tag using different back-calculation models for two species of gobies *Elacatinus evelynae* and *E. prochilos* experiencing different food treatments. Food regime is entered as a fixed factor in the LME models. LBPH: linear body proportional hypothesis, BI: biological intercept, MF: modified Fry, AE: age effect, TVG: time varying growth, p: probability, SE: standard error. Lowest Akaike information criterion (AIC), Bayesian information criterion (BIC) or negative log likelihood (Neg-Log-Lik) indicate models generating the most precise size estimate (in bold). Slope of models is tested against 1 (no bias in size predictions) and slopes smaller than 1 indicate size overestimation (observed size at tag smaller than size-at-tag predicted by back-calculation). Bold font indicates model with smallest bias. NR: non-relevant test. From Wilson et al. (2008).

Species	Statistics	R	LBPH	BI	MF	AE	TVG
E. prochylos	AIC	463	405	426	**390**	752	433
	BIC	479	422	442	**406**	768	450
	Neg-Log-Lik	225	197	207	**189**	370	211
	p Food effect	0.55	0.83	0.88	0.90	0.59	0.85
	Slope	0.050	0.948	0.937	**0.949**	0.610	0.915
	SE slope	0.001	0.007	0.008	0.007	0.022	0.008
	p slope=1	NR	<0.001	<0.001	<0.001	<0.001	<0.001
E. evelynae	AIC	251	208	207	**200**	497	210
	BIC	264	221	220	**212**	509	223
	Neg-Log-Lik	120	98	98	**94**	242	99
	p Food effect	0.39	0.60	0.60	0.50	0.74	0.62
	Slope	0.051	0.944	0.942	**0.948**	-0.019	0.917
	SE slope	0.003	0.011	0.011	0.011	0.008	0.011
	p slope=1	NR	<0.001	<0.001	<0.001	<0.001	<0.001

Morita and Matsuishi (2001) compared the outputs of LSPH, LBPH, MONA-SPH, MONA-BPH, BI, Fraser-Lee, MF and AE models to directly measured size and growth rates of individually tagged white-spotted char (*Salvelinus leucomaenis*). The TVG model was not included in the comparison because the growth effect factor was negligible and not statistically significant. Only the AE model generated unbiased estimates of fish length and growth, but it was also the least precise model, giving lower r^2 values between predicted and observed lengths than any of the other models. It seems possible that the increased dispersion of size estimates produced by the AE model occurred in a relatively random manner, which may have disguised any inherent bias. For example, the AE model predicted a size of ~90–140 mm for a 110 mm fish, while in comparison the MF model narrowed the predicted size to 90–110 mm (see Figure 2 in Morita & Matsuishi 2001). The results of Wilson et al. (2008) confirm that very large errors in predicted size are often produced by the AE model. This result was expected since the AE model is vulnerable to growth effects (as is the case for all proportionality models) and very sensitive to the accuracy and precision of the regression calculated between fish length, otolith radius and fish age. Further illustration of the latter point is provided by Finstad's (2003) study of Arctic charr (*Salvelinus alpinus*). This author included a length by age interaction term in the fish length, otolith radius and age multiple regression. Although this was not a conceptual change to the original AE model, predictions of fish length with and without the interaction term could differ by up to 40% on average (and occasionally much more given a S.D. of ~ 20% around this mean; see Figure 1 of Finstad 2003).

Other studies that have attempted to assess the accuracy and precision of back-calculation models over the last decade (summarised in Table 2) found that biological intercept methods generally performed better or similar to other models (Escot & Granado-Lorencio 1999, Klumb et al. 2001). These studies also highlighted the importance of the shape of the relationship between fish length and otolith radius (particularly where there was curvilinearity) to the relative accuracy and precision of the different models (Smedstad & Holm 1996, Schirripa 2002).

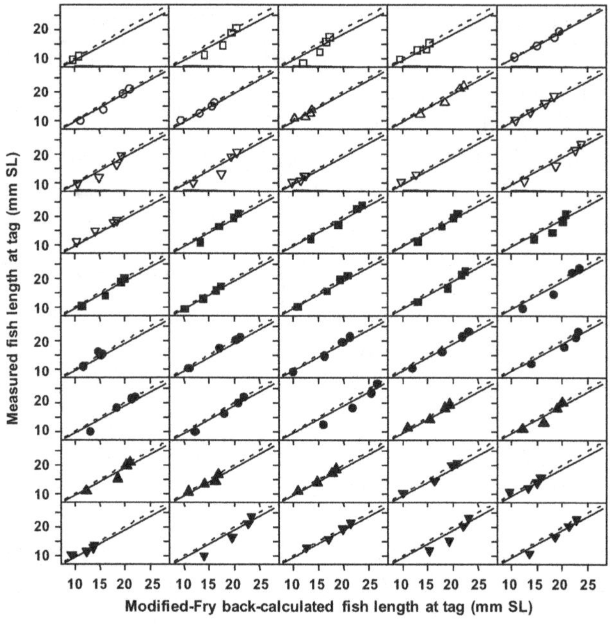

Figure 3. Relationships between observed (multiple internal and external tagging) and predicted (Modified Fry back-calculation model) length-at-age fitted by linear mixed-effects model for two species of tropical gobies. Each plot represents one individual with *Elacatinus evelynae* in open symbols (○) and *E. prochilos* in closed symbols (●). Food regime was entered as a fixed factor in the LME models and treatments are represented as: square = medium food, circle = high food, triangle up = low food, triangle down = varying food. The *solid lines* represent the linear relationships at individual level (i.e., fixed and random effects combined) while the *dashed lines* represent the 1:1 line. *Solid lines* "below" *dashed lines* indicate that observed lengths were smaller than predicted lengths, that is, the back-calculation model overestimated fish length. After Wilson et al. (2008).

Table 2. Summary of studies published in the last decade that sought to identify the most appropriate model for back-calculation. See Appendix 1 for model acronyms.

Study	Models compared	Method	Comparison	Conclusion
Smedstad and Holm (1996)	Fraser-Lee, regression of L on R (BCM 2), Fraser-Lee, symmetrical regression line (modification of BCM 2), MONA-SPH (with log(R) on log(L) regression), MONA-BPH (with log(L) on log(R) regression), LSPH, LBPH	Predictions versus observations from individually tagged fish reared in laboratory and measured repeatedly	Validation at individual level	MONA-BPH (Monarstyrsky-BPH) gave best results
Sirois et al. (1998)	BI, TVG	Predictions versus observations of reared larvae of known age	Validation at population level	Similar results from TVG & BI but TVG slightly more precise
Vigliola et al. (2000)	ABPH, BI, MF, TVG	Predictions vs observations from underwater visual surveys with extrapolated lengths of planktonic larvae.	Validation at population level	MF better, TVG & BI similar, ABPH poor at young ages
Escot and Granado-Lorencio (1999)	Monastyrsky (log(L) on log(R) regression, with intercept), Fraser-Lee (after log_{10} transformation; thus a “non-linear” Fraser-Lee), BI (after log_{10} transformation; thus a “non-linear” BI), simple regression of linear function (after log_{10} transformation; thus a nonlinear model)	Comparisons of back-calculated size at last increment with size at capture	Not a validation	(non-linear) BI gave best results
Morita and Matsuishi (2001)	LSPH, LBPH, MONA-SPH, MONA-BPH, BI, Fraser-Lee, MF and AE	Predictions versus observations from tag-recaptured fish	Validation at individual level	Only AE gave unbiased estimates but AE was less precise than any other model.

Study	Models compared	Method	Comparison	Conclusion
Klumb et al. (2001)	Dahl-Lea, BI, Weisberg[1]	Predictions vs observations from multiple internal tag on reared fish	Validation at population level	Dahl-Lea and BI produced best size estimates
Finstad (2003)	AE, AE with a length by age interaction term	Comparison of models without observation of size	Not a validation	Major differences in predicted size, dependent on interaction term in AE model.
Schirripa (2002)	10 back-calculation models were compared: Dahl-Lea, 5 linear and nonlinear simple regression models, 2 Fraser-Lee (least-square and functional regression), Monarstyrsky (after log transformation), and a derivation of the Weibull function (not described in this chapter)	Data simulated by individual-based bioenergetics model using 4 shapes of fish length – otolith radius relations	Not a validation but a simulation at individual level	Accuracy of the estimated length-at-age directly related to how well the model fitted the otolith radius - fish length relation, implying the L – R function is at least as important as back-calculation model selection.
Wilson et al. (2008)	LBPH, BI, MF, TVG, AE	Predictions versus observations from multiple measurements on reared fish internally and externally tagged	Validation with longitudinal data at individual level	MF best, AE unrealistic, TVG and BI similar; output of all models except AE better length proxies than otolith radius.

1 The Weisberg model (Weisberg & Frie 1987, Weisberg 1993) partitions fish growth into two main effects, growth due to age of the fish and growth due to environmental effects in a given year, in a general additive ANOVA model. This model is rather complex and is not widely used.

2.5 WHICH MODEL?

The studies summarised above confirm that the lack of constraint to biological intercepts by BPH and SPH models results in unrealistic estimates of size, particularly at young ages. Similarly, application of the AE model (an SPH model) produced estimates of size-at-age that varied markedly from reality, although the addition of age in back-calculation models to correct for age effects is a promising avenue for the future improvement of predictions of fish length from otoliths. Thus, the remaining models that can be considered for use based on the evidence above are the MF, TVG and BI. Both Vigliola et al. (2000) and Wilson et al. (2008) found that the MF model tended to produce slightly more precise estimates of size-at-age than the TVG and BI models, and that the outputs of the latter two were essentially equivalent, even where growth rates varied. Given this result, most researchers would be inclined to use the BI rather than the TVG model, since the latter is more complex to compute.

Selection between the modified Fry (MF) and the linear biological intercept (BI) models is rather more difficult, for two reasons. Firstly, the BI model has been used widely for back-calculation since its development by Campana in the early 1990s and most researchers will be comfortable with the calculations involved. The MF model, however, is more recent and there are few published examples of its use. Secondly, calculation of the MF model is more complex than the BI model, and the ideal model should be as simple as possible. There are good reasons, however, to use the MF instead of the BI model in many situations. Vigliola et al. (2000) showed that where there is an allometric L – R relationship, the MF model will produce more precise results than the BI model. The outputs of the MF and BI models will be identical where this relationship is linear, since the BI is a special case of the MF model.

The appropriateness of the 2 models will thus depend on the degree to which the L – R relationship for a species is allometric. Vigliola et al. (2000) concluded that in case of isometry (a linear L – R), the BI model should be used for simplicity, while in case of allometry (a curvilinear L – R) they suggested choosing the MF model due to precision. The results of Wilson et al. (2008) show that this recommendation should be revised. They were able to record the shape of the L – R relationship at an individual level for their study species by repeatedly tagging fish internally (immersion in Alizarin solution) and externally (subcutaneous injection). No allometry was detected using a population level approach (regression analysis of single measurements of L_{capt} and R_{capt} from each fish), and the relationship between these variables appeared isometric for both goby species (Figure 4A). The longitudinal data sets available for each fish, however, showed that on an individual basis there was an allometric L – R relationship (Figure 4B). This result demonstrates that allometric L – R relationships may be undetectable in population analyses (in this case due to low sample sizes), and explains why the MF model performed better despite an apparently linear L – R relationship at population level. It is unlikely that individual longitudinal L – R data will be available in most situations, however, since fish are usually killed when collected. Hence, the most conservative approach will be to use the MF model for routine back-calculations rather than the BI, despite its more complex form.

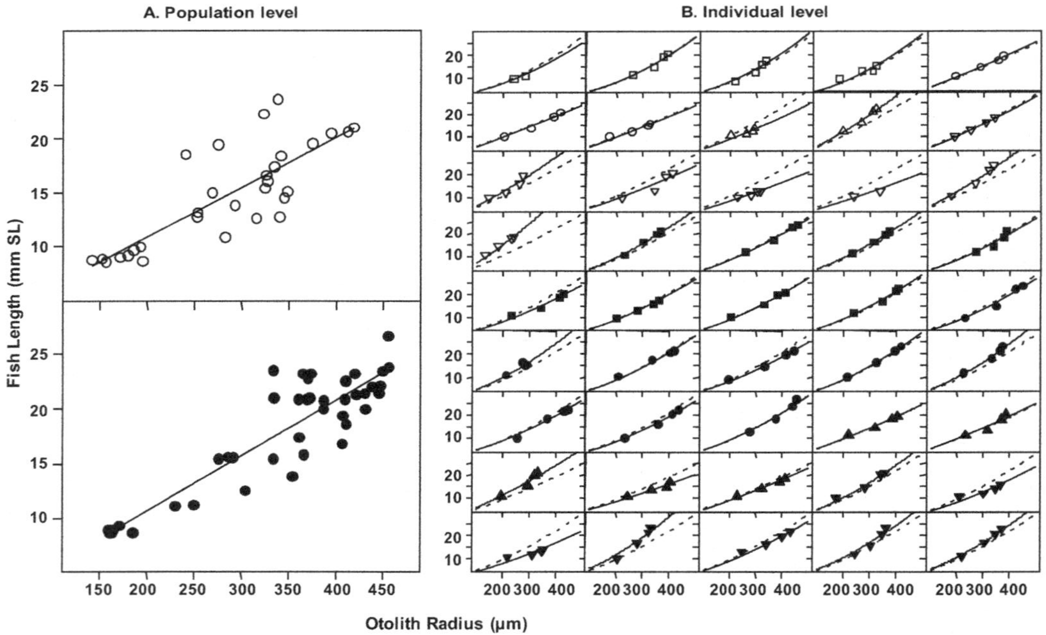

Figure 4. Population (*left*) and individual (*right*) relationships between fish length and otolith radius. *Left panel* (population level): Linear fish length – otolith radius relationship at capture for two species of tropical gobies, *Elacatinus evelynae* (○) and *E. prochilos* (●). *Right panel* (individual level): Allometric fish length – otolith radius relationships fitted by non-linear mixed-effects (NLME) model for the two species of gobies (Each plot represents one individual). Food regime was a fixed factor in the NLME models. Treatments are represented as: ■□ = medium food, ●○= high food, ▲△ = low food, ▼▽ = varying food. *Solid lines* represent allometric relationships at individual level (fixed & random effects combined); *dashed lines* represent allometric relationships at population level (fixed effect). After Wilson et al. (2008).

One of the most desirable features of the BI model is that it does not rely on parameter estimation from a sample of the population. Indeed, back-calculations can be made from an individual fish in the absence of any other fish from the population (Campana 1990). Unfortunately, this is not the case for the MF model where estimates of fish size "a" at otolith formation are required. This parameter will always be very close to fish size at hatching (the biological intercept L_{0p}), however, since otoliths usually form just prior to hatching. For example, average values of body length at otolith formation were a=2.39, 2.42 and 2.45 mm for the three species studied by Vigliola et al. (2000) with a common average length at hatching of L_{0p} = 2.5 mm, which gives ratios a/L_{0p} of 0.96, 0.97 and 0.98, respectively. Vigliola et al. (unpublished data) calculated this ratio for 53 species, mostly coral reef fishes from Florida Keys. They found that 22 (40%) of the species had a significant allometric L – R relationship at the population level and that the average ratio was 0.75, with 95% confidence interval of 0.69–0.80 (range 0.22–0.99). Campana (1990) indicated that back-calculation accuracy of the BI model was relatively insensitive to normal variation around the intercept value, largely because of the small values involved in L_{0p}. This is likely also the case for "a", which is smaller than L_{0p}, so it would be possible to replace "a" in the original MF model by its average value of $0.75L_{0p}$, allowing MF back-calculations from an individual fish in the absence of any other fish from the population. This "experimental MF" model should only be used when there is no data to estimate "a" and can be written as follows:

$$L_i = 0.75L_{0p} + \exp\left(\ln(L_{0p} - 0.75L_{0p}) + \frac{[\ln(L_{cpt} - 0.75L_{0p}) - \ln(L_{0p} - 0.75L_{0p})][\ln(R_i) - \ln(R_{0p})]}{[\ln(R_{cpt}) - \ln(R_{0p})]} \right) \quad (5)$$

Whenever there are linear or allometric L – R relationships, as is the case for most fishes, we recommend using the original MF model or its experimental form (5). This will likely increase precision and reduce bias in back-calculated size-at-age. Furthermore, the MF model is derived from the assumption of proportionality between otolith and somatic growth, a basic assumption that considers back-calculation within the more general context of allometry (Hare & Cowen 1995) for which a large body of literature exists (for reviews see Cock 1966, Gould 1966). When the L – R relationship is not a straight line or an allometric curve (see the study of Tremblay & Giguère 1992 for an example), however, then there is little choice other than using SPH or BPH. In these cases, we recommend including a biological intercept in the f or g function prior to the regression, for example by replacing the intercept a_0 by L_{0p} and R by $(R–R_{0p})$ in the polynomial g7 function of BCM 19 (Appendix 1). These complex SPH or BPH models should be used with extreme caution, however, and validated using longitudinal data at individual level. A detailed example of the recommended back-calculation procedure is provided in Box 1.

Box 1: Recommended back-calculation procedure

Most species display either a linear or an allometric relationship between otolith radius, R, and fish length, L. Where there is insufficient data to examine the shape of the L – R relationship, theoretical considerations, field, and experimental data all indicate that best size estimates are likely from the Modified Fry back-calculation model (MF) (Vigliola et al. 2000). Where L – R relationships are best described by complex functions (e.g., polynomial), we recommend including a biological intercept into the function prior to fitting to data at capture and applying a proportionality hypothesis (SPH or BPH). The MF back-calculation procedure is as follows.

1. Obtain an estimate of fish size at first increment formation (usually hatching) for the study species. This parameter is the biological intercept L_{0p} of the model.
2. Using all samples, calculate the mean radius of first increment in the otolith. This parameter is the otolith radius R_{0p} at the biological intercept.
3. If there is not enough data to examine the shape of the L – R relationship (i.e., a single fish or many fish of similar age and size), then for each individual, back-calculate fish size at age i (L_i) from the longitudinal records of otolith radius at age i (R_i), records of fish length and otolith radius at capture (L_{cpt}, R_{cpt}), knowledge of biological intercept (R_{0p}, L_{0p}, steps 1 & 2) using the experimental MF back-calculation model (otherwise go to step 4):

$$L_i = 0.75L_{0p} + \exp\left(\ln(L_{0p} - 0.75L_{0p}) + \frac{[\ln(L_{cpt} - 0.75L_{0p}) - \ln(L_{0p} - 0.75L_{0p})][\ln(R_i) - \ln(R_{0p})]}{[\ln(R_{cpt}) - \ln(R_{0p})]} \right)$$

4. Using all samples, estimate b, c, d by fitting the following by NL regression:

$$L_{cpt} = L_{0p} - bR_{0p}^{c} + bR_{cpt}^{c} \qquad \text{and} \qquad L_{cpt} = L_{0p} - dR_{0p} + dR_{cpt}$$

   where L_{cpt} is fish length at capture and R_{cpt} is otolith radius at capture.

5. Test c against 1 using estimates of c and standard error SE_c given by non-linear regression (step 4) and calculate $|c - 1|/SE_c$ which follows a t-distribution with n-2 df. If c is significantly different from 1 (allometry) then using estimates of b and c (step 4), calculate $a = L_{0p} - bR_{op}^{c}$. If c is not significantly different from 1 (isometry) then using estimate of d (step 4), calculate $a = L_{0p} - dR_{op}$

   In the original MF model (BCM 14 of Table 1), "a" was calculated from "a_1" and "a_2" obtained from L-on-R and R-on-L regression, respectively. However, Vigliola et al. (2000) showed that differences between a_1 and a_2 were so small (0.08%) that they may be ignored.

6. For each individual, back-calculate fish size at age i (L_i) from the longitudinal records of otolith radius at age i (R_i), records of fish length and otolith radius at capture (L_{cpt}, R_{cpt}), knowledge of biological intercept (R_{0p}, L_{0p}, steps 1 & 2) and estimate of a (steps 4 & 5) using the MF model:

$$L_i = a + \exp\left(\ln\left(L_{0p} - a\right) + \frac{\left[\ln\left(L_{cpt} - a\right) - \ln\left(L_{0p} - a\right)\right]\left[\ln\left(R_i\right) - \ln\left(R_{0p}\right)\right]}{\left[\ln\left(R_{cpt}\right) - \ln\left(R_{0p}\right)\right]} \right)$$

3 Analysis of back-calculated data

Datasets generated from growth back-calculations are longitudinal and autocorrelated, characteristics that make them unsuitable for many statistical analyses (Chambers & Miller 1995). In particular, data that originate from multiple observations per otolith cannot be assumed to be independent. Thus, the longitudinal nature of back-calculated size-at-age data, with repeated measures on each individual, violates the basic assumption of independence that underlies traditional analyses such as univariate regression or ANOVA (Chambers and Miller 1995). Moreover, because statistical hypotheses are accepted or rejected based on the magnitude of the test statistic and the degrees of freedom (df), calculating the df from the total number of non-independent size-at-age data is inappropriate. Chambers and Miller (1995) gave a numerical example for this latter problem with 20 fish collected from each of two populations and 10 increment radii measured on each fish. A traditional regression approach would treat these 200 measurements per population as 200 independent size-at-age observations. Consequently, the hypothesis of a population effect on growth rate would be tested from an F-statistic based on 1,397 df whereas only 40 fish were included in the comparison. This would significantly increase the likelihood of committing a Type 1 error of falsely rejecting the null hypothesis of no difference in size-at-age among populations.

The lack of independence of longitudinal data implies that analysis should occur at the level of individuals, rather than the population. This poses two problems. First, there is often not enough information to fit a growth model by regression to each individual, in addition to the longitudinal and autocorrelated nature of within-individual size-at-age data. Second, it may be unclear how to obtain growth estimates of the population (typically the variable of central interest) from individual analyses, particularly when the growth model is non-linear. We illustrate the issues in Figure 5 where we generated size-at-age data for 10 fish sampled from a simulated population that followed a von Bertalanffy growth model (Box 2, Figure 5A). Individual regressions of a von Bertalanffy growth model could not be calculated for 6 fish as they were too young, with only 2–3 points available for analysis (Figure 5B). Additionally, outcomes of the individual regressions were suspect for at least two other fish. Fish 7 had a growth trajectory unexpectedly higher while fish 4 had a trajectory far lower than the parent population (Figure 5B, L_∞ original population = 30 cm; L_∞ fish 7 = 56 cm; L_∞ fish 4 = 20 cm). A common but flawed approach to obtain an estimate of the growth of the population would be to average growth parameters from individual analyses. Figure 5 shows that this will be unreliable, as the average growth of the 4 fish for which individual regressions could be calculated departed strongly from the trajectory of the original population (Figure 5C). Another common approach would consist of pooling back-calculated size-at-age data of all individuals to collectively derive a mean growth curve by regression. Again, this would be inappropriate as indicated by strong departure of the pooled regression curve from the true population growth curve (Figure 5C). A further risk in the pooling of back-calculated data is that the growth patterns of the oldest individuals will contribute disproportionately, as these older fish will add more points to the data set than young fish. Finally, using only the 10 records of size-at-capture in the regression would not give an appropriate estimate of the true population growth curve (Figure 5C), presumably due to low sample size and the presence of only one old fish in the sample, which is "pooling the trend" for L_∞.

Pinheiro and Bates (2000) recently proposed the use of mixed effects models to analyse longitudinal, autocorrelated, unbalanced data such as those generated by back-calculation. This method fits any linear (linear mixed-effects, LME) or non-linear (NLME) model to longitudinal data with great flexibility in modelling the within-group correlation often present in such data. The term "mixed effects" refers to the fact that both fixed and random effects are modelled. Fixed effects correspond to a population (or factor) level estimate of parameters, while random effects correspond to variability of parameters among individuals relative to the population. Thus, mixed-effects models generate estimates of model parameters at both the individual and population level. Our example of Figure 5 is used to illustrate this point. We fitted a von Bertalanffy growth model using a non-linear mixed-effects model (NLME) for the sample of 10 fish. NLME provided accurate estimates of growth parameters at both the population (Figure 5C) and individual (Figure 5D) levels (see Box 3 for computation of NLME).

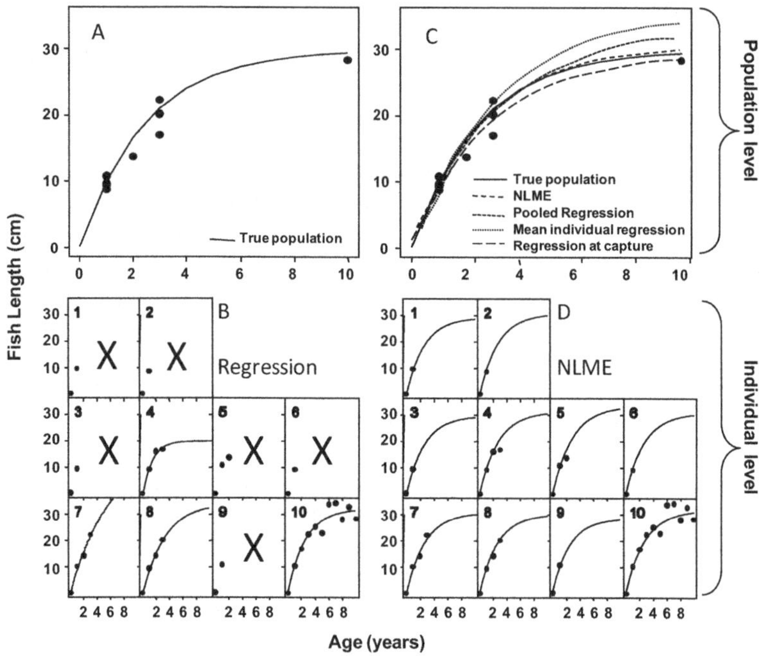

Figure 5. Comparison of regression and mixed effects models. **A**. Ten fish (*points*) sampled from a simulated population of known von Bertanlanffy growth trajectory (*solid line*). Size-at-age was generated for each individual (Box 2) and longitudinal data analysed by non-linear regression (**B**) and non-linear mixed-effects models (NLME, D, Box 3). X indicates individuals for which regressions were not possible due to lack of data. Heteroscedasticity was modelled in NLME by a variance function that was a power of the absolute value of the variance covariate. Plot C compares population level estimates of the growth trajectory by NLME, regression based on pooling size-at-age data of the 10 fish, average individual regression, and regression of size at age of capture of the 10 fish with known growth. Individual ID numbers appear at top of each plot in the lower graphs.

Box 2: Algorithm used to simulate size-at-age data

1. Define a growth function (von Bertalanffy, exponential etc), e.g., $L = L_0 * \exp(K * Age)$, where L = fish length-at-age, L_0 = fish length at hatching, K is the exponential growth rate.
2. Set true population growth parameters μ_p, e.g., $L_0 = 2$ mm and $K = 0.06$ %.d^{-1}.
3. Randomly generate growth parameters for each individual j by: (i) assuming normal distribution of growth parameters with mean μ_p and standard deviation $0.1*\mu_p$ (hence a coefficient of variation of 10%); and (ii) adding a random uniform noise within 2% of the expected value ($\pm 0.02*\mu_p$, which is 5 times smaller than standard deviation). For example:

$L_{0j} \sim N(\mu=2, \sigma=2*0.1)+U(min=-2*0.02, max=2*0.02)$ across j fish;

$K_j \sim N(\mu=0.06, \sigma=0.06*0.1)+U(min=-0.06*0.02, max=0.06*0.02)$ across j fish.

4. Generate size-at-age data for each individual j at each age i. To do this, we first estimated size-at-age from individual growth parameters (points 1–3 above), then introduced a random normal error (mean=0, sd=0.1*size) and a uniform random noise (min=-0.02*size, max=0.02*size). In this example, size of individual j at age i is given by:

$$L_{ji} = L_{0j}*\exp(K_j*i) + \varepsilon_{ji}$$

With:

$$\varepsilon_{ji} \sim N(\mu=0, \sigma=0.1*L_{0j}*\exp(K_j*i)) + U(min=-0.02*L_{0j}*\exp(K_j*i), max=0.02*L_{0j}*\exp(K_j*i))$$

5. Further constraints such as $L_{ij} > 0$ were added in the simulation code. Simulated size-at-age data were analysed by NLME (Box 3).

Box 3: Analysing size-at-age data by NLME: An example in R

1. Set the workspace

Install R software from the internet (http://www.r-project.org/) including required libraries. Create a data file (e.g., in MS Excel™), and save it in .txt format. In the examples below, the data file is named "mydata.txt" & has at least 3 columns: fish identification (id), length-at-age (L) and incremental age (Age). Add further columns for factors such as population (Pop) or time of sampling (Time). ***Caution,*** R is case sensitive. Comments are preceded by # and ignored.

```
# load required libraries
library(grid)
library(lattice)
library(stats)
library(nlme)
```

2. Import data and declare growth functions

```
#Import the data into R under the object name "datgr"
datgr=groupedData(L~Age|id, data=read.table("mydata.txt",header=TRUE, sep="\t"),
labels=list(x="Age",y="Size"), units=list(x="(Years or Days)",y="(cm or mm)"))
#visualise the data
plot(datgr)
#Declare growth functions
LVB=function(x,t0,Lmax,K) {
y=Lmax*(1-exp(-K*(x-t0)))
y
}
EXP=function(x,L0,K) {
y=L0*exp(K*x)
y
}
```

Box 3 (Continued)

3. Fit growth function by NLME

```
# fit a LVB or an exponential growth model by NLME. If the model does not converge, try changing starting values (bold).
LVB.nlme=nlme(L~LVB(Age,t0,Lmax,K),data=datgr,
fixed=list(t0~1,Lmax~1,K~1),
random= t0+Lmax+K~1,
start=list(fixed=c(t0=-0.02,Lmax=30,K=0.4)))
EXP.nlme=nlme(L~EXP(age,L0,K),data=datgr,
fixed=list(L0~1,K~1),
random=L0+K~1,
start=list(fixed=c(L0=2,K=0.06)))
#to see results and plots:
summary(LVB.nlme)
intervals(LVB.nlme)
anova(LVB.nlme)
plot(LVB.nlme)
plot(augPred(LVB.nlme,level=0:1))
coef(LVB.nlme)
```

4. Comparing populations (or levels of other factors)

```
# Compare LVB growth models of two (or more) populations. There must be as many starting values as levels of factors (in this case 2 levels: population 1 and 2). The two models below are inherently identical (type anova(LVB1.nlme,LVBbis.nlme) to verify this). The first model includes an intercept, which is convenient to test the effect of factor population, whereas there is no intercept in the second model, which is convenient to estimate growth parameters for each level of the factor population.
LVB1.nlme=nlme(L~LVB(Age,t0,Lmax,K),data=datgr,
fixed=list(t0~Pop,Lmax~Pop,K~Pop),
random= t0+Lmax+K~1,
start=list(fixed=c(-0.02,0,25,0,0.4,0)))
anova(LVB1.nlme)
LVB1bis.nlme=nlme(L~LVB(Age,t0,Lmax,K),data=datgr,
fixed=list(t0~Pop-1,Lmax~Pop-1,K~Pop-1),
random= t0+Lmax+K~1,
start=list(fixed=c(-0.02,-0.02,30,20,0.4,0.4)))
summary(LVB1bis.nlme)
```

5. Comparing sequential samples (size-selective mortality)

```
#Highlight size-selective mortality effect on exponential growth parameter K with 3 sequential samples (there must be as many starting values as there are samples or levels for factor Time).
#selection on both L0 and K
EXP1.nlme=nlme(L~EXP(age,L0,K),data=datgrl,
fixed=list(L0~Time,K~Time),
random=L0+K~1,
start=list(fixed=c(2,0,0,0.06,0,0)))
#selection on K only
EXP2.nlme=nlme(L~EXP(age,L0,K),data=datgrl,
fixed=list(L0~1,K~Time),
random=L0+K~1,
start=list(fixed=c(2,0.06,0,0)))
```

Box 3 (Continued)

```
#no selection
EXP3.nlme=nlme(L~EXP(age,L0,K),data=datgrl,
fixed=list(L0~1,K~1),
random=L0+K~1,
start=list(fixed=c(2,0.06)))
#most parsimonious model
anova(EXP1.nlme,EXP2.nlme,EXP3.nlme)
#results
EXP2bis.nlme=nlme(L~EXP(age,L0,K),data=datgrl,
fixed=list(L0~1,K~Time-1),
random=L0+K~1,
start=list(fixed=c(2,0.06,0.065,0.075)))
summary(EXP2bis.nlme)
plot(augPred(EXP2bis.nlme,level=0:1))
```

6. Modelling heteroscedastic and/or correlated within-group variance

#The NLME library provides a set of classes of variance functions, the varFunc classes, and a set of classes of correlation structures, the corStruct classes. The former are used to specify within-group variance and thus model heteroscedasticity whereas the latter are used to specify within-group correlation and thus model dependence among observations (see Pinheiro & Bates 2000 for details). For example, heteroscedasticity was modelled by a variance function that is a power of the absolute value of the variance covariate in all simulated examples given in this chapter (Figures 5, 6 & 7). For the size-selection example shown in Figure 7, this gives:

```
EXP2bis.final.nlme=nlme(L~EXP(age,L0,K),data=datgrl,
fixed=list(L0~1,K~Time-1),
random=L0+K~1,
start=list(fixed=c(2,0.06,0.065,0.075)),
weights=varPower())
```

The aim in most studies is not only to estimate growth, but also to compare the growth of 2 or more populations. Mixed-effects models can perform this task easily by incorporating a qualitative co-variable (or factor). Importantly, LME and NLME allow different variances for each level of a factor and hence can be used to flexibly model heteroscedasticity of the within-error group. Figure 6 shows how NLME can estimate the von Bertalanffy growth parameters of two simulated populations that differ only in L_∞ (Pop 1 L_∞ = 30; Pop 2 L_∞ = 20 cm). A von Bertalanffy growth model was fitted by NLME to simulated size-at-age of 10 fish sampled from each of two populations (Boxes 2 and 3). The technique provided not only accurate estimates of growth parameters at both individual and population levels, but also correctly identified that the two populations differed significantly in L_∞.

In the same way that the qualitative factor "population" was inserted into the mixed-effects model in the example above, any other factor (e.g. diet, site, year, time, etc.) or combination of factors (e.g. "diet × population") could also be added to the model. Of all factors whose influence on growth trajectories is likely to be of interest to a researcher, "age" (or "time") is one of the most difficult to analyse. For example, size-selective mortality is a phenomenon that typically removes the smaller, slower growing fish as a cohort ages, so that sequential samples from the same cohort display different growth trajectories (Meekan & Fortier 1996, Sogard 1997, Vigliola & Meekan 2002). Generally, size-at-age data back-calculated from sequential samples are analysed by repeated-measures MANOVA (Chambers & Miller 1995). Although this analysis can successfully demonstrate size-selection, it does not provide any estimate of growth rates, as no growth model is fitted to the data. Furthermore, it restricts the analysis to the youngest fish of each sequential sample. For example, Vigliola and Meekan (2002) were forced to limit the comparison of growth trajectories of settlement stage, 1, 2 and 3 months post-settlement fish to 12 d (a repeated factor at levels 0, 3, 6, 9, 12 d) since the youngest fish collected in the samples was 12 d of age. In contrast, Vigliola et al. (2007) analysed the entire dataset in a NLME model and used individual growth estimates to demonstrate a proximal link between phenotypic and genetic selection.

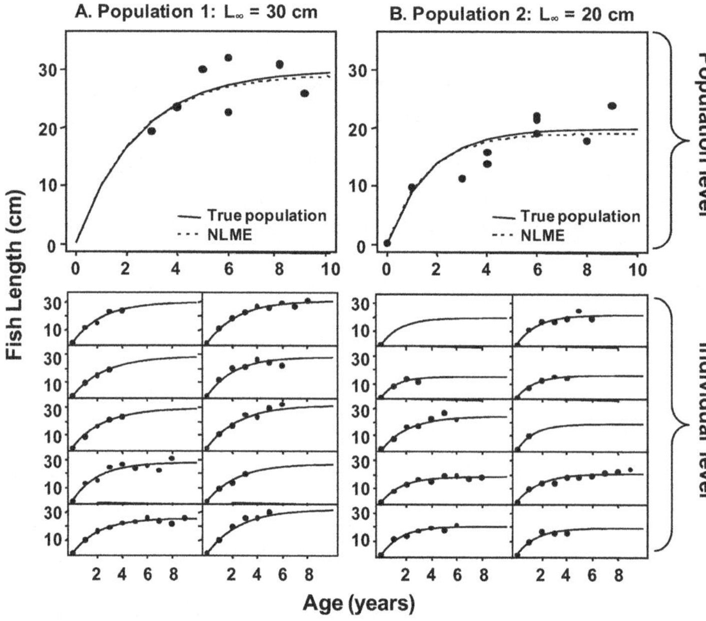

Factor	$df_{Numerator}$	$df_{Denominator}$	F	p
t_0	1	98	0.01	0.91
L_∞	1	98	74.00	< 0.0001
K	1	98	0.66	0.41

Figure 6. Data from two populations of known von Bertanlanffy growth trajectories. The two populations differed only in the value of parameter L_∞. Ten fish were sampled from each population and simulated size-at-age (Box 2) analysed by non-linear mixed-effects models with "population" entered as a fixed factor (Box 3). Heteroscedasticity was modelled by a variance function as a power of the absolute value of the variance covariate. NLME not only provided accurate estimates of the von Bertallanfy model at both population (*upper panels*) and individual levels (*lower panels*), but also provided statistics (*table*) indicating that the two populations only differed in parameter L_∞.

NLME can be a very powerful alternative to repeated-measures MANOVA to study size-selective mortality, while estimating growth parameters and enabling a researcher to determine the growth parameters on which selection acts (Figure 7). A simulated cohort with an exponential growth trajectory ($L = L_0$ exp(K age) with $L_0 = 2$ mm and $K = 0.06$ %.d^{-1}) was sampled so that 10 fish were randomly removed at 10, 20 and 30 days after hatching. No size-selection occurred in the cohort from 0–10d after hatching, but only fish with a growth rate (K) bigger than the 50% quantile survived to 20 d, and only fish with a K bigger than the 90% quantile survived to 30 d. Quantiles were calculated from the random generation of 10,000 individuals using the general algorithm described in Box 2. No selection acted on the size at hatching parameter L_0. Size-at-age data were simulated for the 30 sequentially sampled fish (Box 2) and an exponential growth model fitted by NLME to the entire dataset with "time of sampling" added as a factor (Box 3). NLME analysis successfully estimated growth parameters of each individual and of each sequential sample, while indicating that strong size-selective mortality occurred in this population where fish with higher K survived (Figure 7). This latter result was obtained by identifying the most parsimonious among three possible NLME models. The first included an effect of the factor Time on both L_0 and K; the second only included an effect of Time on K, while the third did not include Time as a covariate (Box 3). The analysis indicated that the second model was the most parsimonious, thus demonstrating that size-selective mortality had indeed occurred and had only acted on growth rate K.

In summary, mixed-effects model are a powerful tool to analyse data from growth back-calculation as they:

- can accommodate longitudinal, unbalanced, auto-correlated data;
- account for individual effects while also estimating group (or factor) effects;
- allow the within-error group to be heteroscedastic and/or correlated;
- are efficient even with small sample sizes;
- can easily integrate factors and covariates; and
- allow researchers to analyse, explore and interpret complex designs.

LME and NLME are programmed in the language S (also used in the software package R). Examples of basic NLME modelling of fish growth are provided in Box 3. Details of LME and NLME can be found in Pinheiro and Bates (2000) and a guideline to programming in S can be found in Venables and Ripley (2002). Implementations of S are available in S-PLUS, a commercial system (http://www.insightful.com), and R, an open source software package (http://www.r-project.org).

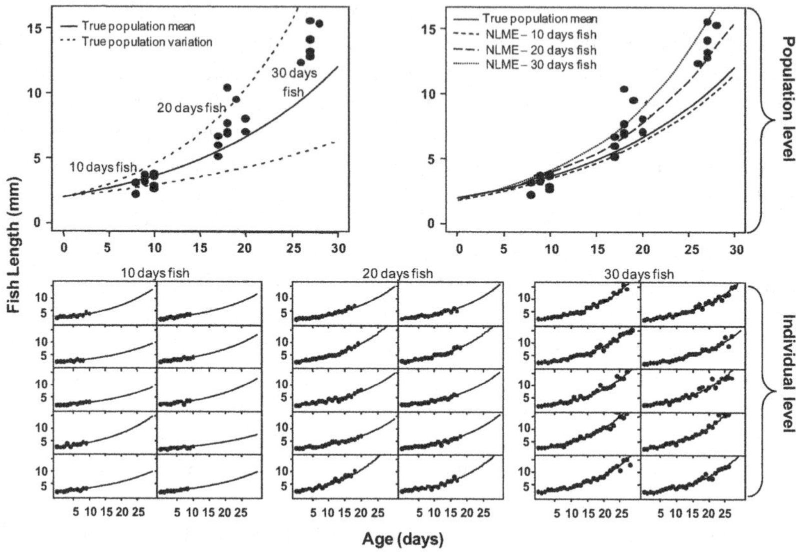

Effect	Model	df	AIC	BIC	L-L	Test	L-L Ratio	p
Selection on K & L_0	1	11	774	822	-376			
Selection on K	2	9	774	814	-378	1 vs 2	3.9	0.1426
No Selection	3	7	794	824	-389	2 vs 3	23.2	< 0.0001

Figure 7. A simulated cohort with an exponential growth trajectory ($L = L_0 \exp(K\ age)$, $L_0 = 2$ mm, $K = 0.06\ \%.d^{-1}$) was sampled so that 10 fish were randomly removed at 10, 20 and 30 days after hatching. No size-selection occurred from 0–10 d after hatching, but only fish with a growth rate (K) bigger than the 50% quantile survived to 20 d, and only fish with a K bigger than the 90% quantile survived to 30 d. No selection acted on the size at hatching parameter L_0. Quantiles were calculated from the random generation of 10,000 individuals using the general algorithm described in Box 2. Size-at-age data were generated from the 30 sequentially sampled fish (see Box 2) and an exponential growth model fitted by NLME to the entire dataset with "Time of sampling" added as a fixed factor (see Box 3). Heteroscedasticity was modelled in the NLME by a variance function, which is a power of the absolute value of the variance covariate. NLME analysis successfully estimated growth parameters of each individual and of each sequential sample. Log likelihood ratio tests (Table embedded in figure) indicated that a model with an effect of the factor Time on both L_0 and K was not better than a model only including an effect of Time on K ($p=0.1426$), while the latter was significantly better ($p<0.0001$) than a third model that did not include Time as a covariable, hence demonstrating that size-selective mortality had indeed occurred and had only acted on growth rate.

4 Conclusions and future directions

The discovery of daily increments within otoliths by Panella in the 1970s has led to a major proliferation in use of back-calculation techniques over the last 30 years. Jones (1992) noted that the number of published studies using daily increments (many of which involved back-calculation) was increasing exponentially. The trend has not abated, so that today even a cursory search of online databases yields hundreds of otolith back-calculation studies published in the last decade. It is now well recognised that the assumptions underlying the technique (periodicity of increment deposition, reliability of ageing and proportionality between otolith and somatic growth) require validation and more than 200 studies have addressed the issue of age validation in the last 10 years. These contrast with the studies that have sought to validate and identify the most appropriate model for back-calculation over the same period: a total of 9 in peer-reviewed journals of which we are aware (Table 2). Of these, 6 attempted a proper validation of the models by comparing direct observations of fish length with estimates provided by back-calculation, while only 3 validated at the individual level. Only one study was based on longitudinal data collected at the individual level under conditions of variable growth.

Is it necessary to validate the output of a back-calculation model every time it is used? Although relative accuracy can only be determined by a comparison of model outputs against independent data sets of growth, this will be logistically difficult to achieve in most situations and virtually impossible for some life history stages such as pelagic larvae. Validation of back-calculation, therefore, will involve considerable effort and will probably only occur in studies designed with this as a central aim. We have shown here, however, that simple criteria can be used to reduce the number of potential back-calculation models to a more manageable short-list. Two recent studies that attempted comprehensive comparisons of these remaining models (Vigliola et al. 2000, Wilson et al. 2008) found that the MF model performed better than others, while one other (Morita & Matsuishi 2001) suggested that the AE model produced the most accurate estimates of size from back-calculation. Experimental studies (Wilson et al. 2008) show that the AE model can generate unrealistic estimates of fish size (as is the case for all proportionality hypothesis and other regression-based models), so we do not recommend use of this model without thorough validation.

Wilson et al. (2008) found that allometric L – R relationships may be undetectable in population level analyses, implying that it may be very difficult to choose between BI and MF models, even when evidence from regression analyses suggests that L – R relationships are isometric (linear). We thus recommend the use of the MF model as a conservative approach in routine back-calculations of fish size at age from otoliths (Box 1).

The development of back-calculation models that can accommodate both isometric and allometric L – R relationships is major advance in the analysis of fish growth from otoliths and other hard parts. These models now need to be validated under a range of environmental conditions and with a variety of life history stages of fish. In particular, validation studies need to focus on the transitions between ontogenetic stages and environments, for two reasons: Firstly, this is where L – R relationships often break down or change; Secondly, transitions such as those between pelagic and reef environments, or between juvenile and adult habitats, are "critical periods" where fish undergo intense periods of mortality (e.g., Doherty et al. 2004, Webster & Almany 2006), and thus are of great interest to ecologists and fisheries biologists due to their influence on population regulation.

Recent advances in statistical methods and the greater availability of NLME models in software packages gives researchers powerful new tools to analyse back-calculated data sets of growth. Like other methods such as repeated-measures MANOVA, these offer a means to include a range of covariates or factors into the model analysis while providing results than can be interpreted in a simple manner. Notably, these models are able to detect processes such as size-selective mortality and to incorporate analysis of the abiotic and biotic correlates of growth within and among different populations. Unlike other methods, however, NLME models account for individual effects while also estimating group (or factor) effects, and can explicitly model heteroscedastic and correlated within-group error. It is important to remember, however, that quality of the output of these new techniques will be determined by the quality of the back-calculated data that is analysed and the growth function that is fitted to data sets. Examples of basic NLME modelling of fish growth are provided in Box 3.

Our call for more validation studies is not new and echoes that of Francis (1990, 1995) and Brothers (1995). In the last decade many more back-calculation models have been proposed, but relatively few studies have attempted to comprehensively assess the ability of these to reproduce real patterns of fish growth. Back-calculation models are theoretical constructs that provide us with a practical means of generating a wealth of useful data, but without rigorous validation we have no idea if they accurately reflect reality. For the most part, this validation remains to be completed.

Appendix 1. Summary of back-calculation models. R: otolith radius; L: fish length; R_i, L_i: radius and length at age i; R_{cpt}, L_{cpt}: radius and length at time when fish were sacrificed; R_0, L_0: radius and length at origin (usually hatching, age=0); R_{0p}, L_{0p}: biologically-determined radius and length at origin ("biological intercept" BI, usually at hatching); f: function such as R=f(L); g: function such as L=g(R); a, b, …: function parameters; BCF: back-calculation function (f or g).

Model & Calculation Details	Abbreviations
Model: $L_i = \frac{R_i}{R_{cpt}} L_{cpt}$	**BCM 1**
Name: Dahl-Lea (Lea 1910), **Function:** Linear w/o intercept; L = bR (or R = L/b) **Hypothesis:** "the scale grows in exact proportion to the length of the individual"; L_i/R_i = constant = b **Geometry:** A straight line passing through (R=0, L=0) & (R_{cpt}; L_{cpt}). **Computation:** Only the directly measured (L_i; R_i; R_{cpt}; L_{cpt}) are required to compute BCM 1. Note that BPH applied on function g1 or SPH applied on function f1 will also result in BCM 1. BCM 3 is identical to BCM 1 if (R_{0p}=0; L_{0p}=0), which may be reasonable when back-calculating size-at-age of adult fish from annual increments.	**DALE** (g1 or f1)

Model & Calculation Details	Abbreviations

Model: $L_i = a + (L_{cpt} - a)\dfrac{R_i}{R_{cpt}}$ **BCM 2**

Name: Fraser-Lee (Fraser 1916, Lee 1920), **FRALE**

Function: Linear w/ intercept; $L = a + bR$ (g2)

Hypothesis: "the growth increment of the scale is, on average…, a constant proportion of the growth increment of the fish"; $\dfrac{(L_i - L_{i-1})}{(R_i - R_{i-1})} = \text{constant} = b$

Geometry: A straight line passing through (R=0; L=a) & (R_{cpt}; L_{cpt}).

Computation: Fraser (1916) and Lee (1920) originally described "a" as the length of the fish at the time of scale formation (and assumed this length was the same for all scales from all fish in a population). Current practice is to set "a" as the intercept of g2 function fitted by "L-on-R" regression (Francis 1990):

1. Fit $L = a + bR$ by L-on-R linear regression to estimate a.
2. Apply BCM 2.

Model: $L_i = L_{cpt} + (R_i - R_{cpt})\dfrac{(L_{cpt} - L_{0p})}{(R_{cpt} - R_{0p})}$ **BCM 3**

Name: Biological Intercept (Campana 1990), or Linear Biological Intercept, **BI, LBI**

Function: Linear w/ intercept; $L = a + bR$ (g2)

Hypothesis: Modification of BCM 2 with replacement of statistically estimated intercept (R=0; L=a) by biologically determined intercept (R_{0p}; L_{0p}).

Geometry: A straight line passing through (R_{0p}; L_{0p}) & (R_{cpt}; L_{cpt}).

Computation: "Generally, I would define the biological intercept of a fish-otolith trajectory as the fish and otolith length corresponding to the initiation of proportionality between fish and otolith growth. …In many cases, the biological intercept could be determined by simple measurements of otolith and fish size in newly-hatched larvae in the laboratory" (Campana 1990).

1. Obtain an estimate of mean fish size at first increment formation (usually size at hatching) for the species. This parameter is the biological intercept L_{0p} of the model
2. Obtain estimate of R_{0p}, mean radius of first otolith increment for the species. R_{0p} may be calculated from the study sample.
3. Apply BCM 3.

Model: $L_i = (a + bR_i)\dfrac{L_{cpt}}{(a + bR_{cpt})}$ **BCM 4**

Name: Linear Body Proportional Hypothesis, **LBPH**

Function: Linear w/ intercept; $L = a + bR$ (g2)

Hypothesis: The BCF is linear and there is "constant proportional deviation from the mean body size" (BPH, Francis 1990).

Geometry: A straight line passing through (R_{cpt}; L_{cpt}) & preserving a constant proportional distance to regression line g2.

Computation:

1. Fit $L = a + bR$ by L-on-R linear regression to estimate a & b.
2. Apply BCM 4.

Model & Calculation Details	Abbreviations
Model: $L_i = L_{0p} + \sum_{j=1}^{i}(W_j + G_e(W_j - W))\frac{(L_{cpt} - L_{0p})}{(R_{cpt} - R_{0p})}$	**BCM 5**
Name: Time-Varying Growth (Sirois et al. 1998),	**TVG**
Function: Linear w/ intercept; $L = a + bR$	(g2)
Hypothesis: Modification of the linear biological intercept model (BCM 3) with inclusion of a growth effect factor in the structure of the model.	
Geometry: Complex curve not necessarily passing through $(R_{0p}; L_{0p})$ or $(R_{cpt}; L_{cpt})$.	
Computation: W_i is the otolith increment width at age i and W is a stage-specific mean increment width. Values of W are calculated for each fish at each developmental stage, e.g., Sirois et al. (1998) calculated W at yolk-sac, preflexion and post-flexion stages for each fish. 1. Obtain a biological intercept $(R_{0p}; L_{0p})$ as for BCM 3. 2. Define developmental stages and calculate average otolith increment for each fish at each stage (one value per fish per stage). 3. Compute the estimated slope of the fish size – otolith size relationship for each fish,: $S = \frac{(L_{cpt} - L_{0p})}{(R_{cpt} - R_{0p})}$. 4. Compute linear growth rate in length for each fish: $G = \frac{(L_{cpt} - L_{0p})}{Age_{cpt}}$. 5. Calculate growth effect factor G_e by S-on-G linear regression, $S = d + G_eG$ using each fish as an independent observation. 6. Apply BCM 5. Note there is one value of W per fish per stage. W should be replaced by W_k for k developmental stage in BCM 5.	
Model: $L_i = \frac{(\frac{R_i}{R_{cpt}}(A + BL_{cpt}) - A)}{B}$	**BCM 6**
Name: Linear Scale Proportional Hypothesis,	**LSPH**
Function: Linear w/ intercept; $R = \frac{(L - a)}{b} = A + BL$	(f2)
Hypothesis: The BCF is linear and there is "constant proportional deviation from the mean scale size" (SPH, Francis 1990).	
Geometry: A straight line passing through $(R_{cpt}; L_{cpt})$ & preserving a constant proportional distance to regression line f2.	
Computation: 1. Fit $R = A + BL$ by R-on-L linear regression to estimate A & B. 2. Apply BCM 6.	

Model & Calculation Details	Abbreviations

Model: $$L_i = -\frac{a}{b} + \left(L_{cpt} + \frac{a}{b} + \frac{c}{b} age_{cpt}\right)\frac{R_i}{R_{cpt}} - \frac{c}{b} age_i$$ **BCM 7**

Name: Age Effect (Morita & Matsuishi 2001), or Age Effect Scale Proportional Hypothesis, **AE, AESPH**

Function: Plane w/ intercept; $R = a + bL + c \times age$ (f3)

Hypothesis: "otolith increases with both fish body length and age following f3" and "the deviation of the otolith length of a fish from the average for that fish length and age is relatively the same throughout the life of a fish (SPH)".

Geometry: A complex trajectory passing through (R_{cpt}; L_{cpt}) and included in a plane passing through (R_{cpt}; L_{cpt}; Age_{cpt}) and preserving a constant proportional distance to regression plane f3.

Computation:

1. Fit $R = a + bL + c \times age$ by R-on-L-and-age multiple linear regression to estimate a, b, c.
2. Apply BCM 7

Model: $$L_i = (A + BR_i + C \times age_i)\frac{L_{cpt}}{(A + BR_{cpt} + C \times age_{cpt})}$$ **BCM 8**

Name: Age Effect Body Proportional Hypothesis, **AEBPH**

Function: Plane w/ intercept; $L = \frac{R\text{-}a\text{-}c \times age}{b} = A + BR + C \times age$ (g3)

Hypothesis: Same as BCM 7 but a BPH rather than a SPH is applied.

Geometry: A complex trajectory passing through (R_{cpt}; L_{cpt}) and included in a plane passing through (R_{cpt}; L_{cpt}; Age_{cpt}) and preserving a constant proportional distance to regression plane g3.

Computation:

1. Fit $L = A + BR + C \times age$ by L-on-R-and-age multiple linear regression to estimate A, B, C.
2. Apply BCM 8.

Model: $$L_i = \left(\frac{R_i}{R_{cpt}}\right)^c L_{cpt}$$ **BCM 9**

Name: Monastyrsky (Bagenal & Tesch 1978), **MONA**

Function: Allometric w/o intercept; $L = bR^c$ (g4)

Hypothesis: The model assumes the BCF is described by an allometric function passing through the origin and the point at capture.

Geometry: An allometric curve passing through (R=0; L=0) & (R_{cpt}; L_{cpt}).

Computation: This model requires an estimate for c. Originally, it seems that c may have been estimated from fit of function g4 by eye (Francis 1990). However, common practice may be to estimate c from L-on-R linear regression of $\ln(L) = B + c \times \ln(R)$ with intercept $B = \ln(b)$.

1. Fit $L = bR^c$ by eye (or regression) to estimate c
2. Apply BCM 9

Model & Calculation Details	Abbreviations
Model: $L_i = \frac{bR_i^c}{bR_{cpt}^c} L_{cpt} = \left(\frac{R_i}{R_{cpt}}\right)^c L_{cpt}$	**BCM 10**
Name: Monastyrsky Body Proportional Hypothesis,	**MONA-BPH**
Function: Allometric w/o intercept; $L = bR^c$	(g4)
Hypothesis: Modification of the Monastyrsky's model (BCM 9) with application of BPH. Morita and Matsuishi (2001) described BCM 10 as a "nonlinear BPH".	
Geometry: Allometric curve through (R=0; L=0) & (R_{cpt}; L_{cpt}).	
Computation: Note an estimate of "b" is not required to compute BCM 10. 1. Fit $L = bR^c$ by L-on-R non-linear regression to estimate c. 2. Apply BCM 10.	
Model: $L_i = \left(\frac{(\frac{R_i}{R_{cpt}} BL_{cpt}^c)}{B}\right)^{\frac{1}{c}} = \left(\frac{R_i}{R_{cpt}}\right)^{\frac{1}{c}} L_{cpt}$	**BCM 11**
Name: Monastyrsky Scale Proportional Hypothesis,	**MONA-SPH**
Function: Allometric w/o intercept; $R = \left(\frac{L}{b}\right)^{\frac{1}{c}} = \left(\frac{1}{b}\right)^{\frac{1}{c}} L^{\frac{1}{c}} = BL^C$	(f4)
Hypothesis: Modification of BCM 9 with application of SPH. Morita and Matsuishi (2001) described BCM 11 as a "nonlinear SPH".	
Geometry: Allometric curve passing through (R=0; L=0) & (R_{cpt}; L_{cpt}).	
Computation: Note that an estimate of B is not required to compute BCM 11. 1. Fit $R = BL^C$ by R-on-L non-linear regression to estimate C. 2. Apply BCM 11	
Model: $L_i = \exp\left(\ln(L_{0p}) + \frac{[\ln(L_{cpt}) - \ln(L_{0p})][\ln(R_i) - \ln(R_0)]}{[\ln(R_{cpt}) - \ln(R_0)]}\right)$	**BCM 12**
Name: Watanabe and Kuroki (1997),	**WAKU**
Function: Allometric w/o intercept; $L = bR^c$	(g4)
Hypothesis: Watanabe and Kuroki (1997) stated that "we assumed that the relationship of i-th otolith ring radius and L on the day of i-th ring formation (L_i) can be expressed by an allometric formula for individual larvae" and "the allometric parameters 'b' and 'c' were calculated for each larvae by solving the two equations; $L_{0p} = bR_0^c$ and $L_{cpt} = bR_{cpt}^c$, where ...R_0 is the measured radius of the first daily ring".	
Geometry: Allometric curve passing through (R=0; L=0) & (R_{cpt}; L_{cpt}).	
Computation: 1. Obtain a biological intercept L_{0p} as for BCM 3. 2. Apply BCM 12 with different measured R_0 (1st increment) per fish.	

Model & Calculation Details	Abbreviations
Model: $L_i = a + \exp\left(\ln(L_0 - a) + \frac{[\ln(L_{cpt} - a) - \ln(L_0 - a)][\ln(R_i) - \ln(R_0)]}{[\ln(R_{cpt}) - \ln(R_0)]}\right)$	**BCM 13**
Name: Fry (1943),	**FRY**
Function: Allometric w/ intercept; $L = a + bR^c$	(g5)
Hypothesis: Assumes the BCF is an allometric function through (R=0; L=a), (R_0; L_0) and (R_{cpt}; L_{cpt}) (but see Francis 1990). Vigliola et al. (2000) showed both allometric and linear BCF were consistent with proportionality between relative growth rates of fish and otoliths".	
Geometry: An allometric curve through (R=0; L=a), (R_0; L_0) & (R_{cpt}; L_{cpt}).	
Computation: Requires estimates for a, R_0 & L_0. Originally, a was chosen to linearize the plot of ln(L-a) against ln(R), and (R_0; L_0) was a point arbitrarily chosen on the curve g5 (Francis 1990). 1. Find a, b, c so that the line $\ln(L - a) = \ln(b) + c \times \ln(R)$ is as close as possible to data, e.g., by numerical optimisation. The function "optim" of R (http://www.r-project.org/) can be used to find the best set of a, b, c that minimise the function $[\ln(L\text{-}a) - \ln(b) - c \times \ln(R)]^2$. 2. Select (arbitrarily) point (R_0; L_0) on the $L = a + bR^c$ curve. 3. Apply BCM 13.	
Model: $L_i = a + \exp\left(\ln(L_{0p} - a) + \frac{[\ln(L_{cpt} - a) - \ln(L_{0p} - a)][\ln(R_i) - \ln(R_{0p})]}{[\ln(R_{cpt}) - \ln(R_{0p})]}\right)$	**BCM 14**
Name: Modified Fry (Vigliola et al. 2000), or Allometric Biological Intercept),	**MF, ABI**
Function: Allometric w/ intercept; $L = a + bR^c$ or $R = \left(\frac{L\text{-}a}{b}\right)^{\frac{1}{c}}$	(g5 or f5)
Hypothesis: Proportionality between relative growth rates of the fish and the otolith (Vigliola et al. 2000). Modification of BCM 13 with biologically-constrained but statistically-estimated "a" and biologically determined intercept (R_{0p}; L_{0p}) instead of (R_0; L_0).	
Geometry: Allometric curve through (R=0; L=a), (R_{0p}; L_{0p}) & (R_{cpt}; L_{cpt}).	
Computation: Requires an estimate for "a", "R_{0p}" and "L_{0p}". 1. Obtain a biological intercept (R_{0p}; L_{0p}) as for BCM 3. 2. Fit $L = L_{0p} - b_1R_{0p}^{c_1} + b_1R^{c_1}$ by L-on-R non-linear regression to estimate "b_1" and "c_1". Then, calculate $a_1 = L_{0p} - b_1R_{0p}^{c_1}$. 3. Fit $R = \left(\frac{L - L_{0p} + b_2R_{0p}^{c_2}}{b_2}\right)^{\frac{1}{c_2}}$ by R-on-L non-linear regression to estimate b_2 and c_2. Then, calculate $a_2 = L_{0p} - b_2R_{0p}^{c_2}$. 4. Calculate $a = \frac{a_1 + a_2}{2}$. 5. Apply BCM 14.	

Model & Calculation Details	Abbreviations

Model: $$L_i = (a + bR_i^c)\frac{L_{cpt}}{(a + bR_{cpt}^c)}$$ **BCM 15**

Name: Fry Body Proportional Hypothesis, or Allometric Body Proportional Hypothesis, **FRY-BPH, ABPH**

Function: $L = a + bR^c$ (g5)

Hypothesis: Modification of the Fry's model (BCM 13) with application of BPH

Geometry: An allometric curve passing through (R_{cpt}; L_{cpt}) and preserving a constant proportional distance to regression curve g5.

Computation:

1. Fit $L = a + bR^c$ by L-on-R non-linear regression to estimate a, b, c.
2. Apply BCM 15.

Model: $$L_i = a + (L_{cpt} - a)\left(\frac{R_i}{R_{cpt}}\right)^c$$ **BCM 16**

Name: Fry Scale Proportional Hypothesis, or Allometric Scale Proportional Hypothesis, **FRY-SPH, ASPH**

Function: $R = \left(\frac{L-a}{b}\right)^{\frac{1}{c}}$ (f5)

Hypothesis: Modification of the Fry's model (BCM 13) with application of SPH.

Geometry: An allometric curve passing through (R_{cpt}; L_{cpt}) and preserving a constant proportional distance to regression curve f5.

Computation: Note that estimate of b is not required in BCM 16.

1. Fit $R = \left(\frac{L-a}{b}\right)^{\frac{1}{c}}$ by R-on-L non-linear regression to estimate a, c.
2. Apply BCM 16.

Model: $$L_i = (a + bR_i + cR_i^2)\frac{L_{cpt}}{(a + bR_{cpt} + cR_{cpt}^2)}$$ **BCM 17**

Name: Quadratic Body Proportional Hypothesis (Francis 1990), **QBPH**

Function: $L = a + bR + cR^2$ (g6)

Hypothesis: Sherriff (1922) first used a quadratic equation but gave no BCM. Francis (1990) gave the BPH formulation of the quadratic function. BCM 17 assumes a quadratic L-on-R BCF and a BPH.

Geometry: A 2nd degree polynomial curve passing through (R_{cpt}; L_{cpt}) and preserving a constant proportional distance to regression curve g6.

Computation:

1. Fit $L = a + bR + cR^2$ by L-on-R non-linear regression to estimate a, b, c.
2. Apply BCM 17.

Model & Calculation Details	Abbreviations
Model: $a + bL_i + cL_i^2 = \frac{R_i}{R_{cpt}}(a + bL_{cpt} + cL_{cpt}^2)$	**BCM 18**
Name: Quadratic Scale Proportional Hypothesis,	**QSPH**
Function: $R = a + bL + cL^2$	(f6)
Hypothesis: Francis (1990) claimed "Thomas (1983) gave a good statement of SPH and applied it with a quadratic body-scale curve" but the BCM was not explicitly developed in Thomas (1983) or in Francis (1990). BCM 18 assumes a quadratic R-on-L BCF and a SPH. **Geometry:** A curve passing through (R_{cpt}; L_{cpt}) and preserving a constant proportional distance to regression curve f6. **Computation:** 1. Fit $R = a + bL + cL^2$ by R-on-L non-linear regression to estimate a, b, c 2. Solve the quadratic equation of BCM 18 by numeric optimisation or using the quadratic solution (for $ax^2 + bx + c = 0$, $x = \frac{-b \pm \sqrt{b^2 - 4ac}}{2a}$).	
Model: $L_i = \frac{a_0 + a_1R_i + a_2R_i^2 + \cdots + a_nR_i^n}{a_0 + a_1R_{cpt} + a_2R_{cpt}^2 + \cdots + a_nR_{cpt}^n} L_{cpt}$	**BCM 19**
Name: Polynomial Body Proportional Hypothesis,	**PBPH**
Function: $L = a_0 + a_1R + a_2R^2 + \cdots + a_nR^n$	(g7)
Hypothesis: Gutreuter (1987) generalized quadratic BCM to a polynomial of arbitrary degree. BCM 19 assumes a polynomial L-on-R BCF and a BPH. **Geometry:** A polynomial curve passing through (R_{cpt}; L_{cpt}) and preserving a constant proportional distance to regression curve g7. **Computation:** 1. Fit $L = a_0 + a_1R + a_2R^2 + \cdots + a_nR^n$ by L-on-R non-linear regression to estimate a_0, a_1 … a_n. 2. Apply BCM 19.	
Model: $a_0 + a_1L_i + a_2L_i^2 + \cdots + a_nL_i^n = \frac{R_i}{R_{cpt}}(a_0 + a_1L_{cpt} + a_2L_{cpt}^2 + \cdots + a_nL_{cpt}^n)$	**BCM 20**
Name: Polynomial Scale Proportional Hypothesis,	**PSPH**
Function: $R = a_0 + a_1L + a_2L^2 + \ldots.. + a_nL^n$	(f7)
Hypothesis: BCM 20 assumes a polynomial R-on-L BCF and a SPH. **Geometry:** A curve passing through (R_{cpt}; L_{cpt}) and preserving a constant proportional distance to regression curve f7. **Computation:** 1. Fit $R = a_0 + a_1L + a_2L^2 + \ldots.. + a_nL^n$ by R-on-L non-linear regression to estimate a_0, a_1 … a_n. 2. Solve BCM 20 (by numeric optimisation or other means).	

Model & Calculation Details	Abbreviations
Model: $L_i = \exp(a + bR_i)\dfrac{L_{cpt}}{\exp(a + bR_{cpt})}$	**BCM 21**
Name: Exponential Body Proportional Hypothesis,	**EBPH**
Function: $L = \exp(a + bR)$	(g8)
Hypothesis: Tremblay and Giguère (1992) used an exponential L-on-R BCF and then used a proportionality method for back-calculation. BCM 21 assumes exponential L-on-R BCF and a BPH.	
Geometry: An exponential curve passing through (R_{cpt}; L_{cpt}) & preserving a constant proportional distance to regression curve g8.	
Computation: Tremblay and Giguère (1992) originally fitted g8 after log transformation ($\ln(L) = a + bR$), which is not BPH per se. 1. Fit $L = \exp(a + bR)$ by L-on-R non-linear regression to estimate a & b. 2. Apply BCM 21.	
Model: $L_i = \exp[a + (\ln(L_{cpt}) - a)\dfrac{R_i}{R_{cpt}}]$	**BCM 22**
Name: Exponential Scale Proportional Hypothesis,	**ESPH**
Function: $R = \dfrac{\ln(L) - a}{b}$	(f8)
Hypothesis: BCM 22 assumes logarithmic R-on-L BCF and a SPH.	
Geometry: An exponential curve passing through (R_{cpt}; L_{cpt}) & preserving a constant proportional distance to regression curve f8.	
Computation: 1. Fit $R = \dfrac{\ln(L) - a}{b}$ by R-on-L non-linear regression to estimate a, b. 2. Apply BCM 22.	

Acknowledgements

We thank Mireille Harmelin-Vivienfor the inspiration for this chapter, the editors and anonymous reviewers for improvements to the various drafts, and Jackie Wilson for use of her work. The work was supported by the Institut de Recherche pour le Développement and the Australian Institute of Marine Science.

References

Bagenal TB, Tesch FW (1978) Age and growth. In: Bagenal TB (Ed) Methods for assessment of fish production in fresh waters, 3rd Edition. Blackwell Scientific Publications, Oxford, UK

Bang A, Gronkjaer P (2005) Otolith size at hatch reveals embryonic oxygen consumption in the zebrafish, *Danio rerio*. Mar Biol 147:1419–1423

Brothers EB (1995) Session II. Overview II. In: Secor DH, Dean JM, Campana SE (Eds) Recent developments in fish otolith research. University of South Carolina Press, Columbia, SC

Campana SE (1990) How reliable are growth back-calculations based on otoliths? Can J Fish Aquat Sci 47:2219–2227

Campana SE (1992) Measurement and interpretation of the microstructure of fish otoliths. In: Stevenson DK, Campana SE (Eds) Otolith microstructure examination and analysis. Can Spec Publ Fish Aquat Sci 117:59–71

Chambers RC, Miller TJ (1995) Statistical analysis of reconstructed life histories from otoliths: special properties of longitudinal data. In: Secor DH, Dean JM, Campana SE (Eds) Recent developments in fish otolith research. University of South Carolina Press, Columbia, SC

Cock AG (1966) Genetical aspects of metrical growth and form in animals. Q Rev Biol 41:131–190

Doherty PJ, Dufour V, Galzin R, Hixon MA, Meekan MG, Planes S (2004) High mortality during settlement is a population bottleneck for a tropical surgeonfish. Ecology 85:2422–2428

Escot C, Granado-Lorencio C (1999) Comparison of four methods of back-calculating growth using otoliths of a European barbel, *Barbus sclateri* (Günther) (Pisces: Cyprinidae). Mar Freshwater Res 50:83–88

Fey DP (2006) The effect of temperature and somatic growth on otolith growth: the discrepancy between two clupeid species from a similar environment. J Fish Biol 69:794–806

Finstad AG (2003) Growth backcalculations based on otoliths incorporating an age effect: adding an interaction term. J Fish Biol 62:1222–1225

Francis RICC (1995) The analysis of otolith data. A mathematician's perspective (what, precisely, is your model?). In: Secor DH, Dean JM, Campana SE (Eds) Recent developments in fish otolith research. University of South Carolina Press, Columbia, SC

Francis RICC (1990) Back-calculation of fish length: a critical review. J Fish Biol 36:883–902

Fraser CMcL (1916) Growth of the spring salmon. Trans Pacif Fish Soc 1915:29–39

Fry FEJ (1943) A method for the calculation of the growth of fishes from scale measurements. Publ Ont Fish Res Lab 61:5–18

Geffen AJ (1992) Validation of otolith increment deposition rate. In: Stevenson DK, Campana SE (Eds) Otolith microstructure examination and analysis. Can Spec Publ Fish Aquat Sci 117:101–113

Gleason TR, Bengtson DA (1996) Size-selective mortality of inland silversides: evidence from otolith microstructure analysis. Trans Am Fish Soc 125:860–873

Gould SJ (1966) Allometry and size in ontogeny and phylogeny. Biol Rev Cambridge Philos Soc 41:587–640

Grimes CB, Isley JJ (1996) Influence of size selective mortality on growth of Gulf menhaden and King mackerel larvae. Trans Am Fish Soc 125:741–752

Gutreuter S (1987) Considerations for estimation and interpretation of annual growth rates. In: Summerfelt RC, Hall GE (Eds) Age and growth of fish. Iowa State University Press, Ames, Iowa, pp 115–126

Hare JA, Cowen RK (1995) Effect of age, growth rate, and ontogeny on the otolith size – fish size relationship in bluefish, *Potamus saltatrix*, and the implications for back-calculation of size in fish early life history stages. Can J Fish Aquat Sci 52:1909–1922

Hovenkamp F (1992) Growth-dependent mortality of larval plaice Pleuronectes platessa in the North Sea. Mar Ecol Progr Ser 82:95–101

Jones CM (1992) Development and application of the otolith increment technique. In: Stevenson DK, Campana SE (Eds) Otolith microstructure examination and analysis. Can Spec Publ Fish Aquat Sci 117:1–11

Klumb RA, Bozek MA, Frie RV (2001) Validation of three back-calculation models by using multiple oxytetracycline marks formed in the otoliths and scales of bluegill × green sunfish hybrids. Can J Fish Aquat Sci 58:352–364

Lea E (1910) On the methods used in the herring investigations. Publ Circ Cons Perm Int Explor Mer 53:7–25

Lee RM (1920) A review of the methods of age and growth determination in fishes by means of scales. Fish Invest Lond Ser 24(2):1–32

Macpherson E (1998) Ontogenetic shifts in habitat use and aggregation in juvenile sparid fishes. J Exp Mar Biol Ecol 220:127–150

Meekan MG (1997) Relationships between otolith and somatic growth of cod larvae (*Gadus morhua*). J Plankton Res 19:167–169

Meekan MG, Dodson JJ, Good SP, Ryan DAJ (1998) Otolith and fish size relationships, measurement error and size selective mortality during the early life of Atlantic salmon (*Salmo salar*). Can J Fish Aquat Sci 55:1666–1673

Meekan MG, Fortier L (1996) Selection for fast growth during the larval life of Atlantic cod Gadus morhua on the Scotian Shelf. Mar Ecol Prog Ser 137:25–37

Morita K, Matsuishi T (2001) A new model of growth back-calculation incorporating age effect based on otoliths. Can J Fish Aquat Sci 58:1805–1811

Mosegaard H, Svedang H, Taberman K (1988) Uncoupling of somatic and otolith growth rates in arctic char (*Salvelinus alpinus*) as an effect of differences in temperature response. Can J Fish Aquat Sci 45:1514–1524

Mugiya Y (1987) Phase difference between calcification and organic matrix formation in the diurnal growth of otoliths in the rainbow trout, *Salmo gairderi*. Fish Bull 85:395–401

Mugiya Y (1990) Long-term effects of hypophysectomy on the growth and calcification of otoliths and scales in the goldfish, *Carassius auratus*. Zool Sci 7:273–279

Pannella G (1971) Fish otoliths: daily growth layers and periodical patterns. Science 173:1124–1126

Pannella G (1974) Otolith growth patterns: an aid in age determination in temperate and tropical fishes. In: Bagenal TB (Ed) Ageing of fish. Unwin Brothers, London

Pinheiro JC, Bates DM (2000) Mixed-effects models in S and S-Plus. Springer-Verlag, New York

Ricker WE (1969) Effects of size-selective mortality and sampling bias on estimates of growth, mortality, production and yield. J Fish Res Board Can 26:479–541

Ricker WE (1975) Computation and interpretation of biological statistics of fish populations. Bull Fish Res Board Can 191:382pp

Ricker WE (1979) Growth rates and models. In: Hoar WS, Randall DJ, Brett JR (Eds) Fish physiology, Vol. 8. Academic Press, Inc Publ, Orlando, FL

Schirripa MJ (2002) An evaluation of back-calculation methodology using simulated otolith data. Fish Bull 100:789–799

Secor DH, Dean JM (1989) Somatic growth effects on the otolith – fish size relationship in young pond-reared striped bass, *Morone saxatilis*. Can J Fish Aquat Sci 46:113–121

Secor DH, Dean JM (1992) Comparison of otolith-based back-calculation methods to determine individual growth histories of larval striped bass, *Morone saxatilis*. Can J Fish Aquat Sci 49:1439–1454

Secor DH, Dean JM, Baldevarona RB (1989) Comparison of otolith growth and somatic growth in larval and juvenile fishes based on otolith length/fish length relationships. Rapp P-V Réun Cons Int Explor Mer 191:431–438

Sherriff CWM (1922) Herring investigations. Report on the mathematical analysis of random samples of herrings. Scient Invest Fishery Bd Scotl 1:25pp

Sirois P, Lecomte F, Dodson JJ (1998) An otolith-based back-calculation method to account for time-varying growth rate in rainbow smelt (*Osmerus mordax*) larvae. Can J Fish Aquat Sci 55:2662–2671

Smedstad OM, Holm JC (1996) Validation of back-calculation formulae for cod otoliths. J Fish Biol 49:973–985

Sogard SM (1997) Size-selective mortality in the juvenile stage of teleost fishes: a review. Bull Mar Sci 41:423–431

Stevenson DK, Campana SE (Ed) (1992) Otolith microstructure examination and analysis. Can Spec Publ Fish Aquat Sci 117:136pp

Templeman W, Squires HJ (1956) Relationship of otolith lengths and weights in the haddock *Melanogrammus aeglefinus* (L.) to the rate of growth of the fish. J Fish Res Board Can 13:467–487

Thomas RM (1983) Back-calculation and time of hyaline ring formation in the otoliths of the pilchard off South West Africa. S Afr J Mar Sci 1:3–18

Tremblay G, Giguère LA (1992) Relation longueur/écaille allométrique chez le saumon atlantique (*Salmo salar*) durant la phase marine. Can J Fish Aquat Sci 49:46–51

Venables WN, Ripley BD (2002) Modern applied statistics with S, 4th Edition. Springer, New York.

Vigliola L, Doherty PJ, Meekan MG, Drown D, Jones ME, Barber PH (2007) Genetic identity determines risk of post-settlement mortality of a marine fish. Ecology 88(5):1263–1277

Vigliola L, Harmelin-Vivien M, Meekan MG (2000) Comparison of techniques of back-calculation of growth and settlement marks from the otoliths of three species of *Diplodus* from the Mediterranean Sea. Can J Fish Aquat Sci 57:1292–1299

Vigliola L, Meekan MG (2002) Size at hatching and planktonic growth determine post-settlement survivorship of a coral reef fish. Oecologia 131:89–93

Watanabe Y, Kuroki T (1997) Asymptotic growth trajectories of larval sardine (*Sardinops melanostictus*) in the coastal waters off western Japan. Mar Biol 127:369–378

Webster MS, Almany GR (2006) The predation gauntlet: early post-settlement mortality in coral reef fishes. Coral Reefs 25:19–22

Weisberg S (1993) Using hard-part increment data to estimate age and environmental effects. Can J Fish Aquat Sci 50:1229–1237

Weisberg S, Frie RV (1987) Linear models for the growth of fish. In: Summerfelt RC, Hall GE (Eds) Age and growth of fish. Iowa State University Press, Ames, Iowa

Whitney RR, Carlander KD (1956) Interpretation of body-scale regression for computing body length of fish. J Wild Manage 20:21–27

Wilson J, Osenberg CW (2002) Experimental and observational patterns of density-dependent settlement and survival in the marine fish *Gobiosoma*. Oecologia 130:205–215

Wilson JA, Vigliola L, Meekan MG (2008) The back-calculation of size from otoliths: validation and comparison of models at an individual level. J Exp Mar Biol Ecol doi:10.1016/j.jembe.2008.09.005

Wright PJ, Metcalfe NB, Thorpe JE (1990). Otolith and somatic growth rate in Atlantic salmon parr, *Salmo salar* L. Evidence against coupling. J Fish Biol 36:241–249

7. OTOLITH MICROSTRUCTURE IN TROPICAL FISH

JACQUES PANFILI[1], JAVIER TOMÁS[1,2], AND BEATRIZ MORALES-NIN[2]

[1] *IRD UR 070, UMR 5119 ECOLAG, Université Montpellier 2, cc 093, Place E. Bataillon, 34095 Montpellier Cedex 5, France*

[2] *CSIC/UIB-IMEDEA, Miguel Marqués 21, 07190 Esporles, Illes Balears, Spain*

1 INTRODUCTION

This chapter focuses on the fine scale microstructure of otoliths, the details of its formation and the biological information stored within the structures. The examination of otolith microstructure can reveal many different aspects of any individual fish's life events, beyond the information stored as daily incremental deposit of the structures and their respective accretion rates. In tropical environments, and specifically in reef areas, the most singular structure within the otolith is the settlement mark corresponding for some species to the change of habitat between the pelagic and demersal environments (e.g., Thorrold & Milicich 1990, Wilson & McCormick 1997, 1999, McCormick et al. 2002). Such specific marks can be relevant for managers if they want information on the presence of a settlement phase for a species or the duration of the pelagic stage. Nevertheless, the treatment of such information often implies the analysis of the daily increments (see Sponaugle, Chapter 4, this volume). Another type of mark that can be found within the otolith structure is related to reproduction (e.g., Massou et al. 2004b) even though only few studies could demonstrate a direct correspondence between reproduction and otolith checks in experimental or natural conditions. This kind of mark can be particularly informative for resource managers as the knowledge of reproduction is one of the key pieces of information for fishery management. The duration of the reproductive period can be extended, particularly in the tropics, meaning that it is often difficult to evaluate the average birth date for cohorts in populations. Otolith reproductive marks, if they exist, are relevant for calculation of birth dates and also to evaluate the number of individual reproductive events and the link between reproductive output and larval supply within specific populations (McIlwain 2002, Sponaugle & Pinkard 2004). Specific variation in the environment can induce a specific mark in the otolith structure which could be useful afterwards as a reference mark to compare individual life history trait responses (Sponaugle & Pinkard 2004).

In the first part of this chapter we provide a detailed description of the microstructures and the processes and factors (endogenous and exogenous) involved in their formation. This is followed by a listing of the potential applications of structure examination, from a patterns related to life-history traits (metamorphosis, settlement, reproduction, migration, etc.) to tropical ecology and fisheries and an explanation of the advantages and disadvantages of the use of otolith microstructures. We then provide a description of the main experimental procedures for otolith manipulation, followed by discussion of the technical aspects of preparing otoliths for examination and analysis.

B.S. Green et al. (eds.), *Tropical Fish Otoliths: Information for Assessment, Management and Ecology*, Reviews: Methods and Technologies in Fish Biology and Fisheries 11,
DOI 10.1007/978-1-4020-5775-5_7, © Springer Science+Business Media B.V. 2009

This chapter mainly links with two other chapters of this book – Sponaugle, Chapter 4 and Thorrold and Swearer, Chapter 8 – but also with others, as the knowledge of structure is the basis for interpretation of all kinds of marks laid down and conserved in otoliths throughout a fish's life. The reader should refer directly to Chapters 3 (Fowler) and 4 (Sponaugle) for discussion of ageing fish using primary increments in otoliths.

2 Formation of otolith microstructures

2.1 DESCRIPTION OF MICROSTRUCTURES

2.1.1 Basic constituents and primary increments

Otoliths are mostly composed of calcium carbonate ($CaCO_3$) precipitated in a protein matrix. The $CaCO_3$ is crystallized mainly in the mineral form of aragonite (Carlström 1963) due to the action of an organic matrix in which acidic amino acids predominate (Degens et al. 1969). The aragonite forms acicular (needle shaped) microcrystals about 0.3 µm long grouped in radial prisms of variable thickness, running from the otolith centre to the external surface (Figure 1). The microcrystals show discontinuities at regular intervals in which the organic matter predominates, although some microcrystals may run across several discontinuities (Morales-Nin 1987) (Figure 1a, c).

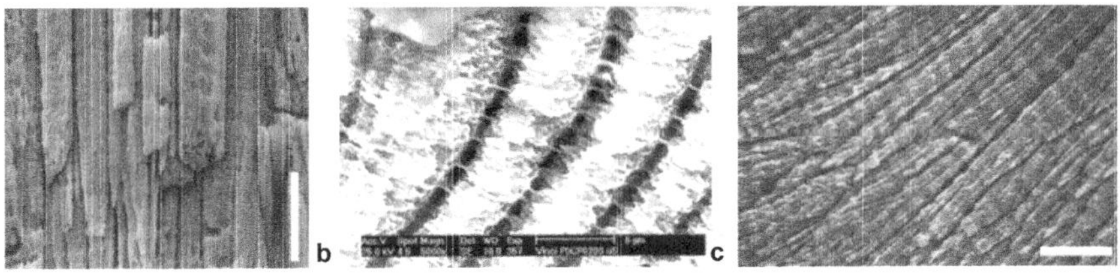

Figure 1. Inorganic constituents of fish otoliths (aragonite crystals within the incremental units) under scanning electron microscopy (SEM) and after acid etching. (**a**) Inorganic constituents within the incremental unit of the otolith of *Stellifer rastrifer*. Scale bar = 7.5 µm (photo B. Morales-Nin). (**b**) Otolith of *Vinciguerria nimbaria*. The radial crystals of aragonite are visible inside the L-zones (carbonate rich zones, light coloured) as well as the bridges formed by water-insoluble proteins inside the D-zones (protein rich zones, *dark coloured*). Scale bar = 5 µm (photo J. Panfili). (**c**) Radial prismatic crystals with primary increments visible on the background (*Lutjanus kasmira*). Scale bar = 10 µm (photo B. Morales-Nin).

The total proportion of the organic matrix is low, between 0.1 and 10% in weight (Borelli et al. 2001), and may vary with otolith growth. For instance, 1 year old *Dicentrarchus labrax* and *Merluccius capensis* had otoliths up to 10 times richer in protein than in older fishes, and there was also a change in the relative proportion of the constituent amino acids with age (Morales-Nin 1986a,c). This organic matter is laid down in the form of a loose reticulum that interpenetrates the crystals and that predominates in the discontinuities, forming thick fibres and layers (Morales-Nin 1987).

The organic matrix is composed of water-soluble proteins (WSP) and water-insoluble proteins (WIP). Both fractions are rich in acidic amino acids. Some of the WSP fractions have calcium-binding capacity and have an important role in mineralisation of the otolith (Asano & Mugiya 1993). These contain a glycoprotein that probably regulates the accretion rate by acting as a calcification inhibitor (Wright et al. 1991). The WIP fraction consists of a single collagen-like protein, termed Otolin-1 (Murayama et al. 2002). The role of WIP is presumably structural (Campana 1999) and might be related to shape control (Figure 2). WSP represent 1.1% by weight of the otolith in *Oreochromis niloticus* whereas total protein accounts for 2.3% (Asano & Mugiya 1993). WSP are composed of a few fractions, including a protein-polysaccharide complex with a moderate calcium-binding capacity and other proteins of low molecular weight that have an intense capacity for calcium binding (Sasagawa & Mugiya 1996).

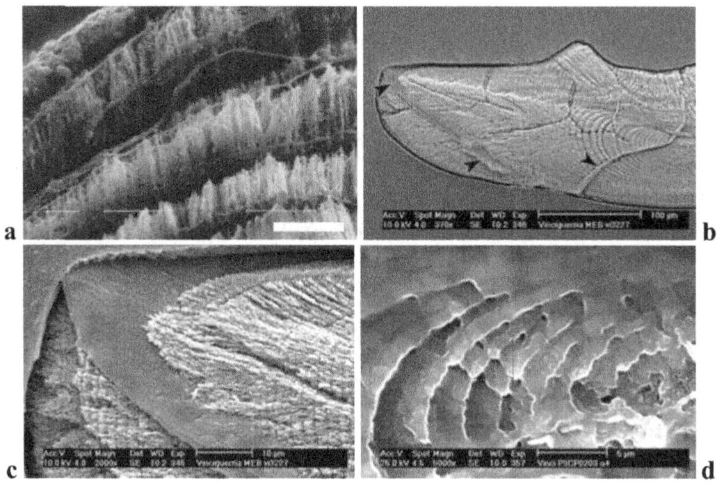

Figure 2. Organic constituents of otoliths seen under SEM, after acid etching. (**a**) Framework of proteins in the microincremental unit of an otolith of *Dicentrarchus labrax*. Scale bar = 10 μm (photo B. Morales-Nin). (**b**) Framework of organic material (*black arrows*) inside the otolith of *Vinciguerria nimbaria*. Scale bar = 100 μm. (**c**) Detail of the image in (**b**) on the ventral edge showing a protein plate crossing the aragonite crystals. Scale bar = 10 μm. (**d**) Water-insoluble proteins in the core area of the otolith of *Vinciguerria nimbaria*. Each protein plate constitutes an element of the primary increment. Scale bar = 5 μm (photos J. Panfili).

Otoliths grow by successive deposition of increments, the so-called primary increments, which constitute bi-partite structures with each composed of one L-zone and one D-zone (Figure 3). The L-zone is a band rich in calcium carbonate crystals, translucent to light and appearing raised in scanning electron microscopy (SEM), with a width varying between 0.4 and 10 μm. The D-zone is a band rich in organic material, opaque to light and appearing as a groove in SEM, with a width smaller than 1 μm (Panfili et al. 2002). The total width of a primary increment ranges from less than 1–12 μm (Pannella 1974) and is thus mainly constituted by the L-zone. The width of the increments depends on the growth rate and the metabolism of the fish (but see Sponaugle, Chapter 4).

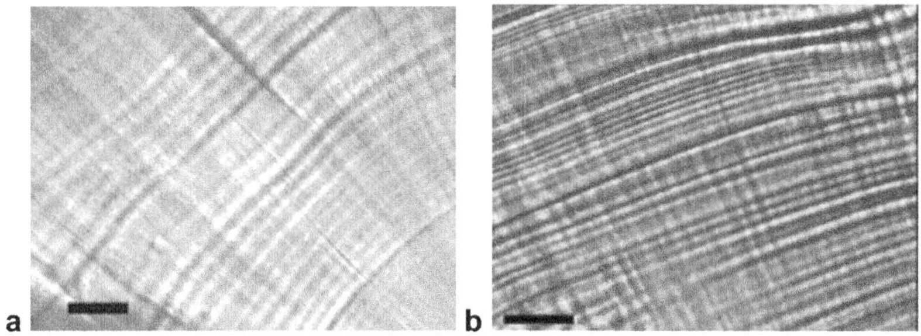

Figure 3. D-zones (protein rich, *dark coloured*) and L-zones (carbonate rich, *light coloured*) of primary increments in the otoliths. (**a**) *Coryphaena hippurus* (scale bar = 10 μm, photo J. Tomás), (**b**) *Ethmalosa fimbriata* (scale bar = 20 μm, photo K. Diop). Note the variations of the L-zone compared to the D-zone in both species.

2.1.2 Initial growth structures

Otolith formation begins with the aggregation of free-floating protein core particles, which are directed to the developing sensory maculae through the action of ciliated cells that line the otosac cavity (Riley et al. 1997). These pre-otoliths are rapidly mineralised forming the primordia or the core, which aggregate to form an optically opaque spot (Figure 4a). X-ray maps have shown that these are formed of non calcium material, as seen in the otoliths of the tropical tilapia *Oreochromis niloticus* (Zhang & Runham 1989) or the tropical *Anguilla* spp. (Arai et al. 2001), as well as other species. The first primary increment surrounding this aggregation delineates an area, generally lentil-shaped, called the core, which includes mineral crystals at this stage (Figure 4).

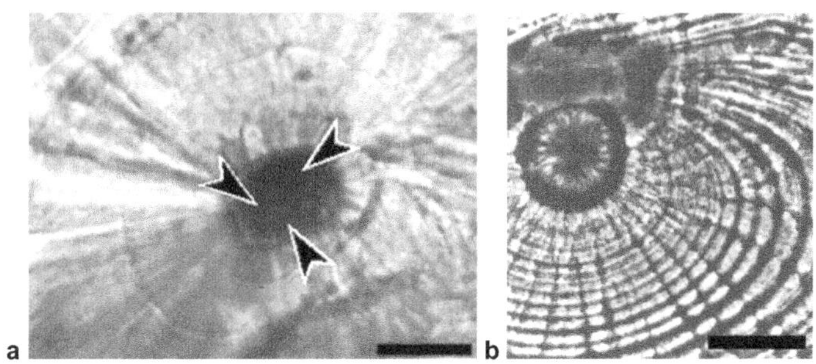

Figure 4. Examples of otolith cores. Scale bar = 10 μm. (**a**) Core area of a *Sarotherodon melanotheron* otolith viewed with a compound microscope. The *arrows* indicate the primordial granules constituting the core (photo K. Diop). (**b**) Central area of a *Vinciguerria nimbaria* otolith viewed under SEM after acid etching. The core, surrounded by a deep depression, is formed by an aggregation of multiple granules (photo L. Marec).

This first increment – also named the first ring, band or check – is generally thought to be formed at hatching and appears like a discontinuity in the otolith (Zhang & Runham 1992b) but this first ring may not have the same physiological basis in all species (Wright et al. 2002b). For example, in the armoured catfish *Hoplosternum littorale*, a few primary increments seem to be laid down prior to the hatching ring (Ponton et al. 2001). In some species, such as tropical clupeids (*Stolothrissa tanganicae* and *Limnothrissa miodon*), otolith growth may initiate increment deposition at yolk sac absorption up to 2–8 days after hatching (Kimura 1995). The biological significance of this first mark is probably species specific and may correspond to several life history events such as yolk-sac absorption, hatching or first feeding (Figure 5) (Wright et al. 2002b). After this, the incremental growth proceeds in a way that has been described as an "onion model" (Figures 4b and 5), with concentric layers of material deposited over the surface of the otolith.

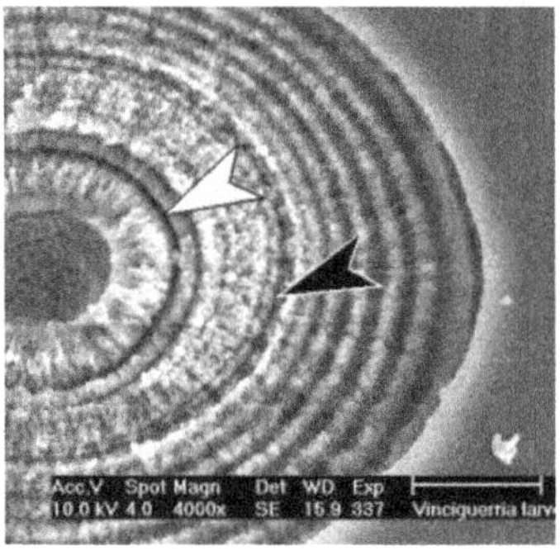

Figure 5. First ring (*white arrow*) and first feeding (*black arrow*) marks interpreted in the central area of an otolith of *Vinciguerria nimbaria* viewed in SEM after acid etching. Scale bar = 5 µm (photo L. Marec & E. Dabas).

2.1.3 Structural discontinuities

Structural discontinuities interrupt the rhythmic growth pattern and can be of different natures. Some are minor, appearing as disruptions in the rhythm. Others are more important and are formed by organic fibres and lamellae, probably due to the cessation of crystal deposition (Figure 6). Once crystal growth is resumed, crystals may have their orientation changed after such a discontinuity. Several possible causes of discontinuities include metamorphosis, settlement, stress or fish reproduction among other changes in life-history characteristics (Victor 1982). Moreover, the discontinuities might determine otolith shape by controlling growth direction and rate by the formation of accessory growth centres, which act as new centres of crystallisation away from the primary core (Figure 7) (Zhang & Runham 1992b). The initiation of their formation is generally concomitant with major developmental, behavioural or ontogenetic changes in the life of the individual. In some species accessory growth centres have been described on the outer surface of adult otoliths resulting in bumps around the outside of the otolith (Wright et al. 2002b, Tomás 2006) or have been observed in the *sulcus acusticus* (Lombarte & Morales-Nin 1995).

The sulcal area, formed in close contact to the *macula acustica*, also is characterised by a major proportion of organic matrix (turning darker brown in charred otoliths) and discontinuities and irregularities both at daily and macro-structural levels (Figure 6a).

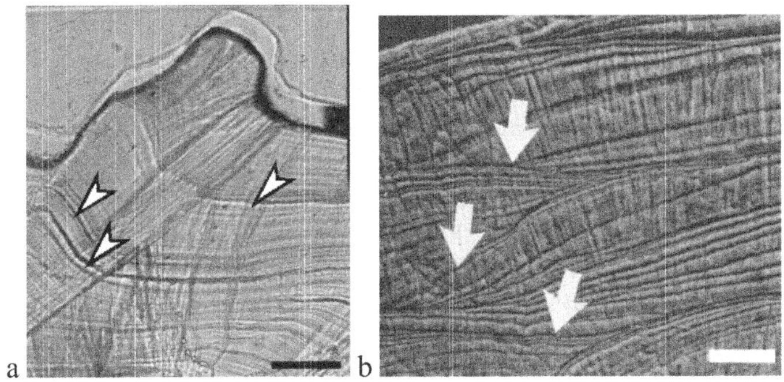

Figure 6. Checks (*arrows*) inside the otolith microstructure. (**a**) Area along the sulcus on the ventral face of an otolith (thin transverse section) of *Oreochromis niloticus* viewed with compound microscopy. Scale bar = 100 μm (photo K. Diop & A. Malam Massou). (**b**) Checks (*arrows*) in the structure of the otolith of *Calamus brachysomus*. Scale bar = 30 μm (photo Morales-Nin 1986b).

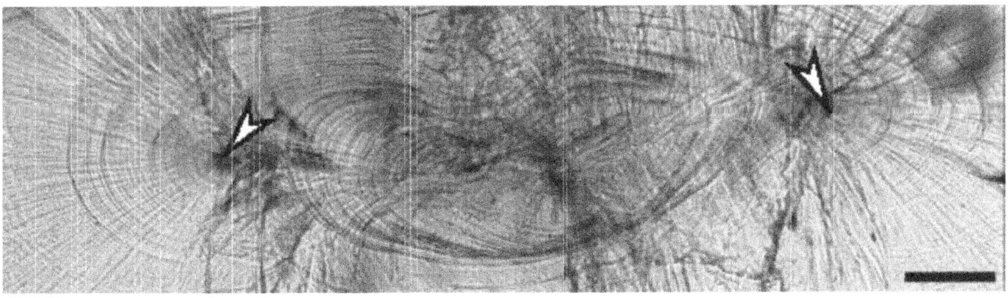

Figure 7. Example of accessory growth centres (arrows) in the central area of the otolith of *Sarotherodon melanotheron*. Thin transverse section observed under compound microscopy. Scale bar = 100 μm (photo K. Diop).

2.1.4 Rhythmical growth patterns

Rhythmical patterns in the deposition of microstructures in the otoliths of tropical fish are the basis of age estimation. Micro-incremental patterns in the otoliths of tropical species vary from sub-daily to daily, lunar and seasonal scales. Daily alternating deposition of calcium and organic matrix results in the formation of daily primary increments predominated by aragonite crystals (L-zone) and a discontinuous organic matrix unit (D-zone) (Watabe et al. 1982, Zhang 1992). This daily rhythm in the formation of the alternating zones is controlled by a circadian variation in plasma chemistry, since during the dark period there is a parallel decline in otolith calcification and total and free plasma calcium concentrations (Mugiya 1987b, Wright et al. 1992). Significant diurnal fluctuations have been recorded in calcium and protein concentrations in the plasma and endolymph of turbot (Edeyer et al. 2000). Incremental growth is general in all ecosystems and the daily nature of the process has been validated for a number of species in the tropics (Sponaugle, Chapter 4). The otolith growth process is a continuous precipitation of aragonite in the endolymph, with incorporation into the otolith structure modulated by neuroproteins secreted at the macula controlling the rate of aragonite crystal growth and possibly crystal size. The effect of this modulation results in a narrow, mineral deficient increment (D-zone) and a wider, mineral-dense increment (L-zone) but also may result in sub-daily rhythmical patterns superimposed on the daily increments (Gauldie & Nelson 1988). Sub-daily increments have been suggested by Pannella (1980) and Campana and Neilson (1985) and described in the otoliths of several species such as *Lepomis sp.* (Taubert & Coble 1977), *Fundulus heteroclitus* (Radtke & Dean 1982) and *Oreochromis niloticus* (Zhang & Runham 1992b, Panfili & Tomás 2001). Note that distinguishing daily and sub-daily increments may be difficult (Taubert & Coble 1977, Sponaugle, Chapter 4).

Lunar patterns in the otolith microstructure have been identified in the juvenile stages of several species inhabiting shallow waters under the influence of tidal rhythms in tropical (Rahman & Cox 2006) and temperate environments (Campana 1984, Geffen & Nash 1995). The influence of lunar rhythms in the recruitment of coral reef species has been sufficiently emphasised (Sponaugle & Pinkard 2004) and the concomitant effect of recruitment and lunar timing may act in phase to alter the otolith microstructural pattern. Nevertheless, lunar rhythms also have been identified and described in the otoliths of pelagic fish inhabiting open oceanic waters, such as *Vinciguerria nimbaria* (Linkowski 1996) or the myctophid *Myctophum asperum* (Hayashi et al. 2001), for which otolith increment widths were narrower around the full moon compared to the new moon.

Despite the widespread assumption of a lack of seasonal environmental variations in tropical environments, tropical regions have marked hydrological seasons, mainly annual, that interact with temperature variations (Morales-Nin & Panfili 2005). These seasonal variations may leave a print in the otolith of tropical fish in the form of opaque and translucent zones which display differences at the microstructure level (Fowler 1995). The seasonal environmental conditions modulate the relative width of primary increments to the point that zones formed at low temperature and high temperature (macrostructures) can be distinguished at low magnification (Morales-Nin & Ralston 1990). Opaque zones are constituted by thick daily microincrements whereas translucent zones consist of thin microincrements, as observed in many coral reef fish (Fowler 1995 and references therein). Finally, the primary increments become narrower as the fish grows older and growth tends to reach an asymptote, leading to possible interruptions in the rhythm of deposition (Ralston & Miyamoto 1983, Hoedt 1992).

2.2 REGULATION OF INCREMENT FORMATION

Otoliths are located within an anatomically closed sac and cells constituting the sac are responsible for synthesising otolith proteins and secreting calcium into the saccular endolymph. The extracellular otolith biomineralisation is complex and requires an orchestrated process bringing proteinaceous and ionic components together in time and space, as proposed in recent models for these processes based on studies of zebrafish (Hughes et al. 2006). Otolith biocalcification is related to endolymph composition

(Romanek & Gauldie 1996), which is heterogeneous inside the otosac probably due to the presence of several types of ionocytes in the membrane surrounding the otolith (Mayer-Gostan et al. 1997). The spatial heterogeneity may be necessary for otolith mineralisation. Payan et al. (1999) proposed that $CaCO_3$ deposition is initiated in the proximal side of the otolith resulting in an heterogeneity of growth along different sides. Increment formation is regulated by both endogenous and exogenous factors.

2.2.1 Endogenous influences on microincrement periodicity

The daily formation of microstructures in the otoliths of fish is under the regulation of an endogenous circadian rhythm. Experimental manipulation of exogenous factors (photoperiod and feeding) showed that these alone did not explain the daily formation of microstructures in *Tilapia mossambica* (Taubert & Coble 1977). Nonetheless, little is known about the endogenous regulation of otolith formation in fish. Early studies ruled out the influence of estradiol (a hormone causing hypercalcemia) upon calcium deposition or resorption in the otolith of the goldfish *Carassius auratus* (Mugiya 1977, 1978). It has been suggested since that otolith growth could be controlled by neuroprotein secretion (Gauldie & Nelson 1988), yet evidence of hormonal regulation of otolith growth has so far only been demonstrated in the otoliths of *Carassius auratus* (Shinobu & Mugiya 1995). It was shown recently in the tropical freshwater green swordtail (*Xiphophorus helleri*) that otolith growth and, more precisely, calcium uptake is neurally regulated, since calcium incorporation into the otolith ceased in fishes that had the vestibular nerve sectioned (Anken et al. 2002).

Changes associated with the development of the embryo also result in noticeable changes in the microstructure of the otolith. Ontogenetic changes usually take the form of: (1) changes in the width of microstructures, as in *Vinciguerria nimbaria* (Figure 5) (Tomás & Panfili 2000), myctophids (Gjøsaeter 1987, Gartner Jr. 1991) or the tropical species *Ambassis vachelli* (Molony 1996); and (2) structural changes associated with accessory primordia and the prismatic growth of the otolith (Figure 7, Zhang & Runham 1992b, Lagardère et al. 1995, Panfili & Tomás 2001). In some cases, the same ontogenetic changes, such as settlement in reef fish, may encompass different changes in the otolith structure at the microincremental level depending on the family. For example, settlement is associated with wider increments in *Scarus rivulatus* and narrower increments in *Ctenochaetus binotatus* (Lou 1993). Ontogenetic trends also may result in the cessation or absence of otolith formation. For example, clupeids only deposit their first microincrement a few days after hatching, possibly coinciding with yolk sac resorption (Kimura 1995).

The sequence of developmental stages (hatching, flexion, metamorphosis, among others) leads to structural changes in the otolith. For example, the formation of the sulcus and increase in otolith concavity (Lagardère et al. 1995) and the appearance of accessory primordia (as new sites of crystal nucleation) redirect the otolith accretion in juvenile (Figure 7) (Panfili & Tomás 2001) and adult (Zhang & Runham 1992b) tilapia *Oreochromis niloticus*. The formation of checks often is associated with metamorphosis (Gjøsaeter 1987). Even reproduction has been shown to leave a mark in the otoliths of certain species (Pannella 1980, Massou et al. 2004b). Therefore, ontogenetic patterns should be identified prior to interpreting variations in microincrement structure potentially caused by exogenous factors.

2.2.2 *Exogenous influences on microincrement periodicity*

Exogenous influences on microincrement periodicity can only be identified and assessed once the ontogenetic (endogenous) patterns of variation of microstructure are known (Molony 1996). This Section lists the major exogenous factors that affect otolith microstructure periodicity in tropical fish species (photoperiod, temperature, food availability, salinity and dissolved oxygen, stress and rearing). The application of using the resulting marks or patterns will be reviewed in the following section on analysis of otolith microstructure.

Otolith microstructure is strongly influenced by alternating light and dark cycles of day and night. Increments were deposited daily in fish reared under a 24 h light-dark periodicity, while microincrements deviated from daily deposition under other light cycles in *Tilapia mossambica* (Taubert & Coble 1977) and the temperate fishes *Fundulus heteroclitus* (Radtke & Dean 1982), *Oncorhynchus mykiss* (Mugiya 1987a) and *Salmo salar* (Wright et al. 1991). Deposition of the aragonite fraction takes place during the light phase for tropical species (Brothers 1981, Ré 1984, Jenkins & Davis 1990, Zhang & Runham 1992c, Tanabe et al. 2003), declining at dusk and resuming at dawn. Nonetheless, other authors have shown the opposite in rainbow trout *Salmo gairdneri* and turbot *Psetta maxima* (Mugiya 1984, Edeyer et al. 2000). It is not clear whether these controversial findings are related to climatic zones, species or seasons. Hence, the timing of formation of the D- and L-zones should be verified for each species where it is important for a particular study. The mechanisms by which photoperiod influences the daily formation of microstructures have yet to be elucidated (Boeuf & Le Bail 1999) and may be related to internal rhythms or other environmental variations related to diel cycle.

Temperature also acts as a primary forcing factor for microstructure formation and appearance in otoliths of fish. The temperature effect on increment periodicity has mostly been studied in temperate species but a few examples are available from tropical environments. A decrease in water temperature from 25°C to 10°C resulted in the cessation of microstructure deposition and the formation of an annulus in the green sunfish, *Lepomis cyanellus* (Taubert & Coble 1977), and a 2–4°C variation at low temperatures (from 8°C to 6 and 4°C) resulted in the absence of microincrement deposition in the otoliths of the European lake whitefish *Coregonus lavaretus* (Klink & Eckmann 1992). Examples of temperature manipulation effects upon otolith microincrement periodicity in tropical species are scarce. Zhang and Runham (1992a) studied the effect of food ration and temperature (18°C and 26°C imposed on 23°C acclimated fish) on the otolith microstructure of tilapia *Oreochromis niloticus*. Results showed that increment deposition persisted in the otoliths of fish kept at low temperatures but increments were less easily discernible. The effects of temperature, reflected in the periodicity of microincrement formation and the width of microstructures, may last for more than one day (Gutiérrez & Morales-Nin 1986). So long as the temperature is high enough to sustain growth (Campana & Neilson 1985), however, a daily temperature fluctuation is not needed for increment formation. For example, daily microstructures were produced in the otoliths of *Tilapia mossambica* held at constant temperature (Taubert & Coble 1977). Microincrements in the otoliths of the Chinese sucker *Myxocyprinus asiaticus*, however, were less discernible in the otoliths of fish reared under constant temperature as opposed to fish that experienced temperature variations between day and night (Song et al. 2008b). The lack of contrast between increments resulted in errors in age estimation for fish reared under constant temperature, at least when the lapillus was used.

Food intake and food deprivation also have an influence on microstructure deposition and increment width and periodicity appear to be sensitive indicators of recent feeding history in fish larvae and juveniles (McCormick & Molony 1992, Molony 1996, Massou et al. 2002). Daily microincrement widths in adult fish decline enough to be not discernible as fish grow older (Hoedt 1992), thus possibly not recording changes caused by food deprivation. Moreover, older fish may have accumulated more reserves and thus be in better condition to cope with food deprivation (Molony & Choat 1990). Hence, the effects of food intake upon otolith microstructure are of main application in the studies of larvae and juveniles (but see comments on adult studies below). In tropical marine waters, which are thought to constitute nutrient poor waters where food can be patchy, it may be of great interest to know whether food deprivation results in changes at the level of otolith microstructure. It has been shown that the feeding regime may lead to the formation of sub-daily or supra-daily microstructures if young fish are fed more or less than once a day (Pannella 1980, Neilson & Geen 1985), although multiple daily feeds did not result in the formation of multiple microstructures in the otoliths of *Chanos chanos* larvae (Tzeng & Yu 1992). Ultimately, food deprivation may result in smaller increment widths but not in the cessation of microincrement deposition (Massou et al. 2002) if the fish has enough body energy reserves (Campana & Neilson 1985). Starvation during 33 days was recorded in the otolith microstructure of the tropical species *Ambassis vachelli* in the form of narrower increment widths, although daily deposition persisted (Molony & Choat 1990). Interestingly, the trends in microincrement widths caused by food deprivation in this species were detectable only after a few days.

The effect of food deprivation on the microstructure of otoliths of adult fish has been studied for the tropical species *Ambassis vachelli* (Molony & Sheaves 1998). Fish exposed to starvation (for a length of time at least 12% of the maximum age of the species) had narrower increments than fed fish. Moreover, increment widths returned to similar widths of fed fish after a period of recovery feeding. Compared to other variables such as lipid content, microincrement widths showed a more conservative pattern of variation indicating that microstructure changes are modulated (probably due to the buffering effect of the endolymph chemistry tending to preserve the functionality of an important organ for hearing) and microstructure will never provide an absolute measurement of the extent of food deprivation. Incidentally, the feeding time during the day does not have any influence on otolith structure. Changes in daily feeding time did not alter the formation of the crystalline part of the microstructure in *Tilapia nilotica* otoliths (Tanaka et al. 1981).

Salinity or dissolved oxygen concentration are also exogenous factors potentially affecting microincrement periodicity but they have scarcely been studied. It is known that salinity induces changes in otolith composition at the microstructure level (Thorrold & Swearer, Chapter 8), yet little is known on how salinity can affect microstructure deposition. Salinity alone does not seem to affect otolith structure but the combination of low food rations and high salinities resulted in the deposition of less than one increment per day in the otoliths of silver perch *Bairdiella chrysoura,* suggesting that salinity could have a physiological effect at low rations (Hales Jr. & Hurley 1991). Oxygen depletion in the water has also been hypothesised to cause disruption of microstructure deposition in *Tilapia sp.* (Pannella 1980) and even resorption of the otolith edge in *Carassius auratus* (Mugiya & Uchimura 1989).

Stress leads to changes in the structure of the otolith in the form of "checks" (sensu Kalish et al. 1995). Stress-induced marks indicate the cessation of otolith growth and appear under transmitted light microscopy as opaque, regular, thin marks (Pannella 1980). It has been suggested that cessation of otolith growth under stress conditions results from disruption in calcium metabolism, although stress is not believed to cause the resorption of otolith calcium (Campana 1983). Stress marks were identified in the otoliths of tilapia *Oreochromis niloticus* associated with fish tagging (Panfili & Tomás 2001) and in juveniles of a subtropical flatfish species, the fringed flounder, *Etropus crossotus* (Reichert et al. 2000), caused by handling the fish to carry out measurements.

Rearing also has been shown to affect otolith structure at the microincrement level. Being a combination of artificial environmental conditions, it may be the actual rearing per se (stress associated with high densities, manipulation of fish, etc.) that affects the microincrement formation but it may also be the relatively homogenous environment provided in most rearing mesocosms. The stress of confinement has been shown to affect the microstructure of tilapia *Oreochromis niloticus* in the form of checks and reduced increment deposition (Massou et al. 2004a). Otolith structure is also affected by fish confinement, as shown by comparing otoliths from laboratory and field caught individuals (Strelcheck et al. 2003). Experimental mesocosms are also unnatural environments where exogenous factors such as photoperiod, temperature and feeding are held constant as independent variables. Various studies have reported that microincrements are more easily discernible in wild fish than in reared fish, in both marine (Al Husaini et al. 2001) and freshwater fish (Vilizzi 1998), to the point that errors in counting microstructures were smaller in field-caught larvae than in reared larvae of *Pagrus major* (Tsuji & Aoyama 1982). The absence of differences in water temperature between night and day led to less discernible increments in the otoliths of reared Chinese sucker compared to fish that experienced diel temperature variations closer to their natural environment (Song et al. 2008a).

3 Analysis of otolith microstructure in tropical fisheries and ecological studies

Otolith microstructure presents a broad spectrum of characteristics that are of immediate application in ecology and fishery management. The first choice to make is which otolith of the three pairs available (sagitta, lapillus, asteriscus) is to be used. The second choice is about the method of analysis, which is highly specific because of the need for a precise examination of the internal structure, requiring time-consuming preparations and sophisticated techniques (compound microscopy, SEM, microprobes, etc.).

3.1 SAGITTA, LAPILLUS OR ASTERISCUS?

All otolith types cannot be employed indiscriminately when studying the structure of the first increments because their formation is not simultaneous, the sagitta and lapillus being formed first (Thorrold 1989). The earliest formed increments in tropical species often are not clearly discernible, depending on the otolith. For example, the asteriscus, with an ambiguous core, renders difficult the identification of the first increment in Cyprinidae (Morioka 2002, Morioka & Matsumoto 2003). The primary increments are often not clearly discernible in smaller lapilli during the first growth phase in the tropical eastern rainbowfish, *Melanotaenia splendida splendida* (Humphrey et al. 2003), probably due to the compression of the growth in smaller otoliths. Some crystallographic and composition differences exist between otoliths, since asterisci are made of the glassy appearing vaterite and sagittae and lapilli are preferentially precipitated in the form of aragonite (Chesney et al. 1998, Campana 1999).

The sagitta is the most commonly used of the three otolith pairs for structural and ultrastructural examination (Morales-Nin & Panfili 2002). This is probably due to the larger size of this otolith compared to lapillus and asteriscus and, therefore, to the larger relative size of its primary increments. There is an exception in the order Ostariophysi, taxa very common in the tropics, in which the asteriscus is bigger than the other two otolith types. Nevertheless, the lapillus has also been preferred in some studies in the tropics of juvenile pelagic species (Morales-Nin et al. 1999, Morioka 2002), juvenile reef species (Allman & Grimes 2002, Zapata & Herron 2002, Nemeth 2005) and even for adult reef fish (e.g., Lou 1993), mainly for ageing purposes and specifically to measure the extent of the planktonic larval duration. Different studies have chosen different otoliths indicating that each new study should carefully select which otolith to use prior to large scale observation of the otolith microstructure (Sponaugle, Chapter 4). Lapilli can be small enough to allow the observation of their internal structure with little or no preparation (Morales-Nin et al. 1999, Morioka 2002, Sponaugle, Chapter 4). Adult fish have thicker otoliths and they are usually sectioned to reveal the primary increments (e.g., in Cichlidae, Zhang & Runham 1992b). In this case (Figure 8), the sagitta is preferred because its increments are wider, whereas the increments of the lapillus are less discernible and the asteriscus structures are more complex (Figure 8). If the comparison between the three otoliths is impossible, the use of the sagitta is recommended because it contains all information from the earliest growth stages.

3.2 PATTERNS RELATED TO LIFE HISTORY EVENTS

The structure and ultrastructure of otoliths are particularly useful to reveal patterns related to life history events such as ontogenetic or environmental changes experienced by individual fish. Investigations often attempt to relate marks with specific events like transition between ontogenetic stages (e.g., metamorphosis, settlement), migration, reproduction or different stresses. The evidence presented below results from experimental rearing in some cases and correlative observations in others. In as much as possible, this Section aims at presenting sound results that invite applying microstructure observation to identifying and analysing metamorphosis, migration, settlement or reproduction in tropical fish species. Otoliths are most powerful, however, when used as a tool combining microstructures, microchemistry and individual growth trajectories to elucidate individual fish life histories.

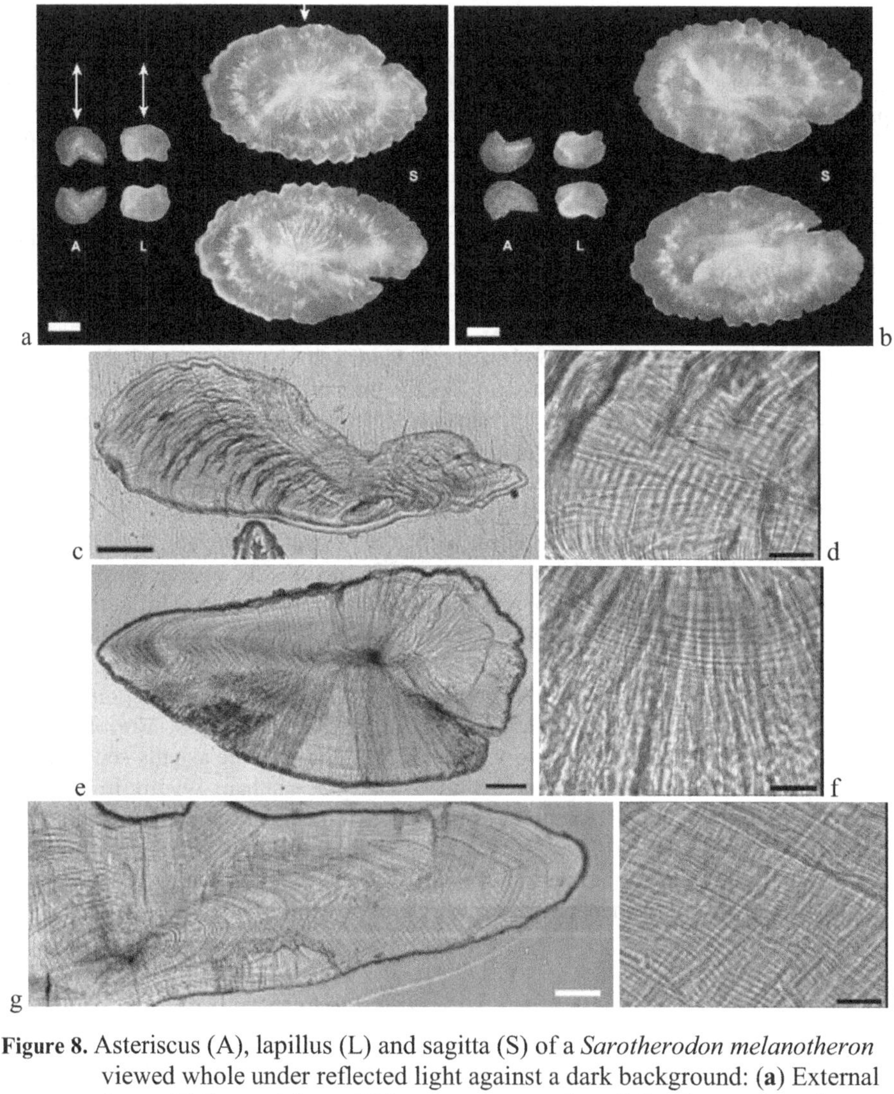

Figure 8. Asteriscus (A), lapillus (L) and sagitta (S) of a *Sarotherodon melanotheron* viewed whole under reflected light against a dark background: (**a**) External faces. (**b**) Internal faces. Thin transverse sections following section plane indicated by arrows in (**a**) of the same asteriscus (**c**, **d**), lapillus (**e**, **f**) and sagitta (**g**, **h**) Scale bars : (**a**, **b**) = 1 mm; (**c**, **e**, **g**) = 100 μm; (**d**, **e**, **f**) = 20 μm (photos K. Diop).

3.2.1 Metamorphosis or environmental changes (migration, settlement)

First feeding can be a critical period in fish from tropical waters as the transitional period between yolk absorption and first feeding is short (Houde 1974). Both hatching and first feeding are frequently recorded in the otolith as distinctive marks (see above). The next major steps for a developing fish can be metamorphosis (i.e., the transition between larval and juvenile life) and then settlement (i.e., final installation in one specific environment), a particularly sensitive stage in tropical areas and especially for reef species (Leis 1991). These developmental stages and environmental conditions are reflected by changes in otolith microstructure which can be used as reference marks.

The metamorphosis stage varies greatly between species but it is generally a strong transition associated with the end of ossification and for that reason it constitutes a tremendous physiological and morphological stress for the fish. Metamorphosis also can be associated with habitat change (Wilson & McCormick 1997, 1999, McCormick et al. 2002) including changes in environmental conditions and behavioural changes, for example in mesopelagic marine species (Myctophidae, Photichthyidae). Metamorphosis in these species leads to a clear change in otolith pattern (Linkowski 1991, Tomás & Panfili 2000) and also to strong morphological, behavioural (from passive pelagic drifting to active diel vertical migration) and habitat transitions (from surface waters to deep waters). Width of daily increments increases with metamorphosis and a distinct shift in habitat (Jenkins 1987, Wilson & McCormick 1999). Metamorphosis in tropical eels appears with a morphological transition from leptocephali and elver to yellow eel stages accompanying movements from marine to inland habitats and produces a marked check on the otolith (Arai et al. 2000a,b, Sugeha et al. 2001). The otolith check induced by metamorphosis will then constitute a reference mark useful for back-calculating the age at metamorphosis for the different species.

Tropical fish species can be involved in either horizontal or vertical migrations. Horizontal migrations often imply changes of habitat and result in changes of environmental conditions, which can mark the otolith microstructure through a corresponding change in growth rate. This kind of migration can also be linked to metamorphosis (e.g., eel species Arai et al. 2000a) or to settlement (Wilson & McCormick 1997) and it is often difficult to distinguish the effects of ontogeny from those of physiology or behavioural changes. Effects of migration across environmental clines may be more pronounced in terms of microchemical otolith composition. For example, the transition between different salinity conditions affects the Sr/Ca ratio, although in eel species metamorphosis changes are also associated with microchemical modifications (Kotake et al. 2003, Thorrold & Swearer, Chapter 8 of this volume). Daily, seasonal or irregular vertical migrations frequently occur among pelagic species in tropical oceanic waters. These can also be associated with metamorphosis and behavioural changes, as is the case for small mesopelagic species. As for horizontal migrations, associated changes in otolith microstructure are mainly reflected in a change in otolith growth rates.

Fish settlement often corresponds with the transition from a pelagic habitat to a demersal or benthic habitat (Wilson & McCormick 1997), as the larvae of many marine and freshwater species are planktonic and, in tropical reef fish, this is often followed by a relatively sedentary, site-attached mode of life. This transition is reflected in both the increment pattern and growth rates in otoliths (Figure 9b) (Wilson & McCormick 1997, 1999). Settlement checks have been documented for many reef species (e.g., Thorrold & Milicich 1990, Danilowicz 1997b, Suthers 1998, Searcy & Sponaugle 2000, Bergenius et al. 2002) but these checks are not found in all tropical species. They are most common in families living close to reefs, such as Acanthuridae, Chaetodontidae, Labridae, Lethrinidae, Lutjanidae, Mullidae, Pomacentridae, Pseudochromidae and Serranidae (Thorrold & Milicich 1990, Risk 1997, Robertson et al. 1999, Wilson & McCormick 1999, Zapata & Herron 2002, Nemeth 2005). In contrast, non-reef species, such as pelagic or estuarine fish, do not present settlement checks in their otoliths, probably because the transition between the larval and the juvenile stages does not correspond to major changes of environment. The transition can correspond to changes in the widths of the primary increments within the otolith (Stequert et al. 1996, Panfili & Tomás 2001) but often not to a check.

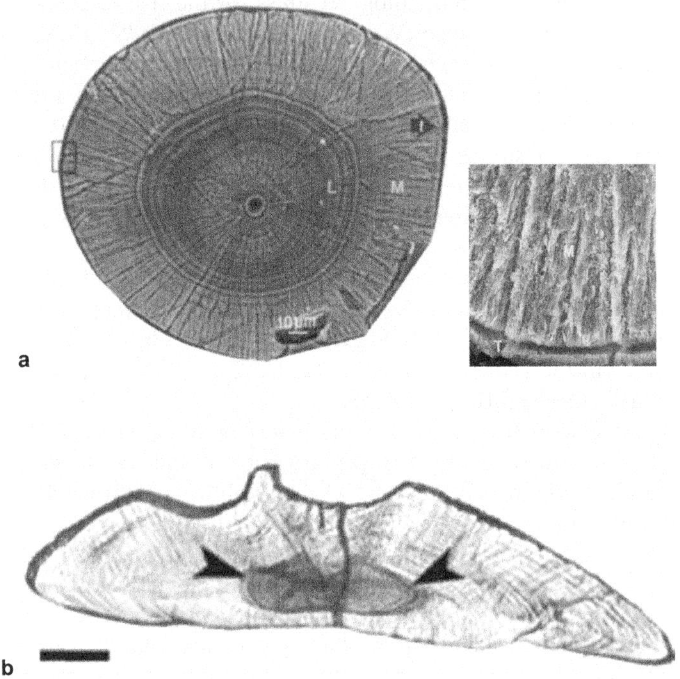

Figure 9. (**a**) Example of metamorphosis zonation on the otolith of *Anguilla marmorata* from Tahiti (French Polynesia). Otolith section viewed in SEM after acid etching. The *right image* shows a detail of the area framed at the *left margin* of the *left image*. L, leptochephalus stage; M, metamorphosis; T and t, transition mark. Scale bar = 10 µm (photos R. Lecomte-Finiger). (**b**) Example of settlement mark (*black arrows*) on the thin transverse section, observed in compound microscopy, of an otolith of *Chaetodon speculum* from the New Caledonian lagoon. Scale bar = 200 µm (photo E. Morize & L. Marec).

Otolith settlement marks in reef species can be grouped into categories based on changes in increment width and optical qualities of the mark (Wilson & McCormick 1999). The most common type corresponds to an "abrupt" settlement mark characterised by a rapid decrease in increment width, whereas other types present more gradual decreases in increment width during the settlement period. Evidence suggests that the increment profile over early development and the increment transitions associated with the settlement event are taxon-specific and may enable late-larval stage fish to be identified to species level (Wilson & McCormick 1999). This microstructural feature is mainly used for the back-calculation of settlement patterns and estimation of larval planktonic durations (Wilson & McCormick 1997). The settlement mark is used for age estimation prior to (e.g., Colin et al. 1997) or after settlement (e.g., Robertson et al. 1999, Rooker et al. 2004, Nemeth 2005), as well as for calculating growth rates during these two periods (Allman & Grimes 2002) or a combination of all types of analyses (Thorrold & Milicich 1990, Suthers 1998, Robertson et al. 1999, Rooker et al. 1999, Searcy & Sponaugle 2000). The planktonic larval duration of each species is then calculated from the otolith (Raventos & Macpherson 2001). Studies on recruitment of fish in the tropics often rely on the measurements of settlement marks. Growth rates of pelagic larvae which have been argued to be one of the principal determinants of recruitment success have also been estimated from otoliths (Bergenius et al. 2002). The identification of settlement location is also possible using the differences in individual growth rates (Danilowicz 1997a) and the spawning dates of arriving recruits can be estimated from back-calculation (Danilowicz 1997b). Body lengths at settlement are also directly back-calculated from otolith settlement checks (Radtke et al. 2001, Vigliola & Meekan 2002, Chapter 6 in this volume).

3.2.2 Reproduction and maturity

Reproduction can slow somatic growth by redirecting energy resources to the reproductive process and the production of gametes instead of whole body metabolism (Roff 1983). As such, reproduction could have an effect on the otolith growth via a slower body growth during this stage. "Spawning checks" or marks on the otolith which coincide with spawning were first described by Williams and Bedford (1974). Nevertheless, experimental studies establishing a precise relationship between reproduction and otolith features are particularly difficult to conduct because spawning should be strictly controlled to establish the exact location of its effect on the otolith. These observations are even more difficult to obtain in the field because the control of spawning or individual maturity is not possible. Massou et al. (2004b) have clearly shown for the first time an effect of reproduction on both somatic and otolith growth in females of the freshwater tilapia *Oreochromis niloticus*. In this study, the otolith had an irregular series of checks during the reproductive period (Figure 10). The responses varied among individuals, however, and there was not a common signal of reproduction on the otolith structure for all fish. The effect could be related to the peculiar behaviour of this mouth-brooding tilapia because of reduction on food intake or oxygen availability. This appearance of checks in the otolith could also be compared with the scale resorption that occurs during the spawning season in salmonids (Persson 1997). Such marks in the otoliths probably could be found in other mouth brooders in marine environments, like apogonids for example, but so far have not been documented. The information given by reproductive checks could allow managers to know the number of reproductive events per individual or within populations but more studies are needed on the characterisation of these checks, at both structural and composition levels.

There are a number of examples of correlative evidence describing checks in otoliths concurrent with change in sex (Walker & McCormick 2004). This study was the first to clearly demonstrate a strong temporal link between sex change from female to male and accelerated growth in an annual species. To our knowledge, however, there is no demonstration in the literature of a direct effect of reproductive activity on otolith microstructures of individuals caught in the field. Research in this direction could be a goal for management because knowledge of reproduction is key for population management.

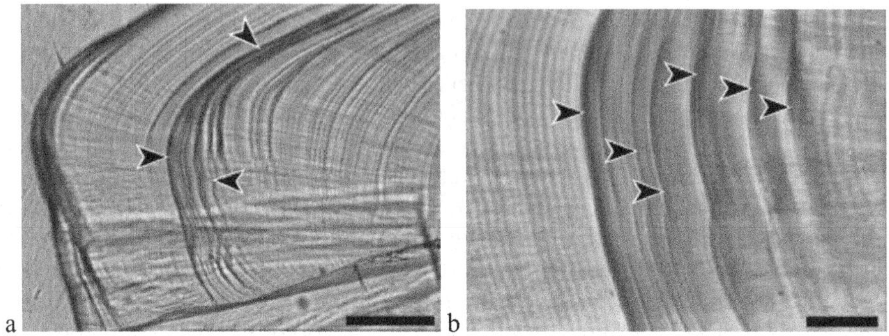

Figure 10. Checks (*arrows*) in the otoliths of *Oreochromis niloticus* females induced by reproduction. Images of the same otolith in thin transverse section observed in compound microscopy. (**a**) Scale bar = 100 μm. (**b**) Detail of (**a**). Scale bar = 20 μm (photos K. Diop & A. Malam Massou).

3.2.3 Stress

Stress of different origins acts on the otolith with a reduction of calcium deposition and produces marks of different intensities. Stress, like reproduction, is a preponderant factor that could lead to changes in the structure and growth of otoliths (Campana & Neilson 1985). There are many sources of stress for fishes, including capture, manipulation in artificial conditions, human or natural environmental pressures, inter- and intra-specific aggression, which, like all environmental changes, can be recorded as checks (discontinuities) inside the structure of the otolith. Pannella (1980) has given a precise but unverified definition of different checks (so called "growth discontinuities") produced by different stresses: "they range from small gaps that grade into interrupted sequences to major unconformities that can be traced continuously along the entire sagitta. The magnitude of the hiatus reflects the importance of the event in the life history of the fish that caused it." He has listed numerous possible causes for growth discontinuities ranging from extreme physiological stresses to starvation, migration or reproduction stresses.

Handling of fish undoubtedly acts as a stress factor during otolith growth. The effect most commonly cited in the literature is that of fish handling for otolith marking (e.g., tetracycline labelling with appropriate dosage). The trauma caused by this marking disturbs the regular deposition of primary increments for 3 to 4 days before the restart of regular deposition (Hoedt 1992, Arai et al. 2000a, Panfili & Tomás 2001, Massou et al. 2002, Massou et al. 2004b). Massou et al. (2004a) have shown that confinement stress could lead to checks in the otolith structure of tilapia *Oreochromis niloticus* (Figure 11). Their results suggest that there is a disruption in the deposition of primary increments during periods of confinement inducing physiological changes like an increase of plasma concentrations of cortisol, a stress hormone, and a reduction of somatic growth. A rapid modification of ambient temperature can also act as a stressor, modifying physiological responses with higher muscle heat shock protein levels and plasma cortisol levels (Nakano et al. 2004). This may reflect significant energetic costs of these stress responses and induce slower growth rates in otoliths.

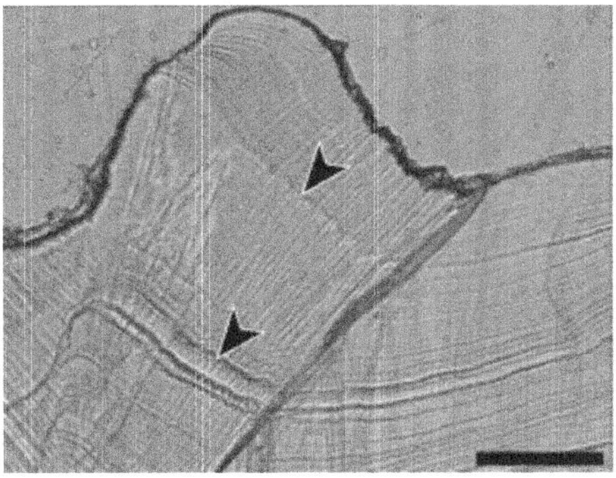

Figure 11. Checks (*arrows*) induced by confinement stress in the otolith of *Oreochromis niloticus*. Image of thin transverse section observed along the sulcus area on the ventral face in compound microscopy. Scale bar = 100 μm (photo K. Diop).

Table 1 summarises specific growth marks associated with life-history traits for tropical species. In conclusion, almost all major life-history or environmental perturbation events encountered by tropical fish may be reflected inside the otolith structure in the form of different marks. Although there are general rules concerning the interpretation of these marks, the responses also seem to be very individual among fish, even for the same species. These marks can be assumed to be reference points in cases such as strong environmental transitions or settlement and metamorphosis. Some difficulties reside in the precise description of these marks in the otolith with the aim of using them as indicators but the information that they contain remains invaluable for managers of natural populations. They can record all steps of developmental stages such as birth, first feeding, metamorphosis, habitat changes and different external stresses. These are taxon-specific and especially useful for back-calculating birth dates, time at settlement or transition between environments.

4 Experimental manipulation of otolith microstructure

The experimental manipulation of otolith microstructure can be undertaken at two levels: (1) the definitive incorporation of new substances or elements inside the structure; and (2) the modification of the growth rate of the structures. The first level includes labelling with fluorescent markers or the incorporation of elements or radionuclides. The second generally uses the manipulation of exogenous factors such as light and temperature to create intrinsic marks. These marks can then be assumed to be references for further analysis.

Table 1. Examples of otolith marks associated with life-history events (excluding hatching and first events after birth) for tropical fish species. Literature review not exhaustive.

L-H event	Otolith mark	Species	References
Metamorphosis	Mark and specific growth rate	*Anguilla celebesensis*	Arai et al. (2000a), Arai et al. (2003)
	Check and specific growth rate	*5 spp. Anguillidae*	Marui et al. (2001)
	Specific growth rate	*Heklotsichthys castelnaui*	Thorrold (1989)
	Metamorphic bands	Labridae (2 species)	Searcy and Sponaugle (2001)
	Accessory primordia	Myctophidae (55 species)	Linkowski (1991)
	Specific growth rate	*Vinciguerria nimbaria*	Tomás and Panfili (2000)
Parturition	Extrusion check	*Sebastes jordani*	Ralston et al. (1996)
Reproduction	Checks	*Oreochromis niloticus*	Massou et al. (2004b)
Settlement	Mark classification, specific growth rate	44 spp. (Great Barrier Reef)	Wilson and McCormick (1999)
	Breakpoint-transition, specific growth rate	38 spp. (Great Barrier Reef)	Brothers et al. (1983)
	Settlement mark, specific growth rate	38 spp. (Great Barrier Reef)	Wilson and McCormick (1997)
	Settlement mark	*Acanthurus triostegus*	McCormick (1999)
	Settlement mark (discontinuity)	Chaetodontidae	Fowler (1989)
	Check	*Chromis atripectoralis*	Thorrold and Milicich (1990)
	Check	*Ctenochaetus binotatus, Scarus rivulatus*	Lou (1993)
	Settlement mark	*Thalassoma bifasciatum, Halichoeres bivittatus*	Victor (1982)
	Settlement mark (check)	*Lentipes concolor*	Radtke et al. (2001)
	Accessory growth centres	*Lota lota*	Fischer (1999)
	Increment settlement transition	Lutjanidae	Zapata and Herron (2002)
	Check	*Lutjanus griseus*	Allman and Grimes (2002)
	Reference mark	Pomacentridae	Wellington and Victor (1989)
	Check	*Pomacentrus coelestis*	Thorrold and Milicich (1990)
	Settlement mark	*Pomacentrus vaiuli*	McIlwain (2002)
	Settlement mark	3 reef spp.	Wellington and Victor (1992)
	Transition mark	*Thalassoma bifasciatum*	Masterson et al. (1997)
	Metamorphic band, transitional mark	*Thalassoma bifasciatum*	Sponaugle and Pinkard (2004), Sponaugle et al. (2006)
Sex change	Check	*Parapercis cylindrica*	Walker and McCormick (2004)
Stress	Specific growth rate	*Abudefduf vaigiensis*	Nakano et al. (2004)
	Check	*Oreochromis niloticus*	Massou et al. (2004a)

4.1 FLUORESCENT LABELLING

The incorporation of a specific fluorescent dye, acting as a marker, into the mineralising surface of the otolith is one of the most common methods used in validation experiments. Several markers have been used but compounds of the tetracycline family have been used mainly for tropical species (Wild & Foreman 1980, Speare 1992, Choat & Axe 1996, Fowler & Short 1998, Thompson et al. 1999, Panfili & Tomás 2001, Grandcourt 2002, Massou et al. 2002, Schwamborn & Ferreira 2002) although calcein (Wilson et al. 1987, Leips et al. 2001) or alizarin complexone (Arai et al. 2000a, Al Husaini et al. 2001, Sugeha et al. 2001, Morioka & Matsumoto 2003, Meifsjord et al. 2006) have also been used. All these markers have the capacity to emit a specific fluorescence under ultraviolet light. The chemical marks are incorporated into otoliths after intra-peritoneal or intra-muscular injection, immersion of the fish in a solution or incorporation of the chemicals into food. Injection doses vary between 25 and 75 mg tetracycline per kg of body weight. The immersion doses and duration of the baths vary with the fluorescent dyes used and fish species (Wright et al. 2002a). The mechanism of incorporation into the otolith structure is rather complex and not very well understood since few experiments have been done to elucidate the incorporation processes. Nevertheless, once the marker is inside the body, its calcium-chelating properties make it become directly integrated into the mineralising surface of the otolith. The non-incorporated fraction is excreted by metabolism after a few minutes or hours (Meunier 1982). This clearly explains the systematic observation of a single fluorescent band after marking, which sometimes can diffuse across a few increments. Multiple incorporations are possible by repeated administration of the dye spaced over a few days or longer (Tsukamoto 1988, Meifsjord et al. 2006).

4.2 THERMAL MANIPULATION

Temperature fluctuations have an immediate effect on otolith accretion rates and thus lead to discernible marks inside the structure. These marks are in the form of changes in the width of increments or as zones more or less dense optically (Volk et al. 1999) that result in incremental patterns distinct from those caused by natural fluctuations. Thermal marking can be applied to mark thousands of fish with minimal effort to create "bar codes" with microstructures. This technique has been used mostly to mark the otoliths of larvae or juveniles but not adults. The idea is to use short-term temperature manipulations to mark juvenile fish otoliths by altering the appearance of D-zones and L- zones in one or more primary increments to produce a clear pattern of events. The production of a visible mark can be done with a temperature depression of only a few degrees for a few minutes or hours (e.g., 3.5°C and 4 h for salmon, Volk et al. 1994). Although this technique has mostly been applied to mark salmonids (Volk et al. 1999), it has been shown that otoliths of fish from non-temperate environments can also be marked with this technique. A change of temperature from 19–21°C to 28°C or 16°C was enough to mark the otoliths of the subtropical species Chinese sucker (*Myxocyprinus asiaticus*) (Song et al. 2008b). The combination of cold and warm temperatures in time lapses of 6–36 h created distinct patterns. Details regarding the facilities necessary to carry out thermal marking can be found in Volk et al. (1999).

4.3 MANIPULATION OF OTOLITH ELEMENTAL COMPOSITION

Otoliths grow by successive deposition of increments, of which the L-zone is a band rich in calcium carbonate crystals in the form of aragonite crystals. The composition of the L-zone can be altered to create a distinct chemical mark. The aragonite group is composed of carbonate minerals with a single divalent cation of radius > 1.00 Å. Carbonates which are isostructural with aragonite include strontianite ($SrCO_3$), cerussite ($PbCO_3$) and witherite ($BaCO_3$) classified by increasing single divalent cation radius. It is possible, therefore, to manipulate the elemental composition of the otolith aragonite by replacing Ca in the crystal lattice by other elements such as Sr, Pb or Ba.

Injections of $SrCl_2$ have been used to chemically mark otoliths of juvenile southern bluefin tuna, *Thunnus maccoyii* (Figure 12) (Clear et al. 2000), whereas strontium chloride baths have been used to mark the otoliths of tropical clupeids (sprat, *Spratelloides delicatulus* and herring, *Herklotsichthys quadrimaculatus*) (Milton et al. 1993) and juvenile barramundi (*Lates calcarifer*) (Milton & Chenery 2001). The Sr mark produced by a Sr-bath was found in the otolith of juvenile barramundi 3 days after the Sr-bath (Milton & Chenery 2001). In all cases, the Sr mark was easily detected using a scanning electron microscope (SEM) fitted with a backscatter detector or Energy Dispersive Spectroscopy (EDS). Nonetheless, caution is needed when marking freshwater fish by immersion as it has been shown that discrimination of strontium against calcium is a function of an osmoregulatory mechanism and increases in waters with low salt concentrations (Rosenthal 1960).

Barium chloride ($BaCl_2$) has been successfully used to mark offspring of coral reef species and Ba isotope ratios in the otolith cores of settlers have permitted larval dispersion to be followed in the environment (Thorrold et al. 2006, Almany et al. 2007). Pb has not been used to mark otoliths, although its proclivity to replace Ca in the crystal lattice has also made it an ideal candidate for marking otoliths and monitoring perturbations such as the exposure of fish to polluted environments (Spencer et al. 2000) or the occupation of estuaries by diadromous or migrating fish (Bath et al. 2000). Alternative chemical species used to mark otoliths are lanthanides such as lanthanum, samarium and cerium (Ennevor & Beames 1993) whose mark in the otoliths of coho salmon (*Oncorhynchus kisutch*) was still detectable 10 months after marking.

Experiments with ^{45}Ca have also demonstrated the use of radio-labelled calcium to mark the otoliths of goldfish *Carassius auratus* (Mugiya et al. 1981, Ichii & Mugiya 1983, Mugiya & Uchimura 1989). Intraperitoneal injection of ^{45}Ca into rainbow trout showed that ^{45}Ca deposition in the otoliths occurred within 3 h (Mugiya 1974). The otoliths of various tropical species (zebra fish, *Danio rerio*, white cloud mountain fish, *Tanichthys albonubes*, and guppy, *Poecilia reticulata*) were marked with similar results using radiostrontium (^{90}Sr), taking advantage of the fact that strontium, fed or injected, accumulates in otoliths as in other calcified tissues (Rosenthal 1957, 1960).

In summary, it is recommended marking the otoliths of tropical fish with tetracycline injection or immersion rather than alizarine, calcein or strontium chloride since it produces a brighter mark than other fluorescent dyes and the detection of the mark is cheaper compared to that of strontium chloride, which requires electron microscopes (Hernaman et al. 2000). Nevertheless, the use of tetracycline constitutes an environmental issue since it does not degrade in the environment and is not suitable for human consumption if accumulated in the fish tissues. The trade-off between these issues should lead the researcher to make a decision regarding which marker to use.

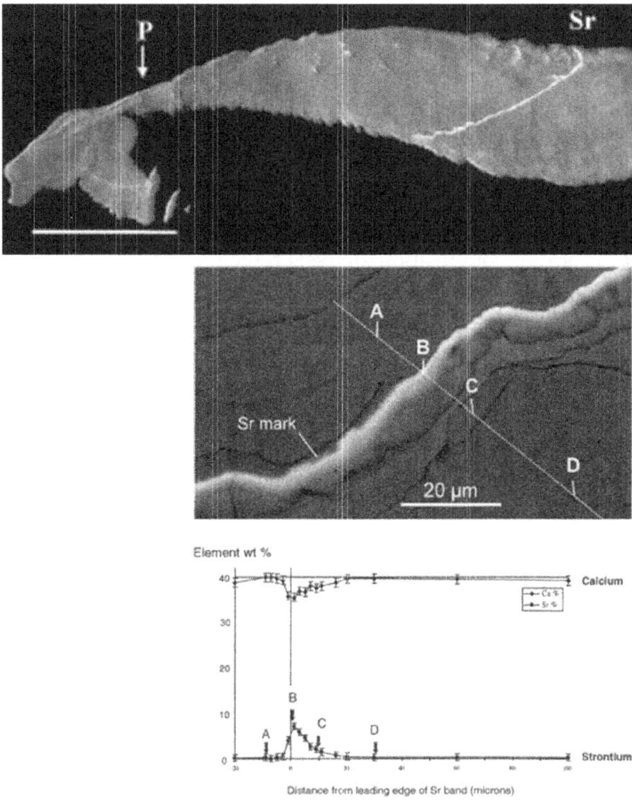

Figure 12. Backscattered SEM image showing a clear strontium mark apparent as a bright band on a section of *Thunnus maccoyii* otolith, along the primordium (P) – post-rostrum (PR) axis (scale bar = 1 mm). The region analysed for the Sr and Ca levels is magnified. The graph indicates the measured variation in Sr and Ca along the transect: points A–D indicate representative positions along the transect to enable comparisons (Source: Clear et al. 2000).

5 Procedural considerations

5.1 TRADE OFFS AND QUALITY ASSURANCE

The study of otolith microstructure demands more effort than any other standard method (e.g. ageing), both in terms of time and cost. The financial investment in materials, equipment and preparation time depends on the type of information that one seeks from microstructural examination. For example, any study of the fine details within the microincrements at high magnifications (above ×2000) requires the use of a SEM because this exceeds the resolution limit of any optical compound microscope. The simple study of life history events, however, mainly relies on observations made with a compound microscope, as details of microincrements can be seen at between ×400 and ×1000 magnifications.

The analysis of the results of experimental otolith manipulation (e.g., fluorescent labelling, thermal manipulation, elemental composition) requires different observation techniques than those used for otolith microstructure examination. Compound microscopes are used for thermal or light-cycle manipulation, UV lamps for fluorescent labelling, and electronic probes for elemental composition. Table 2 lists the comparison of the investment and analysis costs for the different equipment involved in such studies. Techniques used for otolith composition analysis remain the most expensive.

Quality assurance when studying otolith microstructure is directly linked to the instrumentation used. The main constraining factor in compound microscopy is the optical resolution available. The theoretical limit of resolution is about 0.3 μm (Campana 1992). The widths of some microincrements are below this limit and the use of a SEM is thus essential (Morales-Nin 1988). SEM gives the best level of quality assurance provided that the sequence of sample preparation is accurately followed (sectioning, polishing, etching and sample coating).

5.2 SAMPLING DESIGN

The aim of the study will determine the methods to be employed. Only light microscopy is required when the goal is to observe growth micro-increments for age validation, life history events, etc. In some tropical species during certain slow growth periods, however, very thin increments may be laid down on the otolith (Morales-Nin 1988) and when viewed with a compound microscope these would appear as thin translucent bands without structures. It would be necessary in such cases to observe some otoliths of different ages with SEM. This tool is also recommended for microstructural studies and for observation of growth discontinuities to determine if they correspond to periods of several days.

The number of otoliths to be observed will depend on the methods used because SEM time and sample preparation are costly. It is advisable in general to always select one otolith from the same side of the fish and keep the other otolith of the pair for additional observations or as a backup. Time available for the study is also important, as cutting and polishing an otolith might take one hour or more and observation of microstructures is time consuming. Thus, at the beginning of any study, available time and money resources should be considered to define the number of otoliths to be examined. It is advisable to have extra otoliths because frequent losses occur during otolith preparation. It is advisable to examine males and females separately when sexual dimorphism is present or when mature fish are studied. Similarly, when various life phases are studied, it is necessary to increase the number of otoliths examined in the transition sizes (for instance at settlement). The asymptotic otolith growth in adult fish might make it difficult to prepare sections comprising all the increment sequence and extra otoliths might need to be examined to get all required information.

Table 2. Different relative costs of initial investment, analysis and effort (time required) for the preparation and operation of the indicated equipment. X is simply an indicative relative measure and not equal to any monetary value or specific amount of time. ***Legend:*** UV, ultra-violet; SEM, scanning electron microscopy; EDS, energy dispersive microprobe; ICPMS, inductively coupled plasma mass spectrometry; WDS, wavelength dispersive microprobe.

Equipment	Investment	Analysis	Effort	Study
Compound microscopy	X	0	X	Life history events Growth Annulus-age estimates Marginal increment analysis Thermal manipulation Light manipulation
UV light microscopy	XX	X	X	Labelling manipulation
Confocal microscopy	XXX	X	X	3D structure
Raman spectrometry	XXX	XX	XX	3D structure
SEM	XXXX	XX	XXX	Ultrastructure
EDS	XXXXX	XXXX	XXX	Manipulaton of elemental composition
WDS or ICPMS	XXXXXX	XXXXX	XXXXX	Manipulation of elemental composition

5.3 DATA ANALYSIS ISSUES

Microstructural observation is generally descriptive and does not require data analysis. For increment measurement see Vigliola and Meekan (Chapter 6) and for counting Fowler (Chapter 3) and Sponaugle (Chapter 4).

5.4 FUTURE PERSPECTIVES

The microscopes traditionally used, based on light or electrons, are being improved both in their features (optical capacity and quality of the lenses) and with the use of accessory devices such as image analysis systems. These have increased the resolution power and quality of light microscopy, which is now also more user-friendly due to the availability of cameras (digital, with increased resolution). Moreover, PC-based analysis systems help recording, improving, filtering and analysing images, as well as helping in the observation and interpretation of otolith structures displayed on a screen by more than one observer. Thus, new developments concern both the observation devices (microscopes) and the rapidly evolving image treatment systems.

Modern SEM microscopes also now work at ambient or low pressures (Environmental SEM, ESEM), allowing observation of biological samples without metallic coating of the surface. This could be an advantage when otolith microstructure examination is coupled with analysis of chemical composition, since any coating would alter the surface of the otolith. Future studies will certainly use the latest instrumentation (such as Atomic Force Microscopy) to make progress in the analysis of otolith growth at the microincrement level and provide insights into the interaction between aragonite crystals and proteins at the nanoscale.

6 Technical recipes

6.1 TECHNICAL PREPARATION OF OTOLITH MICROSTRUCTURES

Preparation techniques for otoliths are universal and not specific for tropical or temperate species. The examination of otolith microstructure necessarily implies some kind of preparation to reveal its internal structure except if observed with Laser Scanning Microscopy (Lagardère et al. 1995). Preparation usually requires sectioning the otolith along a particular plane. This plane has to be chosen prior to any further manipulation. There are three main planes of sectioning, following the general orientation of the fish: sagittal (vertical longitudinal), transverse and frontal. The sagittal plane can be used for small-sized fish for which both the core and the edge of the otolith can be reached easily. It is more difficult for larger fish with otoliths with a concave-convex or even more complex shape, in which case the transverse section plane is recommended for a standard observation of internal structures.

Whole otoliths usually should be embedded in a suitable block of material prior to sectioning (Box 1) since it is difficult to manipulate the otoliths directly. There are a lot of embedding media available but generally polyester resins or related resins (e.g., epoxy, thermoplastic cement) are used for structure examination. A complete table with the standard mounting materials is available in McCurdy et al. (2002). The internal structure of embedded otoliths can be revealed with two main techniques of sectioning, the first using low speed saws (Box 2) and the second using hand grinding.

The otolith sections can be viewed directly for microstructure examination under light microscopy after appropriate embedding, sectioning and polishing (Box 3). Nevertheless, in order to reveal more precisely the differences in internal crystallisation, acid (Box 4) or proteinase (Box 5) etching is used to accentuate the differentiation between L-zones and D-zones. Etching is also an obligatory step before discriminating zones by staining (Box 6) or examining the structure at high magnification by scanning electron microscopy (SEM). In SEM, the electrons require a high vacuum environment to assure an aligned beam sputtered onto the sample surface and the sample must be electrically conducting to avoid electron clouds on the sample surface that would compromise the quality of observation (Box 7). Consequently, the otoliths must be dried and coated to increase the aragonite conductivity. Techniques to remove the aragonite and preserve the structure of the organic matrix are required when the organic matrix is to be observed. Technical recipes for each step of otolith preparation mentioned above are provided in Boxes 1–7 below (see also Fowler, Chapter 3 and Spanaugle, Chapter 4 in this volume).

7 Observation

7.1 SCANING ELECTRON MICROSCOPY (SEM)

The SEM is a surface topographic examination tool generally used for otolith studies. The instrument offers an over 300-fold increase in the depth of field when compared with the highest quality compound microscope, which is reflected in the superb 3D images it provides at very high magnification (> ×10000). After etching the otolith surface, it reveals the internal crystalline structure as well as the orientation of the crystals and the protein matrix structure (L-zones and D-zones, respectively). Studies of beam-specimen interaction using the SEM can provide useful information about the chemical composition at the specimen surface, as well as the crystallographic, magnetic, and electrical characteristics of the specimen. For example, for microchemical analysis purposes the SEM is coupled to an EDS (Energy Dispersive Spectrometer).

Box 1. Embedding otoliths for sectioning

1. Clean and dry the otolith before embedding.
2. Tag the otolith and embedding mould.
3. Mix carefully one unit of resin (e.g., polyester) by volume or weight and 1% catalyst (e.g., for polyester).
4. Leave the mixture for 2 min to let the largest bubbles escape.
5. Pour a layer of resin into each individual mould (flexible rubber moulds are preferred).
6. Wait for polymerisation that takes at least a few hours (e.g., 4 h for polyester), preferably in a dry oven (30°C). Avoid humid environments for polymerisation.
7. Put the otolith to be embedded over the solid resin layer of the mould.
8. Mix carefully another portion of resin with 1% catalyst.
9. Leave the mixture for 2 min to let the largest bubbles escape.
10. Pour over the otolith until it is covered by the mixture.
11. Turn over the otolith in its mould under the resin to eliminate air bubbles.
12. Orient the otolith precisely in the mould for further preparation (e.g. sectioning).
13. Wait for complete polymerisation over a minimum of several hours (e.g., 24 h for polyester), if possible in a dry oven (30°C), before further manipulation.
14. Remove the embedded otolith from the mould and tag the resin block.

Box 2. Sectioning the otolith

1. It is best to use a slow-speed rotary diamond saw.
2. Use an appropriate liquid media as a cutting fluid (e.g., water is preferred with polyester resin).
3. Mark guidelines for sectioning directly on the resin block holding the embedded otolith.
4. Lock the embedded otolith on the specific support saw and orient it as required to cut in the preferred section plane.
5. Cut the section, being sure not to destroy desired features (e.g., core).
6. If a slice is necessary, adjust the level with the saw micrometer taking into account the thickness of the sectioning tool (e.g. about 300 μm for diamond disc) and make the second cut to produce the section of desired thickness.
7. Dry the section.

Box 3. Grinding and polishing the section

1. Use grinding after sectioning or, for smaller otoliths, use grinding instead of sectioning.
2. Grinding generally is done by hand with wet abrasive paper (grit grades between 120 and 1,200).
3. Polishing follows grinding and generally is done by hand with polishing cloths with different grades of alumina pastes (from 3 µm to 0.3 µm) or diamond powder.
4. Start grinding with a coarser grit paper until the approximate position of the surface is attained and then carefully fine-tune the surface with finer grit paper.
5. Similarly, start polishing with a coarser paste and finish with finer paste.
6. Use random movements when grinding and polishing to avoid systematic distortion of the preparation plane or systematic scratches.
7. Frequently check the state of the ground or polished surface under a microscope to ensure desired features (e.g., the otolith core) are not being damaged or removed.
8. Rinse the preparation in clean water between stages of grinding or polishing.
9. At the end of the process clean the preparation in an ultrasonic bath.

Box 4. Acid etching

1. Prepare acid solution (generally 1–5% EDTA or 1–5% hydrochloric acid).
2. Plunge the preparation to be etched (e.g., otolith section) into the acid solution.
3. Measure the time of acid attack (generally between 1 and 5 min). The speed of etching depends on the temperature of the acid solution, the cooler the solution the slower the reaction.
4. Check the etching process under a binocular or a compound microscope.
5. Thoroughly rinse the etched preparation with distilled water (for a few minutes).
6. Leave the preparation to dry (open air or dry oven at 30°C).
7. Check the etched surface and restart at point 2 (above) if necessary.

Box 5. Proteinase etching (from Shiao et al. 1999)

1. Prepare proteinase K buffer: 10 mg of proteinase K, 0.1 ml of Tri–HCl (pH=8.0, 1 M), 0.5 ml of sodium dodecyl sulphate (SDS), 0.02 ml of NaCl.
2. Digest the otolith section with the solution in a 1.5 ml microtube at 45°C with gentle shaking.
3. Depending on species and temperature, the reaction time varies from a few minutes to a few hours.
4. Check the otolith surface under a compound microscope until optimal etching .
5. Leave the preparation to dry in an oven after digestion.
6. If SDS crystallizes on the surface, use 75% alcohol to dissolve it.

Box 6. Staining (from McCurdy et al. 2002)

1. Prepare staining solution (e.g., 1% Toluidine blue).
2. Plunge the preparation to be stained (e.g., otolith etched section) into the staining solution (or put a drop of stain on the etched section).
3. Measure the time of staining (generally between 1 and 5 min).
4. Rinse the stained preparation carefully with distilled water. Be careful as the stained surface is sensitive to further manipulation.
5. Dry the preparation (open air).
6. Check the stained surface and restart at point 2 if necessary.

Box 7. Surface coating for SEM

1. Dry the sample for 6–8 h at low temperature. Keep it protected from dust.
2. Mount the sample on the SEM sample holder or stub, using carborundum paint, thermoplastic glue or double-sided tape, with surface to be observed uppermost.
3. A thin line of colloidal silver should run from the otolith surface to the stub to avoid electric charges on the otolith during observation.
4. Sputter the otolith surface with gold (100 Angströms) or carbon (10–50 Angströms).
5. Keep dry and dust free until observation.

The electron beam used as "illumination" that is directed onto the specimen surface is accelerated with voltages of 1–30 kV. The interaction between the beam and the specimen generates the image. Specimen-beam interaction, in addition to producing backscattered and secondary electrons, also produces photons, specimen currents, Auger electrons and X-rays, which are characteristic of the elements excited by the beam in the specimen matrix. While in principle any signal generated can be used to produce an image, in practice it is the low-energy secondary electrons released from the sample that are most commonly used. Using surface emitted electrons, rather than those passing through the specimen, as in TEM, surface images revealing some three-dimensional quality can be obtained.

The impression of the 3D surface image is the result of the distribution of light and dark areas. This distribution is accounted for largely by the incident beam which generates more collectable secondary electrons per unit area when it strikes a sharply curved edge or sloping surface than when it hits a flat surface. The contours of biological specimens facilitate the production of such a differential effect. Where the surface is smooth, tilting the sample at an angle to the probe will enhance the desired variation in collected secondary electrons. More subtle effects involving the manner in which structures lying above a primary surface either deflect or absorb the probe electrons also come into play in the process. In any case, the "shadows" seen are a true representation of the 3D character of the specimen surface under study. This is why etching is a pre-requisite of otolith observation under SEM.

Depending on the beam accelerating potential used, which is generally around 15 kV for otolith observation, the ratio between the low-energy secondary to high-energy backscattered electrons will be higher. Higher voltages, although giving more penetration into the sample surface, result in electrical charges due to the low conductivity of aragonite.

7.2 RAMAN SPECTROMETRY

Raman spectrometry can be useful to describe the nature of the otolith layers within the crystal lattice. It measures the Raman effect, which is the inelastic scattering of light by matter. When a photon of visible light interacts with a molecule, it can be elastically scattered (conserving its energy), it can be inelastically scattered losing energy (Stokes scattering) or it can be inelastically scattered gaining energy from the molecule (anti-Stokes scattering). The measurement of such vibrational energy in otolith studies provides information about the metal and carbonate bonds within the crystal lattice of the otolith. A laser excites the sample surface and the Raman effect and the resulting distortion of the crystal lattice are recorded as energy spectra which are characteristic for each polymorph of $CaCO_3$. The volumes excited by the laser can be as small as 20 μm^3. The main use of Raman spectrometry in otolith studies is thus the identification of the $CaCO_3$ polymorph deposited, aragonite, calcite or vaterite (Gauldie et al. 1997, Tomás & Geffen 2003). The readability of the otolith and its chemical composition will vary greatly depending on which polymorph is precipitated (Brown & Severin 1999, Tomás & Geffen 2003) and so it is necessary to verify that the $CaCO_3$ polymorph precipitated is aragonite prior to interpretation.

7.3 CONFOCAL MICROSCOPY

Confocal laser microscopy provides images of the internal structure of samples without preparation for as long as these fluoresce. It can therefore only be used for the observation of microincrements if these are first marked with a fluorescent dye. Confocal microscopy has been applied to detect the internal oxytetracycline mark in the otoliths of larval and juvenile stripped-bass (*Morone saxatilis*) (Secor et al. 1991), to study the ontogenetic changes in the 3D structure of otoliths of various temperate species (Lagardère et al. 1995) and to make observations of the fluorescently labelled saccular membrane of herring (*Clupea harengus*) embryos (Tytler et al. 2002).

References

Al Husaini M, Al Ayoub S, Dashti J (2001) Age validation of nagroor, *Pomadasys kaakan* (Cuvier, 1830) (Family: Haemulidae) in Kuwaiti waters. Fish Res 53:71–81

Allman RJ, Grimes CB (2002) Temporal and spatial dynamics of spawning, settlement, and growth of gray snapper (*Lutjanus griseus*) from the West Florida shelf as determined from otolith microstructures. Fish Bull 100:391–403

Almany GR, Berumen ML, Thorrold SR, Planes S, Jones GP (2007) Local replenishment of coral reef fish populations in a marine reserve. Science 316:742–744

Anken RH, Edelmann E, Rahmann H (2002) Neuronal feedback between brain and inner ear for growth of otoliths in fish. Adv Space Res 30:829–833

Arai T, Limbong D, Otake T, Tsukamoto K (2001) Recruitment mechanisms of tropical eels *Anguilla* spp. and implications for the evolution of oceanic migration in the genus *Anguilla*. Mar Ecol Prog Ser 216:253–264

Arai T, Limbong D, Tsukamoto K (2000a) Validation of otolith daily increments in the tropical eel *Anguilla celebesensis*. Can J Zool 78:1078–1084

Arai T, Miller MJ, Tsukamoto K (2003) Larval duration of the tropical eel *Anguilla celebesensis* from Indonesian and Philippine coasts. Mar Ecol Prog Ser 251:255–261

Arai T, Otake T, Tsukamoto K (2000b) Timing of metamorphosis and larval segregation of the Atlantic eels *Anguilla rostrata* and *A. anguilla*, as revealed by otolith microstructure and microchemistry. Mar Biol 137:39–45

Asano M, Mugiya Y (1993) Biochemical and calcium-binding properties of water soluble proteins isolated from otoliths of the tilapia, *Orechromis niloticus*. Comp Biochem Physiol B 104:201–205

Bath GE, Thorrold SR, Jones CM, Campana SE, McLaren JW, Lam JWH (2000) Strontium and barium uptake in aragonitic otoliths of marine fish. Geochim Cosmochim Acta 64:1705–1714

Bergenius MAJ, Meekan MG, Robertson DR, McCormick MI (2002) Larval growth predicts the recruitment success of a coral reef fish. Oecologia 131:521–525

Boeuf G, Le Bail PY (1999) Does light have an influence on fish growth? Aquaculture 177:129–152

Borelli G, Mayer Gostan N, de Pontual H, Boeuf G, Payan P (2001) Biochemical relationships between endolymph and otolith matrix in the trout (*Oncorhynchus mykiss*) and turbot (*Psetta maxima*). Calcif Tissue Intern 69:356–364

Brothers EB (1981) What can otolith microstructure tell us about daily and subdaily events in the early life history of fish? Rapp PV Reun Cons Intern Explor Mer 178:393–394

Brothers EB, Williams DM, Sale PF (1983) Length of larval life in twelve families of fishes at "One Tree Lagoon", Great Barrier Reef, Australia. Mar Biol 76:319–324

Brown R, Severin KP (1999) Elemental distribution within polymorphic inconnu (*Stenodus leucichthys*) otoliths is affected by crystal structure. Can J Fish Aquat Sci 56:1898–1903

Campana SE (1983) Calcium deposition and otolith check formation during periods of stress in coho salmon, *Oncorhynchus kisutch*. Comp Biochem Physiol A 75:215–220

Campana SE (1984) Interactive effects of age and environmental modifiers on the production of daily growth increments in otoliths of plainfin midshipman, *Porichthys notatus*. Fish Bull 82:165–177

Campana SE (1992) Measurement and interpretation of the microstructure of fish otoliths. In: Stevenson DK, Campana SE (Eds) Otolith microstructure examination and analysis, Vol 117. Can Spec Pub Fish Aquat Sci, Ottawa, Canada, pp 59–71

Campana SE (1999) Chemistry and composition of fish otoliths: pathways, mechanisms and applications. Mar Ecol Prog Ser 188:263–297

Campana SE, Neilson JD (1985) Microstructure of fish otoliths. Can J Fish Aquat Sci 42:1014–1032

Carlström D (1963) A crystallographic study of vertebrate otoliths. Biol Bull 124:441–463

Chesney EJ, McKee BM, Blanchard T, Chan LH (1998) Chemistry of otoliths from juvenile menhaden *Brevoortia patronus*: evaluating strontium, strontium: calcium and strontium isotope ratios as environmental indicators. Mar Ecol Prog Ser 171:261–273

Choat JH, Axe LM (1996) Growth and longevity in acanthurid fishes: an analysis of otolith increments. Mar Ecol Prog Ser 134:15–26

Clear NP, Gunn JS, Rees AJ (2000) Direct validation of annual increments in the otoliths of juvenile southern bluefin tuna, *Thunnus maccoyii*, by means of a large-scale mark-recapture experiment with strontium chloride. Fish Bull 98:25–40

Colin PL, Laroche WA, Brothers EB (1997) Ingress and settlement in the Nassau grouper, *Epinephelus striatus* (Pisces: Serranidae), with relationship to spawning occurrence. Bull Mar Sci 60:656–667

Danilowicz BS (1997a) The effects of age and size on habitat selection during settlement of a damselfish. Environ Biol Fish 50:257–265

Danilowicz BS (1997b) A potential mechanism for episodic recruitment of a coral reef fish. Ecology 78:1415–1423

Degens ET, Deuser WG, Haedrich RL (1969) Molecular structure and composition of fish otoliths. Mar Biol 2:105–113

Edeyer A, de Pontual H, Payan P, Troadec H, Sévère A, Mayer-Gostan N (2000) Daily variations of the saccular endolymph and plasma compositions in the turbot *Psetta maxima*: relationship with the diurnal rhythm in otolith formation. Mar Ecol Prog Ser 192:287–294

Ennevor BC, Beames RM (1993) Use of lanthanide elements to mass mark juvenile salmonids. Can J Fish Aquat Sci 50:1039–1044

Fischer P (1999) Otolith microstructure during the pelagic, settlement and benthic phases in burbot. J Fish Biol 54:1231–1243

Fowler AJ (1989) Description, interpretation and use of the microstructure of otoliths from juvenile butterflyfishes (family Chaetodontidae). Mar Biol 102:167–181

Fowler AJ (1995) Annulus formation in otoliths of coral reef fish – a review. In: Secor DH, Dean JM, Campana SE (Eds) Recent developments in fish otolith research. Belle W. Baruch Library in Marine Science, Hilton Head, South Carolina, USA, pp 45–64

Fowler AJ, Short DA (1998) Validation of age determination from otoliths of the King George whiting *Sillaginodes punctata* (Perciformes). Mar Biol 130:577–587

Gartner JV, Jr. (1991) Life histories of three species of lanternfishes (Pisces: Myctophidae) from the eastern Gulf of Mexico. I – Morphological and microstructural analysis of sagittal otoliths. Mar Biol 111:11–20

Gauldie RW, Nelson DGA (1988) Aragonite twinning and neuroprotein secretion are the cause of daily growth rings in fish otoliths. Comp Biochem Physiol A 90:501–509

Gauldie RW, Sharma SK, Volk E (1997) Micro-Raman spectral study of vaterite and aragonite otoliths of the coho salmon, *Oncorhynchus kisutch*. Comp Biochem Physiol A 118:753–757

Geffen A, Nash RDM (1995) Periodicity of otolith check formation in juvenile plaice, *Pleuronectes platessa* L. In: Secor DH, Dean JM, Campana SE (Eds) Recent developments in fish otolith research, Vol 19. University of South Carolina Press, Columbia, SC, USA, pp 65–73

Gjøsaeter H (1987) Primary growth increments in otoliths of six tropical myctophid species. Biol Oceano 4:359–382

Grandcourt EM (2002) Demographic characteristics of a selection of exploited reef fish from the Seychelles: preliminary study. Mar Freshwater Res 53:123–130

Gutiérrez E, Morales-Nin B (1986) Time series analysis of daily growth in *Dicentrarchus labrax* L. otoliths. J Exp Mar Biol Ecol 103:163–179

Hales LS, Jr, Hurley DH (1991) Validation of daily increment formation in the otoliths of juvenile silver perch, *Bairdiella chrysoura*. Estuaries 14:199–206

Hayashi A, Kawaguchi K, Watanabe H, Ishida M (2001) Daily growth increment formation and its lunar periodicity in otoliths of the myctophid fish *Myctophum asperum* (Pisces: Myctophidae). Fish Sci 67:811–817

Hernaman V, Munday PL, Schläppy ML (2000) Validation of otolith growth-increment periodicity in tropical gobies. Mar Biol 137:715–726

Hoedt FE (1992) Validation of daily growth increments in otoliths from *Thryssa aestuaria* (Ogilby), a tropical anchovy from Northern Australia. Aus J Mar Freshwater Res 43:1043–1050

Houde ED (1974) Effects of temperature and delayed feeding on growth and survival of larvae of three species of subtropical marine fishes. Mar Biol 26:271–285

Hughes I, Thalmann I, Thalmann R, Ornitz DM (2006) Mixing model systems:using zebrafish and mouse inner ear mutants and other organ systems to unravel the mystery of otoconial development. Brain Res 1091:58–74

Humphrey C, Klumpp DW, Pearson RG (2003) Early development and growth of the eastern rainbowfish, *Melanotaenia splendida splendida* (Peters) II. Otolith development, increment validation and larval growth. Mar Freshwater Res 54:105–111

Ichii T, Mugiya Y (1983) Comparative aspects of calcium dynamics in calcified tissues in the goldfish *Carassius auratus*. Bull Jap Soc Sci Fish 49:1039–1044

Jenkins GP (1987) Age and growth of co-occurring larvae of two flounder species, *Rhombosolea tapirina* and *Ammotretis rostratus*. Mar Biol 95:157–166

Jenkins GP, Davis TLO (1990) Age, growth rate, and growth trajectory determined from otolith microstructure of southern bluefin tuna *Thunnus maccoyii* larvae. Mar Ecol Prog Ser 63:93–104

Kalish JM, Beamish RJ, Brothers EB, Casselman JM, Francis C, Mosegaard H, Panfili J, Prince ED, Thresher RE, Wilson CA, Wright PJ (1995) Glossary for otolith studies. In: Secor DH, Dean JM, Campana SE (Eds) Recent developments in fish otolith research, Vol 19. University of South Carolina Press, Columbia, SC, USA, pp 723–729

Kimura S (1995) Growth of the clupeid fishes, *Stolothrissa tanganicae* and *Limnothrissa miodon*, in the Zambian waters of Lake Tanganyika. J Fish Biol 47:569–575

Klink A, Eckmann R (1992) Limits for the detection of daily growth increments in whitefish (*Coregonus lavaretus* L.) larvae. Hydrobiologia 231:99–105

Kotake A, Arai T, Ozawa T, Nojima S, Miller MJ, Tsukamoto K (2003) Variation in migratory history of Japanese eels, *Anguilla japonica*, collected in coastal waters of the Amakusa Islands, Japan, inferred from otolith Sr/Ca ratios. Mar Biol 142:849–854

Lagardère F, Chaumillon G, Amara R, Heineman G, Lago JM (1995) Examination of otolith morphology and microstructure using laser scanning microscopy. In: Secor DH, Dean JM, Campana SE (Eds) Recent developments in fish otolith research. University of South Carolina Press, Columbia, SC, USA, pp 7–26

Leips J, Baril CT, Rodd FH, Reznick DN, Bashey F, Visser GJ, Travis J (2001) The suitability of calcein to mark poeciliid fish and a new method of detection. Trans Am Fish Soc 130:501–507

Leis JM (1991) The pelagic stage of reef fishes: the larval biology of coral reef fishes. In: Sale PF (Ed) The ecology of fishes on coral reefs. Academic Press, San Diego, pp 183–230

Linkowski TB (1991) Otolith microstructure and growth patterns during the early life history of lanternfishes (family Myctophidae). Can J Zool 69:1777–1792

Linkowski TB (1996) Lunar rhythms of vertical migrations coded in otolith microstructure of North Atlantic lanternfishes, genus *Hygophum* (Myctophidae). Mar Biol 124:495–508

Lombarte A, Morales-Nin B (1995) Morphology and ultrastructure of saccular otoliths from 5 species of the genus *Coelorinchus* (Gadiformes, Macrouridae) from the Southeast Atlantic. J Morphol 225:179–192

Lou DC (1993) Growth in juvenile *Scarus rivulatus* and *Ctenochaetus binotatus*: a comparison of families Scaridae and Acanthuridae. J Fish Biol 42:15–23

Marui M, Arai T, Miller MJ, Jellyman DJ, Tsukamoto K (2001) Comparison of early life history between New Zealand temperate eels and Pacific tropical eels revealed by otolith microstructure and microchemistry. Mar Ecol Prog Ser 213:273–284

Massou AM, Le Bail PY, Panfili J, Lae R, Baroiller JF, Mikolasek O, Fontenelle G, Auperin B (2004a) Effects of confinement stress of variable duration on the growth and microincrement deposition in the otoliths of *Oreochromis niloticus* (Cichlidae). J Fish Biol 65:1253–1269

Massou AM, Panfili J, Lae R, Baroiller JF, Mikolasek O, Fontenelle G, Le Bail PY (2002) Effects of different food restrictions on somatic and otolith growth in Nile tilapia reared under controlled conditions. J Fish Biol 60:1093–1104

Massou AM, Panfili J, Le Bail PY, Laë R, Mikolasek O, Fontenelle G, Baroiller JF (2004b) Evidence of perturbations induced by reproduction on somatic growth and microincrement deposition in *Oreochromis niloticus* otoliths. J Fish Biol 64:1–19

Masterson CF, Danilowicz BS, Sale PF (1997) Yearly and inter-island variation in the recruitment dynamics of the bluehead wrasse (*Thalassoma bifasciatum*, Bloch). J Exp Mar Biol Ecol 214:149–166

Mayer-Gostan N, Kossmann H, Watrin A, Payan P, Boeuf G (1997) Distribution of ionocytes in the saccular epithelium of the inner ear of two teleosts (*Oncorhynchus mykiss* and *Scophthalmus maximus*). Cell Tiss Res 289:53–61

McCormick M, Makey L, Dufour V (2002) Comparative study of metamorphosis in tropical reef fishes. Mar Biol 141:841–853

McCormick MI (1999) Delayed metamorphosis of a tropical reef fish (*Acanthurus triostegus*): a field experiment. Mar Ecol Prog Ser 176:25–38

McCormick MI, Molony BW (1992) Effects of feeding history on the growth characteristics of a reef fish at settlement. Mar Biol 114:165–173

McCurdy WJ, Panfili J, Meunier FJ, Geffen AJ, de Pontual H (2002) Preparation of calcified structures. In: Panfili J, de Pontual H, Troadec H, Wright PJ (Eds) Manual of fish sclerochronology. Ifremer-Ird coedition, Brest, France, pp 331–357

McIlwain JL (2002) Link between reproductive output and larval supply of a common damselfish species, with evidence of replenishment from outside the local population. Mar Ecol Prog Ser 236:219–232

Meifsjord J, Midtøy F, Folkvord A (2006) Validation of daily increment deposition in otoliths of juvenile *Limnothrissa miodon* (Clupeidae). J Fish Biol 69:1845–1848

Meunier FJ (1982) Etude expérimentale de l'excrétion de la tétracycline chez la carpe, *Cyprinus carpio* L. (Cyprinidae, Téléostéen). Résultats préliminaires. Cybium 9:53–64

Milton DA, Blaber SJM, Rawlinson NJF (1993) Age and growth of three species of clupeids from Kiribati, tropical central south Pacific. J Fish Biol 43:89–108

Milton DA, Chenery SR (2001) Sources and uptake of trace metals in otoliths of juvenile barramundi (*Lates calcarifer*). J Exp Mar Biol Ecol 264:47–65

Molony BW (1996) Episodes of starvation are recorded in the otoliths of juvenile *Ambassis vachelli* (Chandidae), a tropical estuarine fish. Mar Biol 125:439–446

Molony BW, Choat JH (1990) Otolith increment widths and somatic growth-rate – the presence of a time-lag. J Fish Biol 37:541–551

Molony BW, Sheaves MJ (1998) Otolith increment widths and lipid contents during starvation and recovery feeding in adult *Ambassis vachelli* (Richardson). J Exp Mar Biol Ecol 221:257–276

Morales-Nin B (1986a) Chemical composition of the otoliths of the sea bass (*Dicentrarchus labrax* Linnaeus, 1758) (Pisces, Serranidae). Cybium 10:115–120

Morales-Nin B (1986b) Microestructura de los otolitos de *Calamus brachysomus* Lockington, 1880 (Pisces: Sparidae) al microscopio electrónico de barrido. Invest Pesq 50:479–487

Morales-Nin B (1986c) Structure and composition of *Merluccius capensis* otoliths. S Afr J Mar Sci 4:3–10

Morales-Nin B (1987) Ultrastructure of the organic and inorganic constituents of the otoliths of the sea bass. In: Summerfelt RC, Hall GE (Eds) The age and growth of fish. The Iowa State University Press, Ames, Iowa, USA, pp 331–343

Morales-Nin B (1988) Caution in the use of daily increments for ageing tropical fishes. Fishbyte 6:5–6

Morales-Nin B, Di Stefano M, Potoschi A, Massuti E, Rizzo P, Gancitano S (1999) Differences between the sagitta, lapillus and vertebra in estimating age and growth in juvenile Mediterranean dolphinfish (*Coryphaena hippurus*). Sci Mar 63:327–336

Morales-Nin B, Panfili J (2002) Age estimation. In: Panfili J, de Pontual H, Troadec H, Wright PJ (Eds) Manual of fish sclerochronology. Ifremer-Ird coedition, Brest, France, pp 91–98

Morales-Nin B, Panfili J (2005) Seasonality in the deep sea and tropics revisited: what can otoliths tell us? Mar Freshwater Res 5:585–598

Morales-Nin B, Ralston S (1990) Age and growth of *Lutjanus kasmira* (Forskal) in Hawaiian waters. J Fish Biol 36:191–203

Morioka S (2002) Otolith features and growth of juvenile *Opsaridium microcephalum* (Pisces: Cyprinidae) from the southwestern shoreline of Lake Malawi. Afr Zool 37:165–170

Morioka S, Matsumoto S (2003) Otolith features and utility of lapillus for daily increment analysis in *Opsaridium microcephalum* (Cyprinidae) juveniles collected from Lake Malawi. Ichthyol Res 50:82–85

Mugiya Y (1974) Calcium-45 behavior at the level of the otolithic organs of rainbow trout. Bull Jap Soc Sci Fish 40:457–463

Mugiya Y (1977) Studies on fish scale formation and resorption-II. Effect of estradiol on calcium homeostasis and skeletal tissue resorption in the goldfish, *Carassius auratus*, and the killifish, *Fundulus heteroclitus*. Comp Biochem Physiol A 57:197–202

Mugiya Y (1978) Effects of estradiol-17β on bone and otolith calcification in the goldfish, *Carassius auratus*. Bull Jap Soc Sci Fish 44:1217–1221

Mugiya Y (1984) Diurnal rhythm in otolith formation in the rainbow trout, *Salmo gairdneri*: seasonal reversal of the rhythm in relation to plasma calcium concentrations. Comp Biochem Physiol A 78:289–293

Mugiya Y (1987a) Effects of photoperiods on the formation of otolith increments in the embryonic and larval rainbow trout *Salmo gairdneri*. Nipp Suis Gakk 53:1979–1984

Mugiya Y (1987b) Phase difference between calcification and organic matrix formation in the diurnal growth of otoliths in the rainbow trout, *Salmo gairdneri*. Fish Bull 85:395–401

Mugiya Y, Uchimura T (1989) Otolith resorption induced by anaerobic stress in the goldfish, *Carassius auratus*. J Fish Biol 35:813–818

Mugiya Y, Watabe N, Yamada J, Dean JM, Dunkelberger DG, Shimizu M (1981) Diurnal rhythm in otolith formation in the goldfish, *Carassius auratus*. Comp Biochem Physiol A 68:659–662

Murayama E, Takagi Y, Ohira T, Davis JG, Greene MI, Nagasawa H (2002) Fish otolith contains a unique structural protein, otolin-1. Eur J Biochem 269:688–696

Nakano K, Takemura A, Nakamura S, Nakano Y, Iwama GK (2004) Changes in the cellular and organismal stress responses of the subtropical fish, the Indo-Pacific sergeant, *Abudefduf vaigiensis*, due to the 1997–1998 El Nino Southern Oscillation. Environ Biol Fish 70:321–329

Neilson JD, Geen GH (1985) Effects of feeding regimes and diel temperature cycles on otolith increment formation in juvenile chinook salmon, *Onchorhynchus tschawytscha*. Fish Bull 83:91–101

Nemeth RS (2005) Linking larval history to juvenile demography in the bicolor damselfish *Stegastes partitus* (Perciformes: Pomacentridae). Rev Biol Trop 53(Suppl. 1):155–163

Panfili J, Meunier FJ, Mosegaard H, Troadec H, Wright PJ, Geffen AJ (2002) Glossary. In: Panfili J, de Pontual H, Troadec H, Wright PJ (Eds) Manual of fish sclerochronology. Ifremer-Ird coedition, Brest, France, pp 373–383

Panfili J, Tomás J (2001) Validation of age estimation and back-calculation of fish length based on otolith microstructures in tilapias (Pisces, Cichlidae). Fish Bull 99:139–150

Pannella G (1974) Otoliths growth patterns: an aid in age determination in temperate and tropical fishes. In: Bagenal TB (Ed) The ageing of fish. Unwin Brother's Ltd., London, UK, pp 28–39

Pannella G (1980) Growth pattern in fish sagittae. In: Rhoads DC, Lutz RA (Eds) Skeletal growth of aquatic organisms Biological records of environmental change. Plenum Press, New York, USA and London, UK, pp 519–560

Payan P, Edeyer A, de Pontual H, Borelli G, Boeuf G, Mayer-Gostan N (1999) Chemical composition of saccular endolymph and otolith in fish inner ear: lack of spatial uniformity. Am J Physiol 277:R123–R131

Persson P (1997) Calcium regulation during sexual maturation of female Salmonids: estradiol 17ß and calcified tissues. PhD Thesis, Göteborg University

Ponton D, Mol JH, Panfili J (2001) Use of otolith microincrements for estimating the age and growth rate of young armoured catfish *Hoplosternum littorale*. J Fish Biol 58:1274–1285

Radtke RL, Dean JM (1982) Increment formation in the otoliths of embryos, larvae, and juveniles of the mummichog, *Fundulus heteroclitus*. Fish Bull 80:201–215

Radtke RL, Kinzie RA, Shafer DJ (2001) Temporal and spatial variation in length of larval life and size at settlement of the Hawaiian amphidromous goby *Lentipes concolor*. J Fish Biol 59:928–938

Rahman MJ, Cox IG (2006) Lunar periodicity in growth increment formation in otoliths of hilsa shad (*Tenualosa ilisha*, Clupeidae) in Bangladesh waters. Fish Res 81:342–344

Ralston S, Brothers EB, Roberts DA, Sakuma KM (1996) Accuracy of age estimates for larval *Sebastes jordani*. Fish Bull 94:89–97

Ralston S, Miyamoto GT (1983) Analyzing the width of daily otolith increments to age the Hawaiian snapper, *Pristipomoides filamentosus*. Fish Bull 81:523–535

Raventos N, Macpherson E (2001) Planktonic larval duration and settlement marks on the otoliths of Mediterranean littoral fishes. Mar Biol 138:1115–1120

Ré P (1984) Evidence of daily and hourly growth in pilchard larvae based on otolith growth increments, *Sardina pilchardus* (Walbaum, 1792). Cybium 8:33–38

Reichert MJM, Dean JM, Feller RJ, Grego JM (2000) Somatic growth and otolith growth in juveniles of a small subtropical flatfish, the fringed flounder, *Etropus crossotus*. J Exp Mar Biol Ecol 254:169–188

Riley BB, Zhu C, Janetopoulos C, Aufderheide KJ (1997) A critical period of ear development controlled by distinct populations of ciliated cells in zebrafish. Dev Biol 191:191–201

Risk A (1997) Effects of habitat on the settlement and post settlement success of the ocean surgeonfish *Acanthurus bahianus*. Mar Ecol Prog Ser 161:51–59

Robertson DR, Swearer SE, Kaufmann K, Brothers EB (1999) Settlement vs. environmental dynamics in a pelagic-spawning reef fish at Caribbean Panama. Ecol Monogr 69:195–218

Roff DA (1983) An allocation model of growth and reproduction in fish. Can J Fish Aquat Sci 40:1395–1404

Romanek CS, Gauldie RW (1996) A predictive model of otolith growth in fish based on the chemistry of the endolymph. Comp Biochem Physiol A 114:71–79

Rooker JR, Holt SA, Holt GJ, Fuiman LA (1999) Spatial and temporal variability in growth, mortality, and recruitment potential of postsettlement red drum, *Sciaenops ocellatus*, in a subtropical estuary. Fish Bull 97:581–590

Rooker JR, Landry AM, Geary BW, Harper JA (2004) Assessment of a shell bank and associated substrates as nursery habitat of postsettlement red snapper. Estuar Coast Shelf Sci 59:653–661

Rosenthal HL (1957) Uptake of calcium-45 and strontium-90 from water by freshwater fishes. Science 126:699–700

Rosenthal HL (1960) Accumulation of strontium-90 and calcium-45 by fresh water fishes. Proc Soc Exp Biol Med 104:88–91

Sasagawa T, Mugiya Y (1996) Biochemical properties of water-soluble otolith proteins and the immunobiochemical detection of the proteins in serum and various tissues in the tilapia *Oreochromis niloticus*. Fish Sci 62:970–976

Schwamborn SHL, Ferreira BP (2002) Age structure and growth of the dusky damselfish, *Stegastes fuscus*, from Tamandare reefs, Pernambuco, Brazil. Environ Biol Fish 63:79–88

Searcy SP, Sponaugle S (2000) Variable larval growth in a coral reef fish. Mar Ecol Prog Ser 206:213–226

Searcy SP, Sponaugle S (2001) Selective mortality during the larval-juvenile transition in two coral reef fishes. Ecology 82:2452–2470

Secor DH, White MG, Dean JM (1991) Immersion marking of larval and juvenile hatchery-produced striped bass with oxytetracycline. Trans Am Fish Soc 120:261–266

Shiao JC, Tzeng CS, Leu CL, Chen FC (1999) Enhancing the contrast and visibility of daily growth increments in fish otoliths etched by proteinase K buffer. J Fish Biol 54:302–309

Shinobu N, Mugiya Y (1995) Effects of ovine prolactin, bovine growth-hormone and triiodothyronine on the calcification of otoliths and scales in the hypophysectomized goldfish *Carassius auratus*. Fish Sci 61:960–963

Song Z, Fu Z, Li J, Yue B (2008a) Validation of daily otolith increments in larval and juvenile Chinese sucker *Myxocyprinus asiaticus*. Environ Biol Fish 82:165–171

Song Z, He C, Fu Z, Shen D (2008b) Otolith thermal marking in larval Chinese sucker, *Myxocyprinus asiaticus*. Environ Biol Fish 82:1–7

Speare P (1992) A technique for tetracycline injecting and tagging billfish. Bull Mar Sci 51:197–203

Spencer K, Shafer DJ, Gauldie RW, DeCarlo EH (2000) Stable lead isotope ratios from distinct anthropogenic sources in fish otoliths: a potential nursery ground stock marker. Comp Biochem Physiol A 127:273–284

Sponaugle S, Grorud-Colvert K, Pinkard D (2006) Temperature-mediated variation in early life history traits and recruitment success of the coral reef fish *Thalassoma bifasciatum* in the Florida Keys. Mar Ecol Prog Ser 308:1–15

Sponaugle S, Pinkard D (2004) Lunar cyclic population replenishment of a coral reef fish: shifting patterns following oceanic events. Mar Ecol Prog Ser 267:267–280

Stequert B, Panfili J, Dean JM (1996) Age and growth of yellowfin tuna, *Thunnus albacares*, from the western Indian Ocean, based on otolith microstructure. Fish Bull 94:124–134

Strelcheck AJ, Fitzhugh GR, Coleman FC, Koenig CC (2003) Otolith-fish size relationship in juvenile gag (*Mycteroperca microlepis*) of the eastern Gulf of Mexico: a comparison of growth rates between laboratory and field populations. Fish Res 60:255–265

Sugeha HY, Shinoda A, Marui M, Arai T, Tsukamoto K (2001) Validation of otolith daily increments in the tropical eel *Anguilla marmorata*. Mar Ecol Prog Ser 220:291–294

Suthers IM (1998) Bigger? Fatter? Or is faster growth better? Considerations on condition in larval and juvenile coral-reef fish. Aus J Ecol 23:265–273

Tanabe T, Kayama S, Ogura M, Tanaka S (2003) Daily increment formation in otoliths of juvenile skipjack tuna *Katsuwonus pelamis*. Fish Sci 69:731–737

Tanaka K, Mugiya Y, Yamada J (1981) Effects of photoperiod and feeding on daily growth patterns in otoliths of juvenile *Tilapia nilotica*. Fish Bull 79:459–466

Taubert BD, Coble DW (1977) Daily rings in otoliths of three species of *Lepomis* and *Tilapia mossambica*. J Fish Res Board Can 34:332–340

Thompson BA, Beasley M, Wilson CA (1999) Age distribution and growth of greater amberjack, *Seriola dumerili*, from the north-central Gulf of Mexico. Fish Bull 97:362–371

Thorrold SR (1989) Estimating some early life history parameters in a tropical clupeid, *Heklotsichthys castelnaui*, from daily growth increments in otoliths. Fish Bull 87:73–83

Thorrold SR, Jones GP, Planes S, Hare JA (2006) Transgenerational marking of embryonic otoliths in marine fishes using barium stable isotopes. Can J Fish Aquat Sci 63:1193–1197

Thorrold SR, Milicich MJ (1990) Comparison of larval duration and pre- and post-settlement growth in two species of damselfish, *Chromis atripectoralis* and *Pomacentrus coelestis* (Pisces: Pomacentridae), from the Great Barrier Reef. Mar Biol 105:375–384

Tomás J (2006) The appearance of accessory growth centres in adult whiting *Merlangius merlangius* otoliths. J Fish Biol 69:601–607

Tomás J, Geffen AJ (2003) Morphometry and composition of aragonite and vaterite otoliths of deformed laboratory reared juvenile herring from two populations. J Fish Biol 63:1–19

Tomás J, Panfili J (2000) Otolith microstructure examination and growth patterns of *Vinciguerria nimbaria* (Photichthyidae) in the tropical Atlantic Ocean. Fish Res 46:131–145

Tsuji S, Aoyama T (1982) Daily growth increments observed in otoliths of the larvae of Japanese red sea bream *Pagrus major* (Temminck et Schlegel). Bull Jap Soc Sci Fish 48:1559–1562

Tsukamoto K (1988) Otolith tagging of ayu with fluorescent substances. Nipp Suis Gakk 54:1289–1295

Tytler P, Fox CJ, Folkvord A (2002) Glycoconjugates in the otolithic membrane of herring larvae: a possible framework for encoding the life history recorder in fishes. J Fish Biol 61:39–49

Tzeng WN, Yu SY (1992) Effects of starvation on the formation of daily growth increments in the otoliths of the milkfish, *Chanos chanos* (Forsskål), larvae. J Fish Biol 40:39–48

Victor BC (1982) Daily otolith increments and recruitment in two coral-reef wrasses, *Thalassoma bifasciatum* and *Halichoeres bivittatus*. Mar Biol 71:203–208

Vigliola L, Meekan MG (2002) Size at hatching and planktonic growth determine post-settlement survivorship of a coral reef fish. Oecologia 131:89–93

Vilizzi L (1998) Age, growth and cohort composition of 0+ carp in the River Murray, Australia. J Fish Biol 52:997–1013

Volk EC, Schroder SL, Grimm JJ (1994) Use of a bar code symbology to produce multiple thermally-induced otolith marks. Trans Am Fish Soc 123:811–816

Volk EC, Schroder SL, Grimm JJ (1999) Otolith thermal marking. Fish Res 43:205–219

Walker SPW, McCormick MI (2004) Otolith-check formation and accelerated growth associated with sex change in an annual protogynous tropical fish. Mar Ecol Prog Ser 266:201–212

Watabe N, Tanaka K, Yamada J, Dean JM (1982) Scanning electron microscope observations of the organic matrix in the otolith of the teleost fish *Fundulus heteroclitus* (Linnaeus) and *Tilapia nilotica* (Linnaeus). J Exp Mar Biol Ecol 58:127–134

Wellington GM, Victor BC (1989) Planktonic larval duration of one hundred species of Pacific and Atlantic damselfishes (Pomacentridae). Mar Biol 101:557–567

Wellington GM, Victor BC (1992) Regional differences in duration of the planktonic larval stage of reef fishes in the eastern Pacific Ocean. Mar Biol 113:491–498

Wild A, Foreman TJ (1980) The relationship between otolith increments and time for yellowfin and skipjack tuna marked with tetracycline. IATTC Bull 17:509–560

Williams T, Bedford BC (1974) The use of otoliths for age determination. In: Bagenal TB (Ed) The ageing of fish. Unwin Brothers Ltd., London, UK, pp 114–123

Wilson CA, Beckman DW, Dean JM (1987) Calcein as a fluorescent marker of otoliths of larval and juvenile fish. Trans Am Fish Soc 116:668–670

Wilson DT, McCormick MI (1997) Spatial and temporal validation of settlement-marks in the otoliths of tropical reef fishes. Mar Ecol Prog Ser 153:259–271

Wilson DT, McCormick MI (1999) Microstructure of settlement-marks in the otoliths of tropical reef fishes. Mar Biol 134:29–41

Wright PJ, Panfili J, Folkvord A, Mosegaard H, Meunier FJ (2002a) Direct validation. In: Panfili J, de Pontual H, Troadec H, Wright PJ (Eds) Manual of fish sclerochronology. Ifremer-Ird coedition, Brest, France, pp 114–128

Wright PJ, Panfili J, Morales-Nin B, Geffen AJ (2002b) Otoliths. In: Panfili J, de Pontual H, Troadec H, Wright PJ (Eds) Manual of fish sclerochronology. Ifremer-Ird coedition, Brest, France, pp 31–57

Wright PJ, Rowe D, Thorpe JE (1991) Daily growth increments in the otoliths of Atlantic salmon parr, *Salmo salar* L. and the influence of environmental factors on their periodicity. J Fish Biol 39:103–113

Wright PJ, Talbot C, Thorpe JE (1992) Otolith calcification in Atlantic salmon parr, *Salmo salar* L and its relation to photoperiod and calcium metabolism. J Fish Biol 40:779–790

Zapata FA, Herron PA (2002) Pelagic larval duration and geographic distribution of tropical eastern Pacific snappers (Pisces: Lutjanidae). Mar Ecol Prog Ser 230:295–300

Zhang Z (1992) Ultrastructure of otolith increments and checks in the teleost fish *Oreochromis niloticus*. J Morphol 211:213–220

Zhang Z, Runham NW (1989) Initial development of *Oreochromis niloticus* (Teleostei: Cichlidae) otolith. J Zool (London) 227:465–478

Zhang Z, Runham NW (1992a) Effects of food ration and temperature level on the growth of *Oreochromis niloticus* (L) and their otoliths. J Fish Biol 40:341–349

Zhang Z, Runham NW (1992b) Otolith microstructure pattern in *Oreochromis niloticus* (L.). J Fish Biol 40:325–332

Zhang Z, Runham NW (1992c) Temporal deposition of incremental and discontinuous zones in the otoliths of *Oreochromis niloticus* (L). J Fish Biol 40:333–339

8. OTOLITH CHEMISTRY

SIMON R. THORROLD[1] AND STEPHEN E. SWEARER[2]

[1]*Biology Department MS 50, Woods Hole Oceanographic Institution, Woods Hole MA 02543, USA*

[2]*Department of Zoology, The University of Melbourne, VIC 3010, Australia*

1 Introduction

1.1 GENERAL INTRODUCTION

Age determination of fishes based on periodic growth increments in otoliths has become a routine tool in fisheries science over the last century. Campana and Thorrold (2001) calculated that the ages of over 1 million fish were likely estimated in 2000 by fisheries scientists around the world. We probably have more demographic information on fishes than any other group of organisms on earth, with the exception of humans. The chronological records provided by otoliths are indeed unique. Growth increments in otoliths have now been validated to form on an annual basis in numerous species by many studies (Choat & Robertson 2002). Reef fish biologists were relatively slow to use annual increments in otoliths, perhaps because of the perceived lack of seasonality in tropical environments (Fowler 1995). In contrast, the potential of daily increments in otoliths (Pannella 1971) was recognized almost immediately upon their discovery (Brothers & McFarland 1981, Victor 1983). Otolith microstructure has now been used to determine pelagic larval durations, reconstruct settlement patterns and to investigate the effect of growth rate on larval survival and subsequent settlement (e.g., Bergenius et al. 2002, Searcy & Sponaugle 2001).

Information provided by annual and daily increments clearly has had a significant impact on reef fish ecology (reviewed by Choat & Robertson 2002, Thorrold & Hare 2002). The chemical composition of otoliths is currently poised to make a similar and significant contribution to the field. The use of chemical signatures in otoliths took its inspiration, in large part, from geochemical studies of coral aragonite. Smith et al. (1979) conducted the first study using modern mass spectrometry techniques to derive high-precision analyses of Sr/Ca ratios in coral skeletons. They found a significant correlation between water temperature and Sr/Ca ratios that has subsequently been used to reconstruct sea surface temperatures throughout the Holocene for a number of locations around the world (e.g., Guilderson et al. 1994, McCulloch et al. 1994). Otoliths are also composed primarily of aragonite and most marine fishes are poikilothermic, so Sr/Ca and perhaps other elemental ratios in otoliths may record the physico-chemical properties of water masses inhabited by a fish at any point during its life. Radtke (1985) in a pioneering study combined analysis of daily growth increments with Sr/Ca profiles across the otoliths of recently settled *Dascyllus albisella* with the explicit intention of reconstructing environmental histories during

B.S. Green et al. (eds.), *Tropical Fish Otoliths: Information for Assessment, Management and Ecology*, Reviews: Methods and Technologies in Fish Biology and Fisheries 11,
DOI 10.1007/978-1-4020-5775-5_8,

larval life. Radtke's results were inconclusive but the promise of otolith chemistry to shed light on the larval "black box" was clear.

The potential of otolith chemistry to provide information on larval life history and dispersal pathways was recognised as early as 1985 but only a handful of studies were conducted over the next 15 years. Interest has been revived more recently, however, in the use of geochemical signatures in otoliths of tropical reef fishes. This resurgence has been generated by the need to quantify connectivity among populations and catalysed by developments in analytical instrumentation. Early studies on otolith chemistry used electron probe microanalysis (EPMA) almost exclusively. The instrument was widely available in universities and excellent spatial resolution allowed for individual analyses approaching the level of daily increments in otoliths (Gunn et al. 1992). Unfortunately, the relatively high detection limits of EPMA prevented quantification of a number of elements, including Mg, Mn, Ba and Pb, that proved to be useful water mass tracers in studies of coral geochemistry (Shen et al. 1992). Development of inductively coupled plasma mass spectrometry (ICP-MS) provided an alternative to EPMA with superior detection limits and precision. Plasma source mass spectrometry was introduced in 1980 and has developed since into a mature analytical technique routinely used, for instance, in the certification of standard reference materials (Fassett & Paulsen 1989). ICP-MS was originally conceived for solution-based analysis but laser ablation devices were quickly developed that could be coupled with ICP-MS instruments to provide in situ analyses of solids (including otoliths) at spatial scales approaching that of EPMA. The ability to analyse multiple elements in otoliths at concentrations down to ultra-trace levels has led to new insights into larval dispersal pathways in both tropical and temperate oceans (e.g. Swearer et al. 1999, FitzGerald et al. 2004).

Here, we critically evaluate the use of otolith chemistry in the study of tropical fish ecology. We concentrate on our own experience with plasma source mass spectrometry rather than attempting to provide the reader with a comprehensive review of techniques and applications. We hope to give interested readers enough information to allow them to determine if a particular ecological question lends itself to an otolith chemistry approach. Sample preparation is covered in some depth because it is fundamental to the collection of accurate and precise elemental and isotopic data from otoliths. We then outline issues that a user needs to know in order to accurately interpret ICP-MS data. Operation of ICP-MS instruments will usually be in the capable hands of trained technical staff but the operators will not necessarily have experience with geochemical analyses of biogenic carbonates. We therefore cover topics that include, for instance, the selection of appropriate isotopes and standards. Finally, we provide practical information on the investments in terms of time and money that are required to perform otolith chemistry analyses, along with some recipes that may be useful for researchers new to the field.

1.2 APPLICATIONS

Otoliths are formed during embryogenesis, grow continuously throughout a fish's life (Campana & Neilson 1985), incorporate both minor and trace elements during growth (Campana 1999) and are metabolically inert except under extreme hypoxia (Mugiya &

Uchimura 1989). Geochemical signatures encoded within otoliths, therefore, are natural tags that are continuously recording information throughout the lifetime of a fish. Even with little understanding of the mechanisms governing elemental incorporation, variation in chemistry within otoliths and among individuals and populations can be used to infer differences in a range of ecological characteristics of a species such as migration history, habitat use and spatial distributions. Box 1 provides a summary of the main mechanisms understood to influence elemental composition of otoliths and Box 2 outlines some remaining unresolved issues influencing interpretation of otolith elemental signatures.

Box 1. What determines the elemental composition of tropical fish otoliths?

We have limited understanding of the fundamental mechanisms determining otolith composition and how these mechanisms are influenced by the environment. Nonetheless, the general concepts emerging from studies of temperate species are likely to be relevant to studies in the tropics.

Ambient environment

Physico-chemical properties of ambient water masses are likely to play a major role in the uptake of elements into otoliths. Most studies have investigated the effects of temperature and salinity on otolith Sr/Ca ratios and the evidence points to a dominant effect of ambient element concentration on otolith chemistry, with secondary effects of temperature and salinity (e.g., Bath et al. 2000, Martin et al. 2004). Other studies have revealed more complexity, with interactions amongst elements either facilitating (de Vries et al. 2005) or inhibiting (Okumura & Kitano 1986) deposition. Remarkably, we know virtually nothing about environmental effects on any of the remaining minor and trace elements found in otoliths.

Bioavailability, uptake, and transport of ions within the fish

Elements in free ionic form can be taken up directly from water by fish via chloride cells (also known as ionocytes) in the branchial epithelium of the gills or via the intestine by drinking water (Flik & Verbost 1993). Water has been shown to be dominant source of Ca, Sr and Ba ions deposited in otoliths in at least some marine and freshwater species (Farrell & Campana 1996, Walther & Thorrold 2006). Organically complexed ions or elements present in consumed prey may also be available for uptake if they are broken down into chemical species which are readily absorbed by the intestine. Elemental discrimination also occurs during transport from the blood plasma to the endolymph. The concentration of free ions within the blood plasma is likely related to the availability of metal-binding proteins that may, in turn vary with ontogeny and maturation stage (Kalish 1991).

Processes acting at the crystal surface

Elements can become included in otoliths in several ways upon reaching the otolith growing surface. They can become incorporated directly into the calcium carbonate crystal lattice via substitution for calcium, adsorbed to the crystal surface, trapped in fluid or solid inclusions, or bound to the protein, carbohydrate, or lipid fraction of the organic matrix. The method of incorporation is important because water chemistry is only likely to be related to otolith chemistry for those elements that substitute for Ca in the aragonite lattice. Partitioning of those elements substituting for Ca at the otolith surface may also be influenced by crystal extension rates, although evidence is limited. More generally, models for Ca substitution in inorganic and biogenic aragonite suggest that deposition of these elements is not in thermodynamic equilibrium and there is some evidence for kinetic effects due to surface entrapment phenomena (Gaetani & Cohen 2006). However, the absence of temperature effects on Ba/Ca ratios in otoliths suggests that even non-equilibrium physico-chemical models are unlikely to offer a complete explanation of the Ca substitution processes.

Box 2. Key unresolved issues in otolith chemistry applications

Interpreting otolith chemical signatures undoubtedly remains a significant challenge. Here are some of the key unresolved issues that we believe are hindering our ability to interpret chemical signatures recorded in otoliths.

Diet

Diet may have a variety of direct and indirect confounding effects on otolith chemistry. Direct absorption of ions from food in the intestine may be problematic if the elemental composition of prey is often not similar to concentrations in the surrounding water. This may be particularly apparent for bioaccumulating elements where the trophic transfer of the element results in increasing tissue concentrations with increasing trophic level (e.g., Gray 2002, Long & Wang 2005). Diet can also have indirect effects, by influencing either the transport of metal ions in the plasma or the binding of metals to the organic matrix.

Influence of ontogeny

Some elements show consistent changes in concentration in an otolith with age. There are three likely candidate causes of ontogenetic variation in otolith chemistry, although the exact processes remain unknown. Changes in physiology with development may influence the uptake or transport of ions. A change in otolith precipitation rate as a function of age may increase or decrease the partition coefficient of a particular element. Finally, changes in the aragonite to protein ratio could either increase or decrease the concentration of an element depending on whether it is bound to the crystal lattice or the organic matrix. Future controlled laboratory rearing experiments that expose fish of different ages to a range of different water chemistries are needed to assess the magnitude of ontogenetic influences compared with environmental effects on otolith chemistry.

Genetic effects

Differences in otolith chemistry among individuals within a species or between species may have a genetic basis, although no study has specifically investigated whether genetic variation among individuals is contributing to variation in otolith chemistry. We analysed the chemistry of the larval otoliths of 16 species of coral reef fish from the families: Holocentridae, Haemulidae, Serranidae, Labridae, Pomacentridae, Acanthuridae and Gobiidae, collected from two widely separated localities, US Virgin Islands and Panama, in the Caribbean Sea. Variation in otolith chemistry was dominated by inter-specific differences (Table 1, Figure 2, *S. Swearer* unpublished data), suggesting that there is indeed a genetic component to the uptake or incorporation of elements from the surrounding environment. We now need to establish if more subtle intra-specific genetic differences are also a source of variation in otolith chemistry.

1.2.1 Stock discrimination

Otolith chemistry has been applied to the question of stock discrimination in marine fishes frequently, with the majority of research conducted on temperate species (e.g., Mulligan et al. 1987, Edmonds et al. 1991, 1992, Campana et al. 1994, 1995, Secor & Zdanowicz 1998, Campana et al. 1999, 2000). A few studies, however, including one of the earliest applications of otolith chemistry, have been done on fish populations in the tropics and subtropics (Western Australia – Edmonds et al. 1989, 1999, Caribbean – Patterson et al. 1999, Bangladesh – Milton & Chenery 2001b, Eastern Australia – Begg et al. 1998, Bergenius et al. 2005). The approach rests on the premise that if fish collected from different localities show significant variation in otolith chemistry, whether due to differences in the environment, physiology, or genetics of individuals, there is likely to be little migration among locations. The conclusion is the same regardless of the mechanism(s) generating variation in otolith chemistry: individuals within a location should be treated as discrete populations or separately managed stocks.

Table 1. Estimates of the percentage of the total variation in the log-transformed otolith elemental concentration explained by regional differences (Panama versus St. Croix) and by inter-specific differences within regions based on a nested analysis of variance. Variance percentages were calculated using restricted maximum likelihood estimation.

Element	Factor	df	SS	F	*p*	% Variance
Mg	Region	1	0.2167	3.382	0.075	12.7
	Species(Region)	22	3.226	8.374	<0.0001	47.9
	Error	242	4.238			39.4
Sr	Region	1	0.001	0.004	0.953	0.0
	Species(Region)	22	2.005	34.054	<0.0001	87.3
	Error	242	0.648			12.7
Ba	Region	1	0.377	6.618	0.012	15.6
	Species(Region)	22	1.672	1.639	0.039	10.5
	Error	242	11.218			73.9
Pb	Region	1	0.122	0.540	0.466	0.0
	Species(Region)	22	9.517	3.968	<0.0001	21.7
	Error	242	26.384			78.3

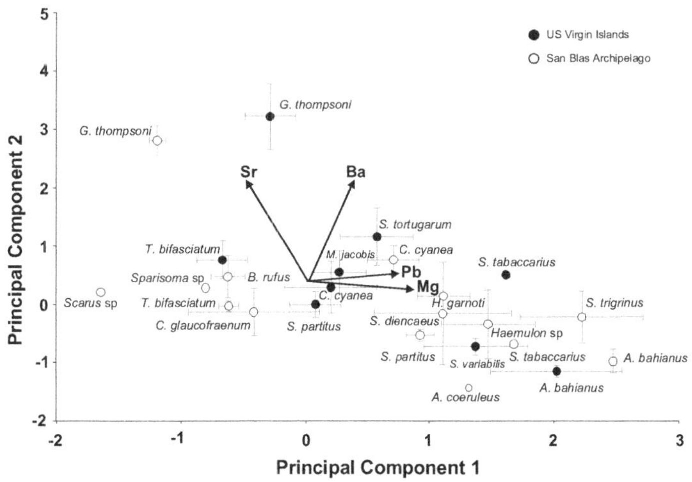

Figure 1. Principal component (PC) analysis of elemental composition of otoliths from newly settled recruits of 16 species of coral reef fish from the U.S. Virgin Islands and the San Blas Archipelago, Panama. Otoliths were embedded in cyanoacrylate glue, ground using diamond lapping film and analysed by laser ablation ICP-MS. Elemental profiles were averaged (±1 SE) over larval growth increments. Principal components 1 and 2 together explained 67.5% of variation in Mg, Sr, Ba and Pb concentrations among samples. Vector length and direction indicate relative weightings of the four elements on each PC.

The majority of studies have used whole-otolith analysis to discriminate among stocks. This approach assumes that the integrated otolith signature is not confounded by either larval dispersal or seasonal adult migration. For instance, stocks may mix during the non-breeding period but migrate to distinct locations to spawn. If the non-breeding period is extensive and the chemical signature associated with that period in the life history is strong, discrete spawning stocks may in fact have very similar otolith chemistries. In contrast, a single stock might contain several discrete foraging areas and these discrete populations may only mix during the spawning season. Consequently, fish collected from different foraging areas will appear to come from distinct stocks when in fact they are part of a single, large spatially fragmented spawning stock. Alternatively, individuals may originate from the same spawning stock but disperse to different locations with distinct chemistries. Subsequent otolith growth will result in differences in whole-otolith chemistry, leading to the false conclusion of apparent stock structure. Thus, direct characterization of the spatial and age-related variation in otolith chemistry using in situ analysis methods (e.g., Campana et al. 1994, Thresher et al. 1994) is necessary to detect stock structure accurately.

The utility of otolith chemistry as a tool for delineating stocks is dependent on the spatial scales over which environmental heterogeneity is sufficient to generate variation in otolith chemistry. Otolith chemistry may be a useful approach if this scale is comparable to or smaller than the scale of movement of fish. Analyses of otolith chemistry, however, will not detect stock structures if scales of variation are considerably larger than the scales over which fish move.

The spatial scales over which fish may show variation in otolith chemistry depends on the system in question. Large, mobile pelagic species that live in the open ocean are likely to experience relatively homogeneous environmental conditions and the scales over which there may be detectable differences are likely to be large (e.g., 1000s of km; Edmonds et al. 1991, Secor & Zdanowicz 1998, Stransky et al. 2005). The spatial scales over which there are detectable differences become smaller as one moves closer to shore, however (e.g., 10s–100s of km, Edmonds et al. 1989, Swearer et al. 1999, Campana et al. 2000, Hamer et al. 2003, Warner et al. 2005) and the influence of land masses on water chemistry (e.g., coastal erosion, rivers, urban and industrial pollution, upwelling, etc.) becomes more apparent. Finally, the scale of spatial resolution can be even finer (e.g., 1–10s of km) for studies of differences in otolith chemistry among estuarine or freshwater fishes (e.g., Mulligan et al. 1987, Edmonds et al. 1992, Gillanders & Kingsford 1996, Thorrold et al. 1998b, Gillanders 2002a, Swearer et al. 2003). Thus, the utility of otolith chemistry to resolve stock structure will depend on both the life history of the species and the system under investigation.

1.2.2 Detection of diadromy

Freshwater and seawater differ dramatically in concentrations of a range elements. Thus, there has been extensive application of otolith chemistry to identify the presence, direction, and frequency of migratory behaviour between rivers and the ocean in both tropical and temperate species (e.g., Radtke et al. 1988, Kalish 1990, Secor 1992, Radtke & Kinzie 1996, Tsukamoto et al. 1998, Secor et al. 2001, Arai et al. 2004, Kraus & Secor 2004b, Elsdon & Gillanders 2005, Woodhead et al. 2005). Most studies have used Sr/Ca ratios as the marker for diadromy, as Sr/Ca ratios are typically an order of magnitude higher in seawater than in freshwater. Kraus and Secor (2004a), however, found evidence that high Sr/Ca ratios can also occur in freshwater systems. Consequently, other tracers such a Ba/Ca ratios (Elsdon & Gillanders 2005) and Sr isotope ratios (Bacon et al. 2004, McCulloch et al. 2005, Woodhead et al. 2005) may prove useful indicators of diadromy involving freshwater systems with anomalously high Sr/Ca ratios. Recent applications of otolith chemistry to investigate diadromy have revealed surprisingly high levels of variation in migratory behaviour (e.g., Tsukamoto et al. 1998, Kraus & Secor 2004b, Elsdon & Gillanders 2005), suggesting that diadromy may be facultative. Future research into the mechanisms that favour migratory versus non-migratory behaviour is needed to evaluate the demographic consequences of such behavioural plasticity.

1.2.3 Detection of philopatry

Philopatry, or the return of adults to the location from which they were born, is not a widely documented life history trait in fishes, although spawning site fidelity in anadromous salmon is a well-known example (Quinn 1993). The use of otolith chemistry to detect philopatry requires that: (1) different natal sites have distinct elemental signatures (e.g., Thorrold et al. 1998a,b, Patterson et al. 2004c); and (2) natal origins of spawning adults can be determined based on geochemical signatures in the otolith cores. Thorrold and co-workers are the only researchers to date to have used otolith chemistry to identify philopatry. Thorrold et al. (2001) collected juvenile weakfish (*Cynoscion regalis*) from five estuaries along the eastern USA to develop ground-truthed geochemical signatures in otoliths of source populations. Spawning adults from this juvenile cohort were subsequently collected from each location 2 years later and the chemistry of the juvenile portion of the otolith characterized to determine the percentage of individuals that had returned to their natal estuary to spawn. Their findings indicated that the majority of adults collected from most estuaries had returned to spawn in their natal estuary.

A variety of fish families with at least some representative species in the tropics migrate to well-defined spawning areas in freshwater (Clupeidae), estuaries (Centropomidae, Mullidae, Sparidae, Sciaenidae) and the open ocean (Anguillidae, Lutjanidae, Serranidae) but we don't know if philopatry is a common feature in the life history of any of these taxa. Many of these families are commercially exploited and such knowledge is likely to be necessary for effective management of those species that are the subject of marine-capture fisheries.

1.2.4 Identification of larval dispersal trajectories and source populations

The great majority of shallow-water fishes have a bipartite life cycle in which demersal adults spawn clutches of benthic or pelagic eggs that hatch into pelagic larvae. The pelagic larval phase can last from weeks to months and the spatial scales over which larvae disperse during this time are not well understood (Warner & Cowen 2002, Swearer et al. 2002). The scales of dispersal for most populations are likely to result in decoupling of local reproduction and recruitment and so local populations are likely to be at least partially demographically open (Roughgarden et al. 1988). Consequently, identifying the source populations of recruiting larvae is integral to understanding the mechanisms that promote persistence of demersal fish populations.

Identifying natal origins is of fundamental importance to understanding the ecology of demersal fishes but application of otolith chemistry to this problem has proved challenging. A variety of approaches has been used to date, listed here in increasing degree of effort and expense. First, whole larval otolith signatures have been compared to identify the presence of discrete larval pools, with inference of their likely origins (e.g., Swearer et al. 1999). Second, otolith cores (i.e., primordia or early larval growth) from recruits have been analysed to determine if there are different sources of larvae for different populations (e.g., Thorrold et al. 1997). Third, post-settlement otolith signatures in recruits or older life stages have been used to characterize the signatures of potential source populations and then these data have been used to interpret the signatures in the cores of otoliths from recruits (e.g., Patterson et al. 2004a). Finally, source population signatures have been characterized by collecting larvae prior to dispersal and using these data to identify the sources of larvae which recruited from that cohort (e.g., Warner et al. 2005, Ruttenberg & Warner 2006, Barbee & Swearer 2007).

Swearer et al. (1999) used solution-based ICP-MS to evaluate the chemical composition of the larval otoliths from recently settled bluehead wrasse, *Thalassoma bifasciatum*. They used a qualitative model to predict the likely signatures associated with local retention of larvae in nearshore waters versus inter-island dispersal in oceanic waters and estimated that a significant proportion of recruitment to St. Croix, USVI, was a result of local production. Retained larvae were prevalent on reefs adjacent to the island wake region where converging currents may have facilitated larval retention (Harlan et al. 2002). Solution-based approaches are reasonably simple but they integrate variability in chemical composition within the otolith. Dispersal may occur primarily during early development when swimming capabilities are at their weakest (e.g., Fisher et al. 2000). Consequently, even dispersing individuals may spend much of the larval period in nearshore waters. Less equivocal results require spatial resolution of elemental signatures within the otolith (e.g., from the core, early and later increments).

Patterson et al. (2004a) applied this approach in their investigation of the spatial scales of variation in elemental signatures among populations of the damselfish, *Pomacentrus coelestis*, along the Great Barrier Reef, Australia. They characterized the chemistry of both the core region and the early post-settlement region of the otolith using laser ablation ICP-MS. Strong differences in chemistry of the post-settlement region of otoliths only occurred between clusters of reefs separated by large distances. They also observed that the chemistry of the core and the post-settlement region were distinctly different, resulting in difficulties in using post-settlement signatures to classify recruits based on their core chemistry. Recent studies have documented high levels of elemental enrichment in otolith cores (e.g., Brophy et al. 2004, Ruttenberg et al. 2005) and there is a high degree of very fine scale heterogeneity in elemental composition within the core region (Barbee & Swearer 2007). The processes generating unique elemental characteristics of otolith cores are unknown but could be a result of maternal effects or differences in crystal mineralogy, element partition coefficients, or mineral/protein ratios.

Unfortunately, ontogenetic variations in element incorporation indicate that determining source populations of recruiting larvae necessitates characterizing natal signatures in otolith cores in larvae prior to dispersal. This conclusion has been substantiated recently by Warner et al. (2005) who determined that a variety of other proxy measures, including adult otoliths, direct seawater and diffusive gradients in thin films that integrate seawater chemistry over time, did not reflect the same pattern of variation among source populations as larval otoliths. Thus, future studies aimed at identifying sources of recruitment in demersal fish populations may be limited to ovoviviparous species or species with benthic egg development from which larvae can be more readily collected prior to the pelagic phase (e.g., Ruttenberg & Warner 2006, Barbee & Swearer 2007).

Returning to the premise of Radtke (1985), that information on larval dispersal may be recorded in the otoliths of coral reef fishes, we analysed Sr/Ca profiles in otoliths of juvenile gray snapper (*Lutjanus griseus*) using a 213 nm laser and an *Element2* single collector sector field ICP-MS. The laser coupled sufficiently well with the otolith material to generate craters of 8 μm in width, allowing high precision estimates of Sr/Ca at intervals of 1–2 days over a larval life of approximately 24 days (Denit & Sponaugle 2004). We found predictable changes in Sr/Ca profiles in otoliths of juveniles from Core Sound, North Carolina and Biscayne Bay, Florida, along the southeast coast of the United States (Figure 2). Results showed a period of relatively low Sr/Ca values during early larval life followed by dramatic increases in the ratio as individuals approached settlement. Waters were approximately 4°C warmer in Biscayne Bay than Core Sound and Sr/Ca was positively correlated with temperature in juvenile gray snapper, based on Sr/Ca ratios at settlement. Bath and co-workers (Bath et al. 2000, Martin et al. 2004) reported a similar result for juvenile spot (*Leiostomus xanthurus*). There were also Sr/Ca variations within each profile, however, that suggested individuals were not spending their larval lives in a single water mass. It is tempting to think that the otolith profiles were driven by ambient temperature, as envisioned by Radtke, given that these fish were likely to have seen little variation in seawater Sr/Ca during their larval lives. If true, this suggests that larval snappers were spending considerable time at depths of > 50 m where water temperatures were below 24°C, and are not within warmer surface waters of the Gulf Stream and adjacent coastal waters. Clearly, it will be necessary to validate the relation between Sr/Ca and temperature for gray snapper before we are able to reconstruct temperature histories of larval snappers. These data nonetheless suggest that significant information on larval transport pathways does reside in the chemistry of daily increments in fish otoliths.

1.2.5 Identification of juvenile nursery areas

Numerous fish species have ontogenetic migrations among discrete habitats during their life history. Considerable debate has focused, for instance, on the role of juvenile nurseries in the demography of fish populations (Beck et al. 2001). Otolith chemistry has been applied, in particular, to assess the importance of estuarine and bay nursery habitats (but see Guido et al. 2004, Chittaro et al. 2005). Several studies have addressed the initial question of whether different nursery areas have distinct elemental signatures (Gillanders & Kingsford 2000, 2003, Hamer et al. 2003, Swearer et al. 2003, Hanson et al. 2004, Kraus & Secor 2004b). Only a handful of studies to date have gone further and attempted to determine the relative importance of different juvenile nurseries as sources of recruitment into adult populations (Thresher et al. 1994, Gillanders & Kingsford 1996, Forrester & Swearer 2002, Gillanders 2002a, Chittaro et al. 2004).

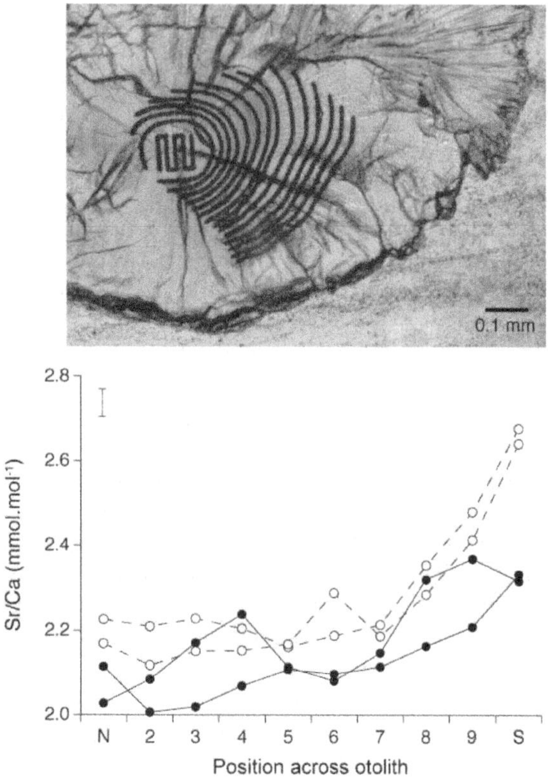

Figure 2. *Top panel*: Sectioned sagittal otolith from a juvenile gray snapper (*Lutjanus griseus*) collected from Biscayne Bay, Florida, showing individual paths along daily increments from which Sr/Ca ratios were quantified using laser ablation ICP-MS. *Bottom panel*: Larval Sr/Ca profiles from the otolith nucleus (N) to the settlement mark (S) of juvenile gray snapper collected from Biscayne Bay and Core Sound, North Carolina. Error bar (± 1 SD, n = 8) from an otolith solution standard run at regular intervals during otolith analyses.

The results of many of these migration studies highlight the advantages of using otolith chemistry to identify juvenile nurseries. Estuaries, in particular, can show tremendous environmental heterogeneity due to variation in mixing dynamics and the physicochemical properties of freshwater inputs. Consequently, many studies have detected fine-scale spatial variation, even within estuaries (e.g., Gillanders & Kingsford 2000), indicating that otolith chemistry can be a powerful tool for identifying critical nurseries. It is surprising, therefore, that few researchers have used otolith chemistry to assess the importance of juvenile nursery habitats, given the importance of sheltered habitats such as mangroves and seagrass beds for many tropical fish species (e.g., Nagelkerken et al. 2000, Halpern 2004, Nagelkerken & van der Velde 2004). Unfortunately, the recent findings of Chittaro et al. (2004, 2005) were not encouraging as they failed to detect strong differences among mangrove and reef habitats, albeit in an oceanic system (Turneffe Atoll, Belize). Studies in other mangrove systems more strongly influenced by freshwater are likely to have greater discriminatory capabilities.

1.2.6 Otoliths as environmental proxies

Unlike most of the applications discussed above, evaluating otoliths as recorders of specific environmental conditions such as water temperature, salinity, or pollution levels requires direct calibration of the relationship between the environmental factor and the elemental concentration in the otolith. One of the most widely researched environmental proxies is the use of Sr/Ca ratios as a measure of ocean temperature (Townsend et al. 1989, Radtke et al. 1990, Hoff & Fuiman 1993, Townsend et al. 1992, Fowler et al. 1995a,b, Hoff & Fuiman 1995, Secor et al. 1995, Townsend et al. 1995, Gallahar & Kingsford 1996, Chesney et al. 1998, Bath et al. 2000, Elsdon & Gillanders 2002, Clarke & Friedland 2004, Elsdon & Gillanders 2004, Martin et al. 2004). The results of these studies indicate that the relationship between water temperature and otolith Sr/Ca is not simple and other factors such as salinity and otolith precipitation rate are clearly influencing factors as well. Consequently, the use of Sr/Ca as a proxy for temperature will depend on the system in question (are, for instance, fish likely to have experienced changes in salinity?) and a better understanding of the mechanisms that cause increases in Sr/Ca ratios with ontogeny (see Campana 1999 for a comprehensive discussion). Only a few studies have investigated the ability of otoliths to record exposure to pollution, either directly in terms of potentially toxic elements such as Cu (Milton et al. 2000), Hg and Pb (Geffen et al. 1998) or indirectly as proxies for exposure along a pollution gradient (Hanson and Zdanowicz 1999). Otolith chemistry was not a good recorder of exposure in any of the studies, with not all elements incorporated in proportion to dissolved concentration in the environment.

1.2.7 Determination of age/timing of metamorphosis

Lastly, otolith chemistry has been used to validate microstructural features in otoliths such as annual increments (e.g., Kalish 2001, Andrews et al. 2005) and metamorphosis and settlement marks (e.g., Arai et al. 1997, Shen & Tzeng 2002). Elemental profiles in otoliths also have been used as an alternative means of estimating age and settlement when these microstructural features are not easily resolvable or are absent (e.g., Campana et al. 1990, 1993, Smith et al. 1995). Most of the applications for age validation and estimation have focused on radiochemical dating (see Campana 1999 for a review) of deep-water species because of the difficulties in resolving annual increments in slow-growing, long-lived species. Otoliths from long-lived tropical species also can be difficult to age as differences in the opacity and width of winter and summer growth increments often are indistinct (e.g., Fowler 1995). Interestingly, radiochemical dating has been used to validate annuli in long-lived tropical fishes in only a few studies (Milton et al. 1995, Baker & Wilson 2001, Baker et al. 2001). Most of the applications of otolith chemistry for validating metamorphosis or settlement marks have focused on diadromous species in the family Gobiidae, where metamorphosis and settlement coincide with a transition from salt to freshwater and consequently a decrease in Sr/Ca ratios (e.g., Radtke et al. 1988, Shen & Tzeng 2002). Metamorphosis and settlement in most demersal fish species, however, result in a transition from pelagic to benthic habitat and thus future studies may detect elemental changes which reflect this transition as well.

2 Methods and Procedures

2.1 SAMPLE PREPARATION

We have witnessed a marked increase in the number of published otolith chemistry studies over the past decade, as instrumentation for elemental analysis has become more sensitive, user friendly, and readily available (Campana 2005). This increase in publication rate, however, has not necessarily been accompanied by an increase in data quality. Careful attention to analytical protocols, beginning when the fish are collected, is required to ensure data quality.

2.1.1 Sample preservation

Both the time between fish capture and the removal of otoliths and the means and duration of fish preservation prior to otolith extraction may influence otolith chemistry. The impact of these factors will be determined by the degree to which the chemistry of the otolith and the surrounding endolymph change post-mortem and whether the preservation mode minimizes these changes. Post-mortem changes in endolymph and otolith chemistry may be analogous to those occurring in the vitreous humour of the human eye. For instance, potassium (K) concentrations in the vitreous humour increase proportionally with time since death. Measuring K concentration in the vitreous humour is one of the methods used by forensic pathologists to estimate time of death (e.g., Lange et al. 1994). Preservation by freezing will presumably slow down the process of decomposition and therefore the rate of increase in ion concentration of the endolymph. The rate of decomposition may be halted entirely if samples are preserved in ethanol but the ethanol itself could act either as a solvent by leaching elements out of the otolith or as a source of contamination. At least two grades of ethanol (ACS and HPLC) appear uncontaminated by either Sr or Ba (Hedges et al. 2004), but some ions appear to be labile and removing this fraction is likely to reduce any confounding effects of sample preservation. Clearly, removing otoliths from fish immediately after capture can eliminate effects of preservation and should be done if logistically feasible.

Five studies have investigated the effects of sample preservation on otolith chemistry. Milton and Chenery (1998) and Proctor and Thresher (1998) compared the effects of immediate removal of otoliths to removal after freezing or preservation of fish in ethanol initiated at different times post-mortem. Both studies detected variation among preservation treatments, particularly for monovalent elements (e.g., Na, K, Cl) and elements with known binding affinities to the organic matrix (e.g., Mg, Zn). The effect of freezing on ions that are substituting for Ca in the aragonite lattice is more uncertain. Brophy et al. (2003) reported significant increases in otolith Mg, Zn, Ba and Pb levels but no effect on Sr after freezing larval herring (*Clupea harengus*). Rooker et al. (2001a), however, found no effect of freezing on Mn, Sr and Ba levels in the otoliths of juvenile *Thunnus atlanticus*. Similarly, Hedges et al. (2004) found no effect of ethanol preservation on either Sr or Ba concentrations. Differences among studies in the effects of particular preservation methods may be related to the ease with which surface contamination can be cleaned from the otoliths after removal. Brophy et al. (2003) suggested that the contamination problems in their study were caused by tissue that remained attached to the small larval herring otoliths.

2.1.2 Otolith removal and storage

The process of removal from the fish and storage prior to analysis can potentially introduce contaminants to the otolith. Otolith removal typically involves the use of knives or scalpels to gain access to the brain cavity and forceps for otolith extraction. Instruments composed of metal alloys may transfer ions upon contact with an otolith. For instance, INOX stainless steel, a common alloy used in the manufacture of fine forceps, contains large quantities of P, S, Si, Cr, Mn, Fe, Ni, and Mo and therefore could be a major source of contamination for these elements. Some studies have employed the use of glass probes, plastic forceps, or Teflon coated forceps as a precaution against such contamination (Table 2). All of these instruments are more chemically inert than metal alloys and can be acid leached to remove metal contaminants prior to use. None of the metal-free materials, however, are used in the manufacture of very fine probes and forceps necessary for the extraction and handling of otoliths from embryos and early-stage larvae.

Once removed, otoliths are subject to contamination from adhering tissue, the walls of the storage container, or atmospheric deposition of particles. Care should be taken therefore to remove any tissue from otolith surfaces. Residual tissue will decompose and release organically bound elements that may contaminate the surface of the otolith. Removal of tissue can be accomplished manually using forceps or small brushes or by placing the otolith in clean water and using ultrasonication to loosen adhering particles of tissue (Table 2). If the otoliths are transferred into the storage container while wet, water can potentially leach ions off the walls of the container and transfer them onto the surface of the otolith as the water evaporates. Thus, care should be taken in assuring the samples are dry and stored in airtight containers that are clean and chemically inert (e.g., plastics, paper envelopes, cardboard depression slides). Dust is a major source of contamination, particularly for Fe, Zn, and Pb, and any surface deposition will likely result in biased estimates of elemental concentrations based on whole-otolith analyses.

2.1.3 Cleaning

The extent of cleaning required depends primarily on how the elemental data will be evaluated. Removal of surface contamination and the subsequent analysis of the composite otolith chemistry including organic, lattice, and labile fractions will be the primary concern for studies using otolith chemistry as a natural tag, where baseline signatures associated with a particular population or stock are assayed. Removal of surface contamination along with matrix and labile fractions is likely to be particularly important for studies focused on reconstructions of dissolved elemental concentrations to estimate temperature histories.

Table 2. Summary of the procedures employed for sample preservation and otolith removal, cleaning and storage in preparation for chemical analysis in ninety-five published studies between 1986 and 2005.

Preservation	References	Removal	References	Cleaning	References	Storage	References
Dissected from fresh specimens	4, 5, 6, 9, 12, 23, 24, 25, 26, 44, 47, 50, 56, 75, 78, 80, 87, 91, 94	Metallic forceps	6, 17, 45, 58, 59	None reported	1, 2, 3, 8, 10, 16, 28, 31, 32, 36, 38, 55, 56, 62	None reported	1, 2, 3, 6, 7, 8, 10, 11, 14, 18, 22, 28, 31, 32, 34, 35, 36, 37, 38, 49, 52, 55, 62, 67, 72, 77, 86, 95
Stored on ice until removed	53, 57, 61, 63, 65, 69, 70, 71, 81	Glass probes	24, 33, 40, 41, 46, 51, 52, 56, 68, 82, 83, 85, 86, 89	Adhering tissue removed with forceps	6, 12, 17, 26, 58, 59, 75, 81	Glass vials	5, 9, 12, 20, 73
Frozen	1, 2, 3, 8, 19, 20, 21, 26, 27, 30, 32, 33, 37, 38, 40, 41, 43, 46, 48, 49, 51, 52, 54, 62, 64, 68, 71, 75, 78, 79, 82, 83, 84, 86, 89, 91, 93, 95	Plastic forceps	12, 43, 78, 85, 86, 91	Adhering tissue removed with glass probes	24, 46, 50, 57, 68, 83, 89	Glass slides	68, 80
Ethanol	2, 3, 10, 15, 22, 26, 28, 31, 34, 45, 67, 74, 76, 93	Teflon coated forceps	49	Adhering tissue removed with plastic forceps	43	Envelopes	16, 19, 23, 25, 42, 50, 87, 90, 94

Table 2. (Continued)

Preservation	References	Removal	References	Cleaning	References	Storage	References
Unknown method	7, 11, 13, 14, 16, 17, 18, 29, 35, 36, 39, 42, 55, 58, 59, 60, 66, 72, 73, 77, 84, 85, 88, 90, 92	Unknown method	1, 2, 3, 4, 5, 7, 8, 9, 10, 11, 13, 14, 15, 16, 18, 19, 20, 21, 22, 23, 25, 26, 27, 28, 29, 30, 31, 32, 34, 35, 36, 37, 38, 39, 42, 44, 47, 48, 50, 53, 54, 55, 57, 60, 61, 62, 63, 64, 65, 66, 67, 69, 70, 71, 72, 73, 74, 75, 76, 77, 79, 80, 81, 84, 87, 88, 90, 92, 93, 94, 95	Adhering tissue removed with unknown method	20, 23, 48, 61, 66, 74, 78, 85, 87, 91, 93	Cardboard depression slides	45
				Rinsed with water	7, 12, 13, 18, 19, 24, 25, 27, 29, 30, 33, 34, 35, 37, 39, 40, 42, 43, 45, 46, 47, 48, 49, 51, 52, 53, 54, 58, 59, 63, 64, 65, 67, 69, 70, 71, 72, 74, 75, 76, 77, 78, 79, 80, 81, 82, 84, 85, 86, 87, 88, 89, 90, 91, 92, 93, 94, 95	Plastic vials	17, 21, 24, 26, 29, 33, 39, 40, 41, 43, 44, 46, 47, 48, 54, 56, 57, 58, 59, 61, 63, 64, 65, 66, 69, 70, 71, 73, 75, 76, 78, 79, 81, 82, 83, 84, 85, 88, 89, 91, 92
				Rinsed with isopropanol	25		
				Immersion in bleach	6, 18, 27, 37, 45, 60, 64, 76, 77		
				Immersion in weak acid	39, 45, 58, 59, 64, 72, 75, 76, 79, 81		

Table 2. (Continued)

Preservation	References	Removal	References	Cleaning	References	Storage	References
Unknown method				Immersion in hydrogen peroxide	39, 40, 46, 51, 52, 58, 59, 75, 79, 80, 81, 82, 83, 89, 93, 95		
				Immersion in hot water bath	45, 64, 76		
				Immersion in hexane and methanol	92		
				Scrubbed and rinsed in clean water	4, 11, 14, 17, 21, 26, 44, 47, 73, 90		
				Ultrasonication in water	6, 40, 41, 44, 45, 46, 47, 49, 51, 52, 67, 69, 77, 83, 87, 88, 89, 92, 93		
				Ultrasonication in ethanol	22		
				Unknown method	9, 15, 17		

[1]Gauldie et al. (1986), [2]Mulligan et al. (1987), [3]Radtke et al. (1988), [4]Edmonds et al. (1989), [5]Grady et al. (1989), [6]Kalish (1989), [7]Radtke (1989), [8]Townsend et al. (1989), [9]Kalish (1990), [10]Radtke et al. (1990), [11]Edmonds et al. (1991), [12]Kalish (1991), [13]Mugiya et al. (1991), [14]Edmonds et al. (1992), [15]Gallahar and Kingsford (1992), [16]Gauldie et al. (1992), [17]Gunn et al. (1992), [18]Secor (1992), [19]Hoff and Fuiman (1993), [20]Campana et al. (1994), [21]Thresher et al. (1994), [22]Arai et al. (1995), [23]Campana et al. (1995), [24]Fowler et al. (1995a), [25]Hoff and Fuiman (1995), [26]Proctor et al. (1995), [27]Secor et al. (1995), [28]Townsend et al. (1995), [29]Dove et al. (1996), [30]Farrell and Campana (1996), [31]Gallahar and Kingsford (1996), [32]Gauldie (1996), [33]Gillanders and Kingsford (1996), [34]Milton et al. (1997), [35]Mugiya and Satoh (1997), [36]Thorrold et al. (1997), [37]Chesney et al. (1998), [38]Geffen et al. (1998), [39]Secor and Zdanowicz (1998), [40]Thorrold et al. (1998b), [41]Thorrold et al. (1998a), [42]Edmonds et al. (1999), [43]Hanson and Zdanowicz (1999), [44]Patterson et al. (1999), [45]Swearer et al. (1999), [46]Bath et al. (2000), [47]Campana et al. (2000), [48]Gillanders and Kingsford (2000), [49]Kennedy et al. (2000), [50]Milton et al. (2000), [51]Thorrold and Shuttleworth (2000), [52]Wells et al. (2000), [53]Gillanders (2001), [54]Gillanders et al. (2001), [55]Limburg (2001), [56]Milton and Chenery (2001a), [57]Milton and Chenery (2001b), [58]Rooker et al. (2001a), [59]Rooker et al. (2001b), [60]Secor et al. (2001), [61]Thorrold et al. (2001), [62]Brophy and Danilowicz (2002), [63]Elsdon and Gillanders (2002), [64]Forrester and Swearer (2002), [65]Gillanders (2002a,b), [66]Secor et al. (2002), [67]Arai et al. (2003), [68]Brophy et al. (2003), [69]Elsdon and Gillanders (2003), [70]Gillanders and Kingsford (2003), [71]Hamer et al. (2003), [72]Milton and Chenery (2003), [73]Morris et al. (2003), [74]de Pontual et al. (2003), [75]Rooker et al. (2003), [76]Swearer et al. (2003), [77]Zlokovitz et al. (2003), [78]Brophy et al. (2004), [79]Buckel et al. (2004), [80]Chittaro et al. (2004), [81]Hanson et al. (2004), [82]Hedges et al. (2004), [83]Martin et al. (2004), [84]Patterson et al. (2004c), [85]Patterson et al. (2004a), [86]Patterson et al. (2004b), [87]Bergenius et al. (2005), [88]Humphreys et al. (2005), [89]Martin and Thorrold (2005), [90]McCulloch et al. (2005), [91]Miller and Shanks (2004), [92]Munro et al. (2005), [93]Ruttenberg et al. (2005), [94]Stransky et al. (2005), [95]Warner et al. (2005)

The majority of otolith chemistry studies to date have performed limited sample cleaning, often simply rinsing otoliths in high-purity water, presumably to remove adhering tissue and other surface contaminants (Table 2). Few studies have adopted more rigorous cleaning, particularly the use of solvents such as bleach (NaOCl) and hydrogen peroxide (H_2O_2), to remove organic matrix-bound fractions. This is surprising given the wide use of organic solvents in geochemistry and the testing of the efficacy of different cleaning procedures (e.g., Love & Woronow 1991, Gaffey & Bronnimann 1993, Boiseau & Juillet-Leclerc 1997, Stoll et al. 2001, Martin & Lea 2002). As mentioned previously, failure to remove organic material or labile fractions often results in variable and biased estimates that can obscure finer patterns of variation within the structure (e.g., Boiseau & Juillet-Leclerc 1997) or result in inaccurate estimates of environmental conditions such as ocean temperature (e.g., Stoll et al. 2001).

We examined the effects of a variety of different cleaning solutions and the duration of immersion in the solutions on the chemistry of otoliths measured by LA-ICP-MS (S. Swearer, unpublished data). Both of the common solvents used for organic matrix removal in biogenic calcium carbonates (NaOCl and H_2O_2) were tested at two different concentrations (Figure 3). The H_2O_2 solution was buffered with NaOH to raise the pH in order to minimize crystal dissolution. Proteinase K buffer, a non-specific proteinase that has been used for etching otoliths for SEM (e.g., Shiao et al. 1999), was also tested but over a shorter period than the other two solutions. Preliminary trials determined that proteinase K is very effective at removing inter-crystalline proteins but compromised the structural integrity of the otolith with longer immersion times.

There was a strong effect of reagent for all elements except Sr (Table 3). NaOCl generally was a source of contamination, resulting in significantly higher concentrations of Mg, Mn, and Ba compared to the other reagents. This conclusion was supported by the trend towards increasing concentrations with increasing duration of immersion. Proteinase K was the most effective at removing Mn and this effect was quite rapid as there was no difference in concentration among durations. Buffered H_2O_2 appeared to remove organically bound elements without being a source of contamination for other elements, with the 24 h immersion in a 15% solution most effective. Higher concentrations and longer duration treatments resulted in loss of structural integrity of the otolith that introduced both considerable variability into the assays and significant Ba contamination in the cracks that formed. In summary, Ba and Sr appeared almost exclusively bound to the crystal lattice; cleaning had little effect on their concentrations but cleaning had a significant effect on Mg and Mn levels, indicating that organically-bound fractions contributed significantly to total Mg and Mn in the otoliths.

2.1.4 Preparation for solution-based analysis

There is very little additional preparation required for whole-otolith analysis by dissolution once the otoliths have been removed and cleaned. Efforts must be taken to minimize contamination, however, as with other preparation steps. For solution-based analysis, otoliths are typically dissolved in ultrapure HNO_3 within acid leached plastic vials within a HEPA filtered laminar flow cabinet to reduce the risk of dust particles being deposited into the vial. Acid leaching of vials in a strong acid prior to sample dissolution is critical as elements can be effectively removed from the plastic by the excess H^+ ions present in the sample solution.

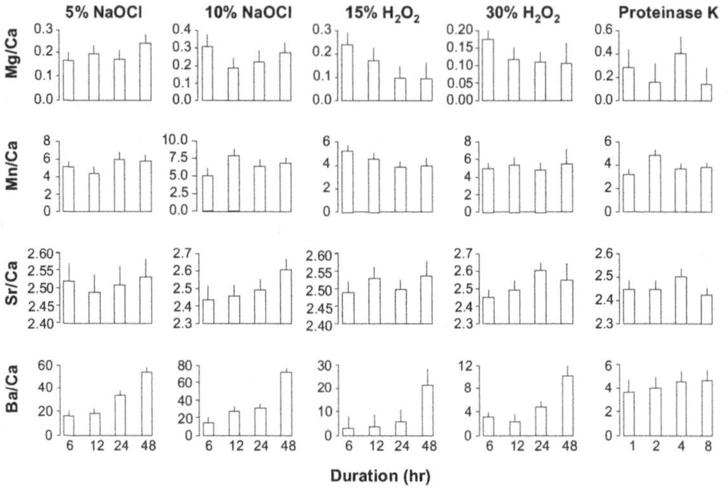

Figure 3. Effects of cleaning reagent and immersion duration on elemental composition of juvenile kelp bass, *Paralabrax clathtratus*, otoliths. Samples were embedded in epoxy resin, ground with diamond lapping film, polished and randomly assigned to one of the 20 treatments (5 reagents × 4 durations). All samples were rinsed with 18 MΩ water after treatment, dried in a class 100 laminar flow bench, and analysed by laser ablation ICP-MS. Elemental ratios (vertical axes) are in $mmol.mol^{-1}$ for Mg/Ca and Sr/Ca and in $\mu mol.mol^{-1}$ for Mn/Ca adn Ba/Ca. Error bars ±1 SE.

Table 3. Effects of cleaning reagent (R) and cleaning duration (D) on log-transformed elemental composition of otoliths from kelp bass (*Paralabrax clathratus*) recruits based on two-way analyses of variance and post-hoc Tukey HSD tests.

Element	Factor	df	SS	*F*	*p*	Tukey HSD
Mg	R	4	0.927	5.35	0.0006	5%, 10% NaOCl > 15%, 30% H_2O_2
	D	3	0.341	2.62	0.0552	
	RxD	12	0.534	1.03	0.4311	
	Error	116	4.206			
Mn	R	4	0.740	6.39	0.0001	10% NaOCl > Proteinase K
	D	3	0.117	1.34	0.2647	
	RxD	12	0.283	0.82	0.6350	
	Error	114	2.752			
Sr	R	4	0.003	1.87	0.1220	
	D	3	0.001	1.41	0.2459	
	RxD	12	0.003	0.66	0.7872	
	Error	117	0.034			
Ba	R	4	21.267	122.2	<0.0001	5%, 10% NaOCl > rest
	D	3	1.932	14.8	<0.0001	6 h, 12 h < rest
	RxD	12	1.333	2.5	0.0056	
	Error	118	4.309			

2.1.5 *Preparation for in situ analysis*

The preparation of otoliths for *in situ* microprobe analysis is more involved than for solution-based assays. Laser ablation analyses require the surface of the otolith to contain all growth layers and so samples must be sectioned and polished. Significant contamination during the preparation process can be introduced from the embedding compound, the lubricant, the saw blades, or the polishing compound. We analysed the chemical composition of several different types of otolith mounting and embedding compounds to assess the contamination potential of each material (Figure 4). The compounds varied tremendously in the level of elemental contamination, with both a polyester resin and a thermoplastic cement having levels of contamination 1–2 orders of magnitude greater than a cyanoacrylate glue and an epoxy resin. Concentrations within the more contaminated compounds were within the range of median otolith concentrations reported in the literature (Campana 1999). Fragments of embedding compound can also accumulate in cracks in the surface of the otolith and along the interface between the otolith and the embedding compound during sectioning and polishing. For instance, Proctor and Thresher (1998) reported elevated levels of contaminating elements associated with cracks from in situ analysis of otolith chemistry using micro-PIXE.

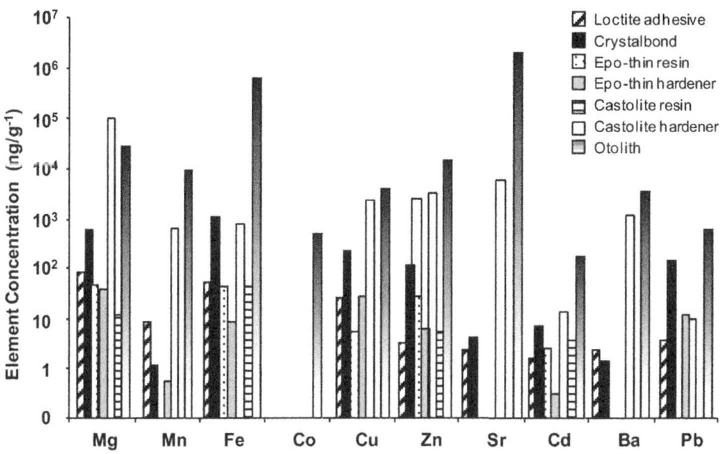

Figure 4. Elemental composition of four types of embedding compounds: cyanoacrylate adhesive (Loctite), thermoplastic cement (Crystalbond) epoxy resin (Buehler Epo-thin resin and hardener), and polyester resin (Buehler Castolite resin and hardener), in comparison to median otolith concentrations from Campana (1999). Compounds were digested in HNO_3 in heated Teflon crucibles. Subsamples of digests were then analysed by standard addition ICP-MS.

The choice of polishing media introduces another potential source of contamination in otolith chemistry studies. Polishing films are generally preferable to pastes because the polishing compound (e.g., SiC, AlO, diamond) is embedded in the film and doesn't coat the sample but even films can be sources of contamination. We analysed several different types of polishing films to determine if abraded particulates from the film could also accumulate within surface imperfections (Figure 5; S. Swearer, unpublished data). We observed that the films were potentially major sources of contamination, as qualitative measures of concentrations of several elements, particularly Mg and Ba, were several orders of magnitude higher in films than in otoliths. The difference in the performance of the films was principally a result of how the polishing compound was attached to the film. The abrasive compound is traditionally slurry-coated onto the film. This method was used in both the AlO and DFP diamond films and it clearly resulted in high rates of contamination. In contrast, the 3M™ imperial diamond lapping film used a resin bonding system that was apparently much cleaner. Finally, ultrapure water should be used if a lubricant is required when sectioning or polishing otoliths for elemental analysis.

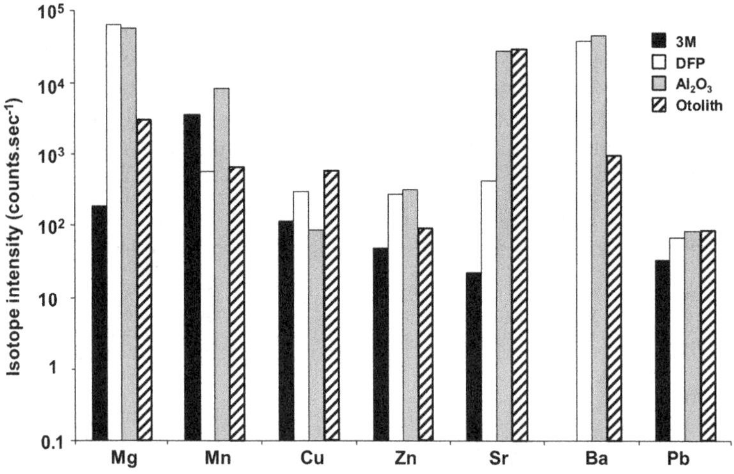

Figure 5. Background corrected intensities from laser ablation ICP-MS analysis of 3M diamond lapping film, DFP diamond lapping film and Al_2O_3 lapping film in comparison to reef fish otoliths analysed using identical laser and instrument settings. Note that the analyses assume similar ablation efficiencies among lapping films and otoliths due to the absence of a suitable internal standard.

The amount of handling required for in situ analysis will almost undoubtedly result in some sample contamination even when care is taken to minimize sources of contamination during sample preparation. Consequently, samples will need to be cleaned after polishing. Samples should be cleaned using clean water, or even a weak acid, in combination with ultrasonication as both particulate contamination on the otolith surface and ion adsorption within the otolith matrix are likely.

2.2 QUANTIFYING ELEMENTAL SIGNATURES IN OTOLITHS

The use of plasma source mass spectrometry to quantify the elemental composition of otoliths opened up the possibility of analysing a number of elements that were present in otoliths at trace to ultra-trace levels and therefore below the detection limits of the electron microprobe. Campana and Gagne (1995) provided the first systematic examination of otolith composition using ICP-MS. They reported a total of 22 elements in the otoliths of adult cod (*Gadus morhua*), with significant differences in multivariate elemental composition among individuals from different locations. Many of these elements, however, were unlikely to have been reliably quantified in otolith samples using the instrumentation available at the time. More recent studies have concentrated efforts on a smaller number of elements that are both relatively easy to accurately measure using ICP-MS and form divalent cations that are likely to substitute for Ca in the aragonite lattice, including Mg, Mn, Sr, Ba and Pb.

2.2.1 Selection of isotopes for analysis

An important criterion when selecting elements for quantification using ICP-MS is that each element must have at least one isotope free from isobaric interferences. The presence of significant interferences can be checked for elements that are not mono-isotopic by measuring isotopic ratios in representative samples. Significant departures from natural ratios indicate that one or both of the isotopes expressed in the ratio are compromised by isobaric interferences. If, however, the ratios are close to natural values, then it is likely that the both isotopes are suitable for quantification. High resolution ICP-MS can be particularly useful when checking isotopes for interferences. We ran an otolith standard reference material (Yoshinaga et al. 2000) in medium resolution (r = 4,000) using a Thermo Elemental *Element2* ICP-MS to check for interferences on Mg, Ca, Mn, Fe, Ni Co, Cu and Zn isotopes that have been commonly used in otolith chemistry studies. Analysis of a sample blank identified a number of molecular interferences resulting either from the Ar plasma gas or the 2% HNO_3 solution, including a small interference on ^{55}Mn and significant interferences on ^{57}Fe, ^{58}Ni, ^{59}Co, and ^{68}Zn (Figure 6). Results from the blank-corrected otolith SRM run found significant isobaric interferences, presumably from Ca molecules, on ^{24}Mg, ^{57}Fe, ^{58}Ni, ^{60}Ni, ^{59}Co, and ^{64}Zn. Indeed, ion counts on masses 57, 58, 59 and 62 were almost entirely the result of molecular interferences rather than the Fe, Ni and Co isotopes of interest. Doubly-charged ^{48}Ca ions were again found to represent a significant interference on ^{24}Mg, as noted by Thorrold and Shuttleworth (2000).

There are several ways to minimise effects of molecular interferences in otolith analyses using ICP-MS. Molecular interferences originating from the Ar sample gas can be eliminated by simple blank subtraction. For instance, an Ar molecule causing an interference on ^{55}Mn is effectively removed by subtracting counts from a blank solution (Figure 6). Although many molecular interferences are resolved by sector field ICP-MS instruments operating at resolutions of 4,000 (Figure 6), there is a significant loss of sensitivity. This is a problem for elements such as Fe, Co and Ni that appear to be present in otoliths at ultra-trace levels. Another option is to remove Ca from otolith samples before ICP-MS analysis using Ca-chelating columns or $CaFl_2$ precipitation (e.g., Arslan & Paulson 2003, Arslan 2005, Arslan & Secor 2005). These approaches also allow for pre-concentration of the analyte during Ca removal and may be particularly useful for quantifying elements that are below detection limits using standard acid dissolution techniques. There is currently no way to remove Ca online during laser ablation assays, and in situ analyses of elements with significant molecular interferences from the sample matrix will need to use sector field instruments operating at high resolution (Thorrold & Shuttleworth 2000).

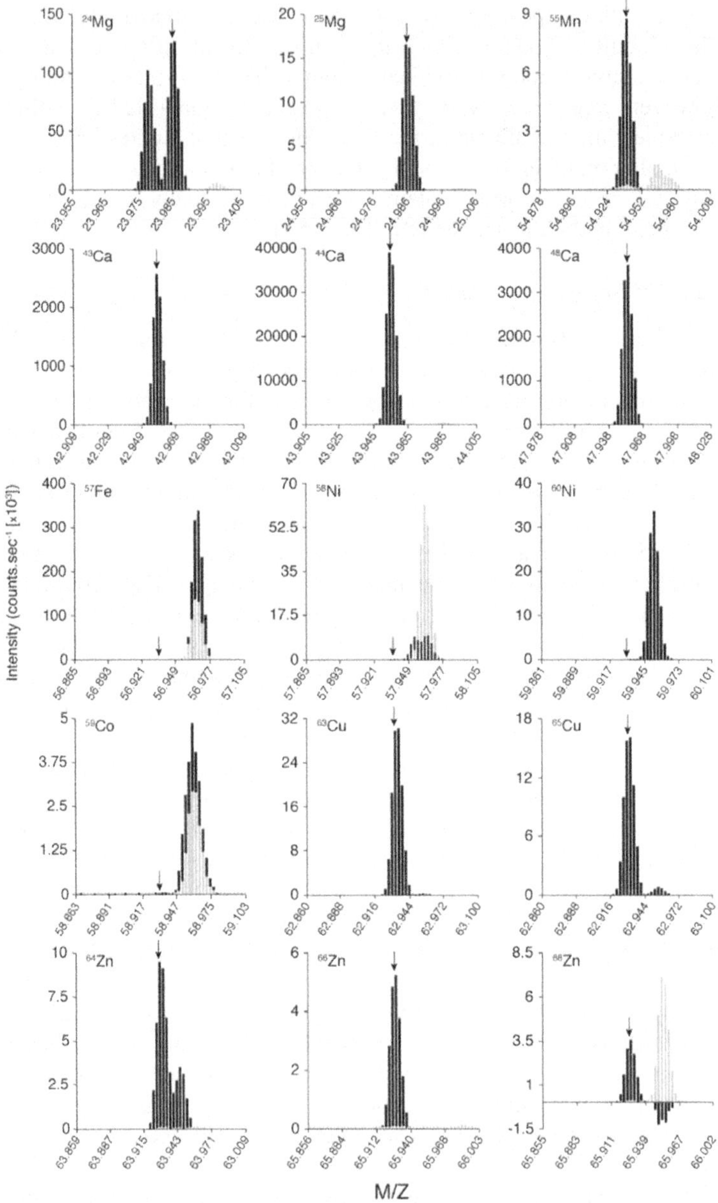

Figure 6. Count rate (intensity) as a function of mass/charge ratio (*m/z*) from medium resolution ($r = 4{,}000$) scans of isotopes potentially useful for analyses of otolith chemistry using a high resolution *Element2* ICP-MS. *Shaded bars* are data from a 2% HNO_3 blank solution and *solid bars* are blank-corrected data from a dissolved otolith reference material (Yoshinaga et al. 1999, 2000). *Arrows* mark the peak center of the isotope of interest.

2.2.2 *Quantitative analysis of otolith composition*

Several standardization approaches are available when analysing dissolved otoliths by ICP-MS. Most studies have used a combination of external and internal standardization to determine elemental concentrations (e.g., Campana & Gagne 1995, Fowler et al. 1995a, Gillanders & Kingsford 1996, Patterson et al. 1999, Swearer et al. 2003, Dorval et al. 2005). Signal intensities of all analyte isotopes are measured in standards with elemental concentrations that cover the concentration range likely to be found in the samples. A calibration curve is then established to calculate the concentration of the elements in samples of unknown composition. Ideally standards are matched to the matrix of samples in terms of both acid and Ca concentrations to account for potential matrix effects.

Changes in instrument sensitivity during an analysis session are unavoidable with ICP-MS measurements. Internal standardization is usually implemented to account for this instrument drift. All data are normalized to one or more isotopes that have usually been spiked into samples, standards and blanks at a known concentration before analysis. The technique can also account for matrix effects in some instances (Horlick & Montaser 1998). Correction for instrument drift using internal standardization is only useful if mass-dependent effects on mass bias drift are small. Similarly, the method only compensates for matrix differences among samples if the internal standard and the isotope of interest experience the same matrix effect. Failures to account for the effect of Ca concentrations in otolith assays using ICP-MS has typically occurred when otoliths were dissolved using a predetermined amount of nitric acid rather than by altering acid volumes to provide for matched Ca levels among samples (e.g., Patterson et al. 1999, Gillanders et al. 2001, Dorval et al. 2005). Matrix-dependent errors will act to magnify within-location variances if there are no significant among-location differences in otolith mass. Systematic variation in otolith mass, however, may lead to false claims of differences in otolith composition among locations that are an artefact of matrix effects during the analysis. We therefore suggest that researchers ensure that Ca concentrations of sample solutions and standards are matched to minimize matrix effects as much as possible.

Isotope dilution (ID) mass spectrometry is the method of choice for certification of standard reference materials and the most accurate quantification method available for analysis of dissolved otoliths (Fassett & Paulson 1989, Yoshinaga et al. 2000, Sturgeon et al. 2005). ID mass spectrometry was first used for otolith assays by Campana et al. (1995) but has been applied in relatively few studies of otolith composition (Thorrold et al. 1998a,b, Bath et al. 2000, Campana et al. 2000, Humphreys et al. 2005). The technique is based on addition of a known amount of enriched isotope to a sample. The isotope ratio of the sample is measured using mass spectrometry after equilibration of the spiked isotope and the concentration of the element (C_x) in the sample calculated using the following formula (Fassett & Paulson 1989):

$$C_x \quad \frac{C_s W_s}{W_x} \quad \frac{A_s \quad R_m B_s}{R_m B_x \quad A_x}$$

where C_s refers to the concentration of the element in the spike, A_x and B_x are the atom fractions of isotopes A and B in the sample, A_s and B_s are the atom fractions of the isotopes A and B in the spike, W_x and W_s are the weights of the sample and spike, respectively, and R_m is the measured ratio of isotope A to isotope B.

Systematic errors in ID mass spectrometry can occur during sample preparation and measurement of isotope ratios. Equilibration of the spike and sample is easily achieved because otoliths dissolve rapidly in concentrated nitric acid. Isotope dilution lends itself particularly well to otolith assays using ICP-MS because isotopic ratios of a number of elements can be measured during the same analysis with good sensitivity, accuracy and precision. A major advantage of isotope dilution over other quantification strategies used in ICP-MS analyses is that the effect of instrument drift on concentration estimates is minimized because both the spiked and unspiked isotope of any element will be similarly influenced by changes in sensitivity and mass bias. Matrix effects are also eliminated and so [Ca] levels need not be matched among samples. Finally, ID is particularly useful when using off-line chemistry because recovery of target elements need not be quantitative, providing the spike is equilibrated in the sample before separation. ID ICP-MS cannot be used to quantify concentrations of mono-isotopic elements such as Mn in otoliths, however, or elements in which only one isotope is free of isobaric interferences.

Many analyses of biogenic carbonates require assays of elemental ratios to Ca rather than quantitative estimates of elemental composition. For instance, Mg/Ca and Sr/Ca ratios in foraminiferal calcite and coral aragonite are routinely used for paleo-reconstructions of sea surface temperatures (e.g., Guilderson et al. 1994, Lea et al. 1999). Physicochemical properties of ambient water masses are similarly likely to be recorded by Ca ratios rather than elemental concentrations in otoliths (e.g., Bath et al. 2000). Rosenthal et al. (1999) introduced a technique for precision estimates of elemental ratios to Ca (Me/Ca) in carbonates using sector field single collector ICP-MS. The approach is elegant in its simplicity, requiring only a single matrix-matched external standard from which elemental mass bias is measured and then applied to unknown samples. Mass bias ($C_{Me,Ca}$) between an element (Me) and Ca is quantified by measuring the elemental ratio in a matrix-matched standard ($M_{Me/Ca}$) of known elemental composition ($S_{Me/Ca}$) as follows:

$$C_{Me/Ca} \quad \frac{S_{Me/Ca}}{M_{Me/Ca}}$$

The mass bias factor is then used to correct the measured elemental ratio in an unknown sample to give the true Me/Ca value. Rosenthal et al. (1999) reported small matrix effects for some, but not all, elements found in concentrations typical of calcitic marine foraminifera. We examined matrix effects in otolith aragonite using a similar technique with an *Element2* ICP-MS operating in low resolution. A laboratory otolith standard was initially diluted to [Ca] values of 6, 12, 40, 100, 150 and 200 $\mu g \cdot g^{-1}$. We then determined elemental mass bias by measuring an otolith reference material (Yoshinaga et al. 1999, 2000) at a single Ca concentration of 40 ppm. We found significant ($>2\sigma$) matrix effects for all elements except Sr (Figure 7). In general there was an inverse, non-linear, relation between [Ca] and Me/Ca ratios, with matrix effects becoming pronounced at [Ca] levels below 40 ppm. Magnesium values were a complex polynomial function of [Ca]. An obvious solution to minimize matrix effects is to ensure that [Ca] values are similar in both samples and standards, as we advocated earlier for traditional external standardization. This is usually possible with otolith assays because sample sizes are sufficiently large that they can be weighed accurately before dissolution. Rosenthal et al. (1999) reported precisions of better that 1% for multi-elemental analyses and 0.1% for Sr/Ca ratios. We examined external reproducibility of the technique in otolith samples by running a powdered otolith standard a total of 32 times over 5 days of analysis (Figure 8). The results were comparable with those of Rosenthal and co-workers, with RSDs ranging from 2% (Pb/Ca) to 0.2% for Sr/Ca and Ba/Ca ratios. Moreover, each analysis used approximately 5 μg of otolith material, making the approach particularly useful for small sample sizes.

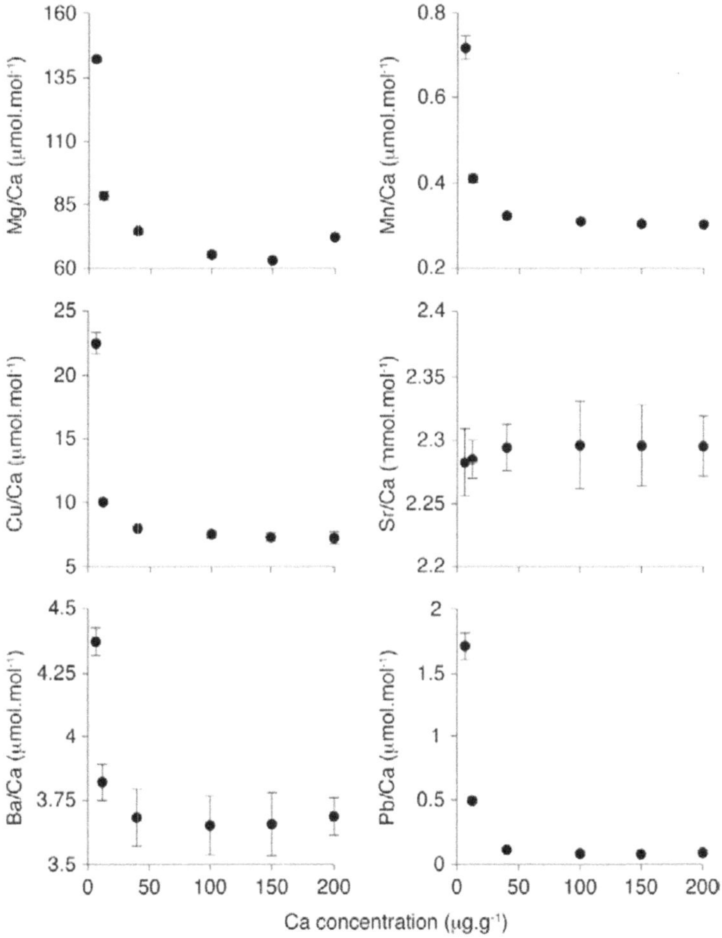

Figure 7. Mean (n = 3, ± 1 SD) concentrations of six elements, expressed as a ratio to Ca, in a standard consisting of powdered otolith material digested in HNO_3 and diluted to varying Ca concentrations. Elemental mass bias was quantified by assaying an otolith CRM (Yoshinaga et al. 1999, 2000) at a Ca concentration of 40 μg·g^{-1}.

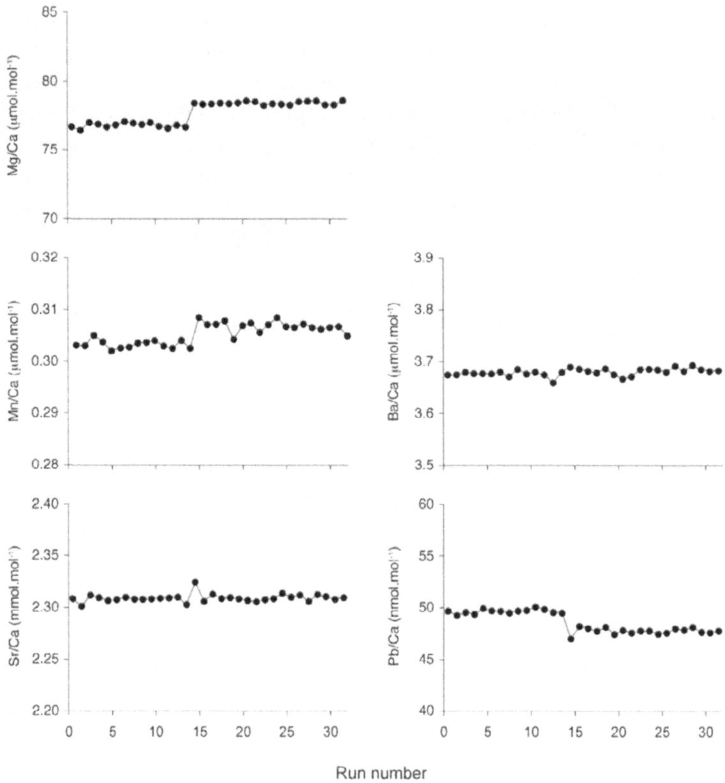

Figure 8. External reproducibility ($n = 32$) of five elements in a dissolved otolith standard with a Ca concentration of 40 $\mu g \cdot g^{-1}$ measured at regular intervals over a 3 day period. Elemental mass bias was quantified by assaying a 40 $\mu g \cdot g^{-1}$otolith CRM solution (Yoshinaga et al. 2000) immediately before running the lab standard.

In summary, the calibration strategy employed when analysing dissolved otoliths will be determined by the questions being addressed in a study. Isotope dilution is the logical choice if absolute concentrations of elements are required, providing there are suitable isotopes available for the analysis. We recommend the approach outlined by Rosenthal et al. (1999), however, if elemental ratios to Ca are sufficient. In either case, it is essential that some quality control procedure is also used during analyses. Otolith researchers are particularly fortunate because two standard reference materials, both consisting of otolith powders, are readily available (Yoshinaga et al. 1999, 2000, Sturgeon et al. 2005). These standards are ideal for external determination of elemental mass bias and for quality assurance. A number of elements, including Mg, Ca, Sr and Ba, are certified in both materials, while Mn and Pb values are provided in at least one of the standards. Taken together, this list represents most elements likely to be of interest to fish biologists using otolith chemistry. Finally, we note that there have been very few inter-laboratory comparisons of otolith chemistry (Campana et al. 1997, Secor et al. 2002). The availability of two SRMs, however, will make comparisons of otolith chemistry among studies significantly easier if researchers consistently analyse the SRMs and report the appropriate results in their publications.

2.2.3 Laser sampling

There is little doubt that the popularity of ICP-MS analyses for otolith chemistry applications stems in large part from the ability to perform in situ analyses by laser ablation. Gray (1985) first coupled laser technology with a plasma source mass spectrometer and since then laser ablation ICP-MS has gained wide acceptance within the analytical chemistry community for semi-quantitative elemental analyses in a number of applications (Durrant & Ward 2005).

A fundamental characteristic of all systems is the operating wavelength of the laser. Gray's (1985) initial experiments were conducted with a 694 nm ruby laser but more recent studies have typically used neodymium:yttrium aluminium garnet (Nd:YAG) lasers. Campana et al. (1994) produced the first elemental data collected from otoliths using a solid state Nd:YAG laser operating at its fundamental wavelength (1,064 nm) coupled with a quadrupole ICP-MS. It was quickly realized, however, that infrared lasers did not couple particularly well with otoliths (e.g., Geffen et al. 1998) and attention switched to lasers operating at frequency-tripled (335 nm) or frequency-quadrupled (266 nm) wavelengths (e.g., Campana et al. 1997, Milton et al. 1997, Thorrold et al. 1997). A further advancement was made by coupling high resolution sector field ICP-MS instruments with 266 nm laser systems to provide high precision elemental ratios (Thorrold & Shuttleworth, 2000). Several research groups have used this combination to analyse elemental composition of fish otoliths (Elsdon & Gillanders 2002, 2003, FitzGerald et al. 2004, Ruttenberg et al. 2005, Warner et al. 2005) and gastropod shells and statoliths (e.g., Zacherl et al. 2003). The trend to lower wavelength lasers that couple better with opaque and transparent materials continued with the development of frequency-quintripled Nd;YAG lasers operating at 213 nm (Jeffries et al. 1998) and ArF excimer lasers operating at 193 nm (Günther et al. 1997). Recent studies using these lasers to examine elemental composition of fish otoliths include research on both coral reef (Patterson et al. 2004b,c, Patterson & Swearer 2007) and temperate fish species (Brophy et al. 2003, 2004, Walther & Thorrold 2006) using 193 nm and 213 nm systems, respectively.

The performance of laser ablation ICP-MS instruments is influenced by several parameters that are under the control of the operator. For instance, the efficiency of particle transport from the sample to the ICP-MS is dependent upon the carrier gas. Efficiency is of the order of only 25–50% using Ar but ablation in He has been shown to increase particle transport to approximately 90% (Eggins et al. 1998). The majority of laser ablation ICP-MS studies of otolith chemistry have used a dry plasma where the aerosol is removed from the sample by a carrier gas and injected directly into the plasma. Operating in dry plasma mode has the advantage of reducing the formation of oxides and other molecules that may generate interferences on isotopes of interest. Instrument blanks are generally low because they are limited primarily by contamination in the Ar and He gases. Several groups have mixed the laser aerosol with dilute HNO_3 from a self-aspirating nebulizer in a spray chamber (Figure 9) before the plasma (e.g., Günther & Heinrich 1999, Thorrold et al. 2001, Zacherl et al. 2003, Warner et al. 2005). A wet plasma approach allows for liquid standards to be used for semi-quantitative analyses of elemental ratios, usually at the expense of higher oxide levels and blanks.

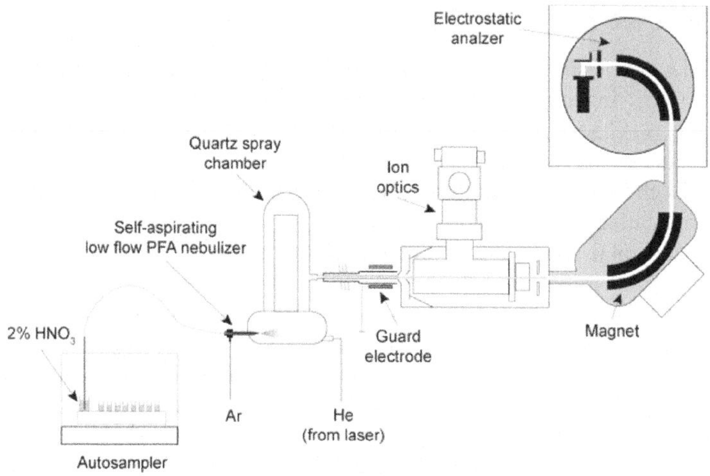

Figure 9. Schematic diagram (not to scale) of a double focusing magnetic sector inductively coupled plasma mass spectrometer showing a sample introduction system used for laser ablation analysis of biogenic carbonates using a wet plasma mode.

An ongoing problem with laser ablation ICP-MS analyses is the lack of suitable solid standards. Most otolith studies continue to use NIST glasses to convert raw instrument counts to elemental concentrations or ratios. The silicate matrix of the glasses is very different to that of otolith aragonite, however, and many elements in the glasses are found at different concentrations than they occur in otoliths. Campana and co-workers (Campana et al. 1997, Thorrold et al. 1997) constructed a series of glass bead standards that consisted of a 1:1 ratio of ground otolith powder and Li tetraborate flux and spiked with different levels of minor and trace elements. The elemental concentrations in the beads were closer to otoliths than the NIST glasses but the high tetraborate matrix was obviously quite different to that of biogenic aragonite. Elsdon and Gillanders (2002) analysed a pressed pellet made from powdered otolith material at the beginning and end of each laser session to calculate an average elemental mass bias for that day. Several researchers have used liquid standards to monitor elemental mass bias while operating in wet plasma mode in attempts to calibrate instrumental mass bias more regularly. The approach is similar to that used for quantification of elemental ratios in solutions (see above) but in this case the sample material comes from a laser aerosol rather than a solution. It is possible to use dissolved otolith reference material as a calibration standard because plasma conditions are identical for both laser samples and solution standards. These solution standards can be run frequently without opening the laser cell and, therefore, variability in elemental mass bias of the instrument can be constrained very accurately (e.g. Swart et al. 2002). Instrument mass bias can be accurately measured with this approach but a difficulty is that elemental fractionation during the ablation process is ignored. It is therefore difficult to compare among studies that have used different laser wavelengths and operating parameters.

2.3 DATA INTERPRETATION

Analyses of otolith chemistry have largely focused on either environmental reconstruction of the physico-chemical properties of water masses or on the use of geochemical signatures in otoliths as natural tags of natal origins. Examples of the use of otoliths to reconstruct water mass properties include studies that have inferred dispersal pathways of larvae from otolith chemistry (Thorrold et al. 1997, Swearer et al. 1999, Patterson et al. 2005) and research that has tracked movements of migratory fishes from marine systems to estuarine and riverine habitats (e.g., Limburg 1995, Secor & Piccoli 1996, Elsdon & Gillanders 2005). The technique assumes that otolith chemistry reflects dissolved ambient concentrations as modified by temperature and, therefore, elemental profiles across otoliths reflect environmental conditions experienced by individual fish. Assignment of individual movement trajectories is then made based either on ground-truthed signatures from otoliths known to have been deposited within a given water mass or, more frequently, estimated from assumptions on the influence of water chemistry on otolith composition. For instance, movements from freshwater to marine environments are often assumed to generate otolith Sr/Ca profiles that transition from relatively low values characteristic of river waters to higher values associated with seawater (Kraus & Secor 2004a).

Statistical analyses of elemental profiles are complicated by the observation that points within an otolith are not independent and, therefore, traditional Analysis of Variance (ANOVA) approaches are not valid. Longitudinal analyses that account for autocorrelations within individual profiles allow for more rigorous testing of movement patterns. Repeated measures ANOVA has been used to test hypotheses concerning variations in elemental profiles across otoliths (Thorrold et al. 1997, FitzGerald et al. 2004). Repeated measures ANOVA, however, accommodates only a single response variable, whereas ICP-MS analyses of otolith composition often generate multivariate elemental signatures. Sandin et al. (2005) described a statistical approach that used maximum likelihood methods to test specific hypotheses concerning dispersal trajectories of coral reef fishes based on multivariate elemental profiles in otoliths. Sandin and co-workers acknowledged that some of the assumptions of their model were unlikely to have been met but their technique clearly is a significant improvement over less quantitative methods for assigning dispersal trajectories of individual fish based on multivariate elemental profiles across otoliths.

Multivariate geochemical signatures in otoliths have proved useful as natural tags of natal origins or stock membership in fish (e.g., Campana et al. 1999, Thorrold et al. 2001). The technique doesn't rest on the assumption that otolith chemistry records physico-chemical properties of ambient water but requires significant geographic variation in otolith chemistry. The mechanisms generating these variations are not important providing that the signatures of likely source groups can be ground-truthed in order to classify the origins or stock composition of an unknown sample. Multivariate ANOVA is typically used to determine if geochemical signatures in otoliths differ significantly among locations. The signatures can be visualized in reduced dimensional space using canonical discriminant analysis (CDA). Confidence intervals on the canonical group means can then provide a useful way of performing a posteriori pairwise tests among groups (e.g., Thorrold et al. 1998b).

Determining population affinities or spawning site fidelity based on geochemical signatures in otoliths can be accomplished on a sample of unknown group composition, or alternatively individual fish can be assigned to a stock or to their natal spawning location. Maximum likelihood analysis is currently the method of choice for estimating stock proportions from a mixed-stock sample (Campana et al. 2000, Thorrold et al. 2001). A ground-truthed data set consisting of geochemical signatures in otoliths from fish of known group membership is used to train a discriminant function that is then used to estimate the proportion of fish from each of the groups present in the unknown sample. Similar approaches have been used to determine the stock composition of unknown samples based on genetic (Millar 1990) and morphometric markers (DeVries et al. 2002).

It is more difficult to ascertain the natal origin or stock identity of an unknown individual. A number of studies have used discriminant function analysis (DFA) to determine the accuracy with which ground-truthed otolith signatures can assign individuals to potential groups using cross-validation (e.g., Campana et al. 1995, Thorrold et al. 1997, 1998b). A jacknife "leave-*n*-out" strategy avoids the circularity of classifying a fish using a function that was trained with data from the fish that is subsequently considered an unknown. Researchers have often failed to realize, however, that discriminant analyses are sensitive to prior probabilities of group membership that must be chosen before classifying unknown individuals (DeVries et al. 2002). These probabilities are, of course, the group proportions in the sample that one is trying to estimate from the analysis. Thorrold et al. (1998b) compared the accuracy of DFA and neural networks to classify juvenile weakfish (*Cynoscion regalis*) to natal estuaries based on geochemical signatures in otoliths. They found that neural networks were able to identify the natal estuary of the juveniles more accurately than DFA. More importantly, neural networks alleviate the need for prior probabilities and, therefore, can be used to classify origins of individual fish in samples of unknown origin.

2.4 FUTURE DIRECTIONS

Plasma source mass spectrometry has matured as an analytical technique in inorganic chemistry over the last 10 years, although there remain several active areas of research that may prove to be useful for elemental and isotopic analyses of otolith chemistry. Spatial analysis of otolith elemental composition using laser ablation is often hindered by the need for sequential analysis of isotopes with single collector quadrupole or magnetic sector mass spectrometers. Time-of-flight mass spectrometry (TOF-MS) is well suited to elemental analysis of transient signals from lasers because detection of ions is simultaneous over a large mass range (Vázquez et al. 2002). Simultaneous measurements of isotopes minimize multiplicative errors associated with sample introduction and plasma fluctuations that commonly limit instrument precision with single collector plasma source mass spectrometry (Willie et al. 2005). The major problem with plasma source TOF-MS has been poor sensitivity compared to conventional ICP-MS (Vázquez et al. 2002, Bings et al. 2004). TOF-MS appears well suited to multi-elemental analysis of otoliths, if ion transmission can be improved.

Researchers continue to develop lasers that reduce elemental fractionation during ablation of solids. Commercial laser units currently available employ lasers with pulse durations of 6–30 ns. Femtosecond (fs) lasers will, in theory, reduce heating in solid samples during ablation compared to a nanosecond laser because the photon relaxation time in a solid is on the order of 100 fs (Russo et al. 2002). Stoichiometric conversion of ions produced by fs laser ablation is therefore possible if elemental fractionation is dominated by thermal effects at the sample surface (Bian et al. 2005). Russo and colleagues examined performance of an 800 nm fs laser coupled to an ICP-MS. They noted improvements in crater morphology over nanosecond laser systems but fractionation behaviour during ablation of NIST glasses was similar to that of nanosecond UV lasers. It remains to be seen if improvements in elemental fractionation occur with UV fs lasers similar to that documented with nanosecond lasers. In the future, however, we may see fs lasers coupled to plasma source TOF-MS instruments producing elemental data in otoliths at spatial resolutions approaching that of the smallest daily increments.

There is little doubt that the instrumentation available to otolith researchers will become more sophisticated in the future. We may have already reached the stage, however, where progress is not limited by technology but rather by metabolic controls on the degree to which otolith composition reflects ambient water chemistry. This realization has, in turn, begun to focus more attention on isotope systems that may be less susceptible to biological effects. For instance, C and O isotopes have been used as natural tags of natal origins in estuarine systems (Thorrold et al. 1998b, 2001, Hanson et al. 2004). Similarly, Sr isotopes in otoliths have been used to determine movement patterns of salmon in freshwater systems (Kennedy et al. 1997) and to investigate the prevalence of marine migrations in anadromous species (Hobbs et al. 2005, McCulloch et al. 2005). In situ analyses of Sr isotopes in otoliths has been possible due to the development of multiple collector ICP-MS instruments capable of high precision isotope ratio measurements (Halliday et al. 1998). First generation multiple collector ICP-MS systems coupled with laser systems were significantly less precise than thermal ionisation mass spectrometry (Christensen et al. 1995, Outridge et al. 2002, Milton & Chenery 2003). Newer instruments perform comparably to TIMS, however, but with significant advantages in terms of sample preparation and throughput (Barnett-Johnson et al. 2005, Hobbs et al. 2005, McCulloch et al. 2005, Woodhead et al. 2005).

Strontium isotopes are typically not fractionated by biological processes in nature (Blum et al. 2000) and most of the Sr deposited in fish otoliths comes from the ambient water (Farrell & Campana 1996, Walther & Thorrold 2006). Sr isotope ratios in otoliths therefore accurately reflect those of the environment. Marine $^{87}Sr/^{86}Sr$ ratios are constant in seawater because of the long residence time of Sr in the world's oceans. Ambient Sr isotope values in freshwater systems depend on the bedrock geology of the drainage basins that are almost always different from seawater values and can vary significantly even among adjacent streams. Otolith $^{87}Sr/^{86}Sr$ values are potentially useful for identifying residency in specific brackish or freshwater habitats for those reef fish species that migrate to nearshore nursery areas as juveniles. Other isotope systems, including S and Pb, may be useful as natural tracers in otoliths. For instance, Weber et al. (2002) used natural variations in S isotope ratios in otoliths to distinguish between hatchery and naturally-spawned chinook salmon (*Onchorhynchus tshawytscha)*. Spencer et al. (2000) presented data that suggested Pb isotopes in otoliths may be a useful tracer of habitat residency in coral reef fishes collected from inshore habitats in

Kaneohe Bay, Hawaii. Lead isotope signatures from the Kaneohe Bay watershed were significantly different from those in adjacent ocean waters.

A need for more accurate temperature proxies in biogenic carbonates has led to considerable interest in new isotope systems displaying temperature dependent fractionation. For example, Ca isotopes have been shown to fractionate with temperature in foraminifera skeletons, resulting in a new method for reconstructing paleotemperatures (Nägler et al. 2000, Gussone et al. 2004). The use of Ca isotope fractionation to determine water temperatures is particularly exciting because, unlike oxygen isotopes, the approach should be insensitive to salinity effects in the oceans (e.g., Rosenheim et al. 2005). New experimental studies (Marriot et al. 2004, Gussone et al. 2005) have suggested that Ca isotope temperature dependence in biogenic aragonite and calcite is small (~5 ppm·amu^{-1}·°C^{-2}). However, the precision of Ca isotope ratio measurements using TIMS (Gussone et al. 2004) or multiple collector ICP-MS (Wieser et al. 2004) instruments is approximately 50 ppm and therefore useful temperature reconstructions based on Ca isotope fractionation in otolith aragonite appear unlikely at this stage. Nonetheless, we believe that, as in inorganic geochemistry (Halliday et al. 1998), significant advances in otolith chemistry will lie in the study of isotope systems that are not subject to biological fractionation. Such isotopes will likely record physicochemical properties of ambient environments with more fidelity than the elemental tracers commonly used in studies of otolith chemistry.

3 Technical information

3.1 SAMPLE PREPARATION

The following section provides details of the laboratory supplies, equipment, and facilities necessary for preparing otoliths for either solution-based or probe-based analysis, estimates of the time investments involved and protocols for how to prepare your own samples. We do not provide protocols for sectioning and polishing of otoliths as these are discussed in detail in Chapters 3 and 4.

3.1.1 Time investment

In our experience, sample preparation is the rate-limiting step in generating reliable data on otolith chemistry. A proficient researcher should be able to extract and mechanically remove adhering tissue at a rate of 5–10 fish.h^{-1}. We then transfer the otoliths to racks that can hold 30 samples for sonication. It takes approximately 30 min to rinse and sonicate one rack of samples. Samples will not be fully dry, however, for a further 12–24 h. A proficient researcher should be capable of cleaning 90 samples at a time with organic solvents, if necessary. The whole cleaning cycle takes ~30 h to complete. Cleaned otoliths have to be mounted, polished, and cleaned again for probe-based analyses. Decontaminating polished thin sections also requires ~30 h to complete and we typically will clean up to 60 samples (five 12-well tissue culture trays) in one cleaning run.

3.1.2 Material investment

Most of the supplies and equipment required to prepare samples for chemical analysis of otoliths are typically found in laboratories that routinely section otoliths for microstructure analysis. There are a few specific equipment items worth mentioning in the context of preparing otoliths for probe-based analysis.

One of the distinctions between preparing polished thin sections for chemical versus microstructure analysis is the thickness of the section. Most probes either only characterize the chemical composition of the surface of the section or, in the case of laser ablation, a few microns to tens of microns deep, depending on the laser settings. Thus, otoliths must be polished to the core to ensure accurate characterization of natal elemental signatures. We use a SBT polishing wheel and multilap sample polishing mount to achieve high precision polishes to the core without overgrinding. Samples are mounted for polishing on 18 mm acrylic discs using epoxy resin, which are attached to 18 mm round mounting stubs using a soft wax. Each metal stub attaches to individual micrometer pistons by magnetism. The multilap fixture is a tripod mount with 6 pistons. Each piston can be adjusted to precisely and independently control the amount of material removed from the surface of each sample. The fixture is rotated in place by an arm positioned above the lapping film on the polishing wheel. The cores of the mounting stubs are hollow, so samples can be removed easily and viewed under transmitted light with a compound microscope to assess the distance from the polished surface to the core (i.e., the amount of material to be removed). The ease of viewing samples, the precise control over sample removal and the ability to polish multiple samples simultaneously are all advantages over more traditional polishing methods.

3.1.3 Sample preparation protocols

The following protocols for preparing otoliths for either solution-based or probe-based chemical analysis involve the use of hazardous chemicals and appropriate operational health and safety guidelines must be followed at all times. It is the responsibility of the researcher to ensure that such guidelines are in place and that all individuals have been trained in the safe handling and use of these hazardous chemicals.

Sample removal and initial cleaning

Otoliths should be removed using non-metallic forceps or glass probes and then transferred to an acid-leached glass microscope slide or depression slide with a drop of ultrapure water. Adhering tissue can then be scraped off the otolith surface using forceps or probes. The cleaned otolith is then transferred to an acid-washed storage container and stored dry.

Mechanical removal of adhering tissue is reasonably effective but there will be residual organics coating the sample that may be a source of contamination and reduce the binding efficiency of the resin to the sample if the otolith is to be sectioned. This material can be cleaned from the otoliths by placing one otolith in a 0.5 ml microcentrifuge tube and filling the tube with 500 μL of hydrogen peroxide, buffered with NaOH (a 1:1 mixture of 30% hydrogen peroxide and 0.1 N NaOH, both high purity grade). Mount vials in a rack and partially submerge them in an ultrasonic water bath for 5 min, then transfer the racks to a laminar flow hood. Centrifuge the samples after ~12 h to make sure the otoliths are on the bottom of the vial, aspirate off as much of the cleaning solution as possible and add another 500 μL of cleaning solution. Aspirate off the cleaning solution after 24 h of cleaning and then rinse the samples in the vials twice with ultrapure water (sonicate for 5 min during each rinse). Then add ~500 μL of 0.001 N ultrapure HNO_3 and sonicate again for 5 min. Aspirate off the acid and follow with two more ultrapure water rinses (sonicating for 5 min again during each rinse). Aspirate off as much water as possible after the final rinse and leave the vials uncapped in a laminar flow hood until dry. Cap and store the otoliths until embedding, if they are to be sectioned, or transfer otoliths to acid-leached vials and store until dissolution for solution-based analysis.

Cleaning sectioned samples

Otoliths must be embedded, sectioned and polished to expose growth layers of interest for chemical profiling using probe-based analyses. Sectioning and polishing is also required for age and growth analysis so we refer the reader to Chapters 3 and 4 for detailed descriptions of these procedures. We recommend using epoxy resins as embedding compounds, 3M™ imperial diamond lapping films for polishing, and ultrapure water as a lubricant to minimise contamination during otolith preparation.

Otolith sections can be cleaned by placing mounted samples in individual wells of a 12-well tissue culture tray or a larger container, depending on the size of sample mounts, after rinsing the containers five times in ultrapure water prior to use. Add buffered hydrogen peroxide (see above) until wells are half full. Sonicate samples in an ultrasonic bath for 5 min (tissue culture trays can be "floated" on top of two stacked microcentrifuge racks in the ultrasonic bath), then leave uncovered in a laminar flow hood until dry. As with cleaning of unmounted otoliths, change the cleaning solution once during the 24 h period by aspirating off the cleaning solution and filling the wells half full again with fresh cleaning solution and sonicating for another 5 min.

Confirm that no otoliths have become dislodged from the resin after 24 h and then aspirate off the cleaning solution and wash samples twice with ultrapure water by filling up wells and then aspirating away the water (do not sonicate samples at this stage as there is some risk of otoliths becoming dislodged). Follow the water rinses with an acid leach. Add enough 0.001 N ultrapure $HN0_3$ to cover the discs completely and leach for 5 min. It is a good idea to add 18 Mohm water to the wells to dilute the acid before aspirating off the acid. Then aspirate off the diluted acid solution and follow with 2 more 18 Mohm water rinses (do not sonicate samples during rinses). Aspirate off as much water as possible after the final rinse and leave uncovered in a laminar flow hood until completely dry, after which the samples can be covered and stored indefinitely until chemically analysed.

3.2 SAMPLE ANALYSIS

3.2.1 Time and material investment

There are two distinct time components to the analysis of otolith chemistry using plasma source mass spectrometry. Initially, time needs to be invested in developing a familiarity with the instrumentation to be used. It is possible to get reliable data from ICP-MS analysis of otolith chemistry without understanding all the details of ICP mass spectrometry providing the samples are run by an experienced operator. Nonetheless, we have found it helpful to have at least some familiarity with the particular instrument being used. We find that students become reasonably proficient operators of the instrumentation in our respective labs in approximately 6 months.

Sample processing is relatively quick using ICP-MS in either solution or laser mode once the instrument has been optimized. Data acquisition for a multi-element method will normally take 1–2 min and so it is routine to be able to run at least 50 samples, along with blanks and standards, in a normal day of operation. It is difficult to assign a cost of running an individual sample as this will largely depend on the pricing structure of the facility.

3.2.2 Supplies and equipment

Most supplies and equipment needed for analyses of otolith chemistry, including the ICP-MS and laser ablation unit, will be provided by the ICP-MS facility. The facility will also in all likelihood supply expendables such as torches, injectors and sample cones that are specific to the instrument that they operate, along with gases required to operate the ICP-MS. We have found that it can be beneficial for researchers to purchase smaller items, however, including nebulizers and spray chambers, which are particularly useful for otolith analyses and may not be available at commercial labs.

Sample introduction of dissolved otoliths is usually achieved with a nebulizer that produces a wet aerosol that is then injected into a spray chamber before entering the mass spectrometer (Figure 9). We prefer low flow self-aspirating nebulizers, although some nebulizers require that a peristaltic pump be used to transport the sample solution. Sensitivity of the low flow nebulizers is surprisingly good despite the low sample volumes because droplet sizes in the resulting aerosol are typically small and relatively uniform, leading to high transport efficiency (Aeschilman et al. 2003). The choice of spray chambers will also influence instrument sensitivity and signal stability. Spray chambers are designed to remove large particles produced by nebulizers, leading to higher transport efficiency, lower formation of molecular interferences and better instrument precision (Montaser et al. 1998). Scott-type double pass spray chambers are perhaps the most common spray chamber used in ICP-MS. The reduced internal volume of cyclonic spray chambers offers lower memory effects and faster rinse out times than Scott-type chambers without sacrificing precision. Finally, there are at least two commercial desolvating nebulizer systems available that both increase sensitivity of many elements and significantly reduce the formation of oxides and hydrides. These nebulizers, while relatively expensive, may be useful for analysing elements found at ultra-trace concentrations in otoliths.

3.2.3 Instrument performance

Researchers hopefully will have an experienced technician who will be responsible for daily tuning and optimizing of the mass spectrometer. Nonetheless, it is advisable to track instrument performance during sessions. Our experience suggests that at least three parameters, sensitivity, stability and oxide levels, should be routinely monitored during otolith chemistry assays. Instrument sensitivity depends on a number of factors, including type of instrument and sample introduction system, and is quantified by measuring a standard solution (e.g., ^{115}In) of known concentration. Sensitivity of the *Element2* ICP-MS equipped with a 20 μL·min^{-1} low flow nebulizer and a cyclonic spray chamber is typically 1–2 million counts/ppb ^{115}In in our labs. Signal stability with this sample introduction system is ~1% (RSD) for a single 90 s analysis. Finally, oxide formation levels can be quantified by measuring a single isotope of an element (typically Ce or U) and an oxide of the same isotope and expressed as a ratio (i.e., CeO^+/Ce^+). We typically find oxide levels (UO^+/U^+) of 2–5% using our standard sample introduction system but much lower levels can be achieved using a desolvating nebulizer. These levels are particularly important when measuring isotopes with significant oxide interferences.

High or noisy background levels can also affect data quality in ICP-MS analyses of otolith chemistry. Instrument backgrounds can be assessed by running ultrapure solutions (typically 2% HNO_3) or by measuring gas blanks if operating in dry plasma mode. These blanks are typically minimized by regular maintenance of sample introduction systems and by using high purity Ar and He gases when running the ICP-MS. Procedural blanks are assessed by following the exact procedure used when preparing a sample for analysis without the sample itself. Sample vials, pipette tips and chemical reagents can all contain significant contamination if not cleaned appropriately or used in a suitably clean environment. Most researchers have prepared otoliths in class 100 clean hoods if they have not had access to clean rooms used for trace element analysis. Reagents used in the cleaning and digestion procedure should be as high a purity as is available. In any case, it is important to check for procedural contamination problems *before* digesting otoliths for analysis.

There are, of course, several other factors that may influence data quality when conducting elemental analyses using ICP-MS, including mass calibration, peak shape, and the condition of cones, injectors and torches. These factors, however, are typically not influenced by settings that are altered on a daily basis during instrument tuning and optimization. Certainly, the chances of a researcher producing accurate and precise measurements of otolith chemistry are enhanced when sensitivity, stability and oxide levels are maintained within normal operating limits and these parameters should always be checked before starting sample analyses. Similarly, every effort should be made to maintain low procedural and instrument blanks.

4 Conclusions

The field of otolith chemistry is still in its infancy. The earliest otolith chemical analyses post-date the first rigorous geochemical studies by more than three decades and the growing pains that the field of otolith chemistry are currently experiencing are reminiscent of similar trends in the evolution of geochemical tracer studies in marine carbonates (Lea 2003). Thus, there is room for optimism: our understanding of the processes involved in elemental incorporation will continue to improve as new studies contribute to a growing body of literature. Certainly, an improved mechanistic understanding of elemental partitioning during deposition is likely to lead to more accurate environmental reconstructions based on otolith chemistry, as has occurred in studies of aragonitic coral skeletons (Gaetani and Cohen 2006). New techniques have also been developed that use the otolith as a storage device for unique transgenerational isotope labels rather than as a record of natural environmental variations (Thorrold et al. 2006). For instance, mass-marking of larvae using enriched Ba isotopes recently has revealed significant local replenishment of coral reef fishes to a marine reserve in Papua New Guinea (Almany et al. 2007). Ultimately, we believe that otolith chemistry is poised to make significant contributions to the conservation and sustainable management of coral reef fishes because spatial management techniques, including the use of marine protected areas, require information on geographic structure and connectivity of reef fish populations. We hope that studies of otolith chemistry, perhaps in conjunction with coupled biophysical models (e.g., Cowen et al. 2006) or genetic markers (e.g. Jones et al. 2005), will generate innovative management strategies that will significantly improve the resilience of coral reef fish populations to natural and anthropogenic disturbances.

Acknowledgements

Support for the preparation of this manuscript was provided by NSF grants OCE-0134998, 0215905 and 0424688 to SRT, and ARC grants DP0345511 and DP0342784 to SES. We thank Larry Ball and Stan Hart for numerous discussions on plasma source mass spectrometry, and Jennifer FitzGerald for help with data analysis and graphics.

References

Aeschliman DB, Bajic SJ, Baldwin DP, Houk RS (2003) High-speed digital photographic study of an inductively coupled plasma during laser ablation: comparison of dried solution aerosols from a microconcentric nebulizer and solid particles from laser ablation. J Anal At Spectrom 18:1008–1014

Almany GR, Berumen ML, Thorrold SR, Planes S, Jones GP (2007) Local replenishment of coral reef fish populations in a marine reserve. Science 316:742–744

Andrews AH, Burton EJ, Kerr LA, Cailliet GM, Coale KH, Lundstrom CC, Brown TA (2005) Bomb radiocarbon and lead-radium disequilibria in otoliths of bocaccio rockfish (*Sebastes paucispinis*): a determination of age and longevity for a difficult-to-age fish. Mar Freshwater Res 56:517–528

Arai N, Sakamoto W, Maeda K (1995) Analysis of trace elements in otoliths of Red Sea bream *Pagrus major*. Fish Sci 61:43–47

Arai T, Kotake A, Lokman PM, Miller MJ, Tsukamoto KK (2004) Evidence of different habitat use by New Zealand freshwater eels *Anguilla australis* and *A. dieffenbachii*, as revealed by otolith microchemistry. Mar Ecol Prog Ser 266:213–225

Arai T, Otake T, Tsukamoto K (1997) Drastic changes in otolith microstructure and microchemistry accompanying the onset of metamorphosis in the Japanese eel *Anguilla japonica*. Mar Ecol Prog Ser 161:17–22

Arai T, Sato H, Ishii T, Tsukamoto K (2003) Alkaline earth metal and Mn distribution in otoliths of *Anguilla* spp. glass eels and elvers. Fish Sci 69:421–423

Arslan Z (2005) Analysis of fish otoliths by electrothermal vaporization inductively coupled plasma mass spectrometry: aspects of precipitating otolith calcium with hydrofluoric acid for trace element determination. Talanta 65:1326–1334

Arslan Z, Paulson AJ (2003) Solid phase extraction for analysis of biogenic carbonates by electrothermal vaporization inductively coupled plasma mass spectrometry: an investigation of rare earth element signatures in otolith microchemistry. Anal Chim Acta 476:1–13

Arslan Z, Secor DH (2005) Analysis of trace transition elements and heavy metals in fish otoliths as tracers of habitat use by American eels in the Hudson River estuary. Estuaries 28:382–393

Bacon CR, Weber PK, Larsen KA, Reisenbichler R, Fitzpatrick JA, Wodden JL (2004) Migration and rearing histories of chinook salmon (*Onchorhyncus tshawytscha*) determined by ion microprobe Sr isotope and Sr/Ca transects of otoliths. Can J Fish Aquat Sci 61:2425–2439

Baker MS, Wilson CA (2001) Use of bomb radiocarbon to validate otolith section ages of red snapper *Lutjanus campechanus* from the northern Gulf of Mexico. Limnol Oceanogr 46:1819–1824

Baker MS, Wilson CA, VanGent DL (2001) Testing assumptions of otolith radiometric aging with two long-lived fishes from the northern Gulf of Mexico. Can J Fish Aquat Sci 58:1244–1252

Barbee NC, Swearer SE (2007) Characterizing natal source population signatures in the diadromous fish *Galaxias maculatus*, using embryonic otolith chemistry. Mar Ecol Prog Ser 343:273–282

Barnett-Johnson R, Ramos FC, Grimes CB, MacFarlane RB (2005) Validation of Sr isotopes in otoliths by laser ablation multicollector inductively coupled plasma mass spectrometry (LA-MC-ICPMS): opening avenues in fisheries science applications. Can J Fish Aquat Sci 62:2425–2430

Bath GE, Thorrold SR, Jones CM, Campana SE, McLaren JW, Lam JWH (2000) Strontium and barium uptake in aragonitic otoliths of marine fish. Geochim Cosmochim Acta 64:1705–1714

Beck MW, Heck KL, Able KW, Childers DL, et al. (2001) The identification, conservation, and management of estuarine and marine nurseries for fish and invertebrates. Bioscience 51:633–641

Begg GA, Cappo M, Cameron DS, Boyle S, Sellin MJ (1998) Stock discrimination of school mackerel, *Scomberomorus queenslandicus*, and spotted mackerel, *Scomberomorus munroi*, in coastal waters of eastern Australia by analysis of minor and trace elements in whole otoliths. Fish Bull (Wash DC) 96:653–666

Bergenius MAJ, Mapstone BD, Begg GA, Murchie CD (2005) The use of otolith chemistry to determine stock structure of three epinepheline serranid coral reef fishes on the Great Barrier Reef, Australia. Fish Res 72:253–270

Bergenius MAJ, Meekan MG, Robertson DR, McCormick MI (2002) Larval growth predicts the recruitment success of a coral reef fish. Oecologia 131:521–525

Bian QZ, Koch J, Lindner H, Bberndt H, Hergenroder R, Niemax K (2005) Non-matrix matched calibration using near-IR femtosecond laser ablation inductively coupled plasma optical emission spectrometry. J Anal At Spectrom 20:736–740

Bings NH, Bogaerts A, Broekaert JAC (2004) Atomic spectroscopy. Anal Chem 76:3313–3336

Blum JD, Taliaferro EH, Weisse MT, Holmes RT (2000) Changes in Sr/Ca, Ba/Ca and ^{87}Sr/^{86}Sr ratios between trophic levels in two forest ecosystems in the northeastern USA. Biogeochemistry 49:87–101

Boiseau M, Juillet-Leclerc A (1997) H_2O_2 treatment of recent coral aragonite: oxygen and carbon isotopic implications. Chem Geol 143:171–180

Brophy D, Danilowicz BS (2002) Tracing populations of Atlantic herring (*Clupea harengus* L.) in the Irish and Celtic Seas using otolith microstructure. ICES J Mar Sci 59:1305–1313

Brophy D, Danilowicz BS, Jeffries TE (2003) The detection of elements in larval otoliths from Atlantic herring using laser ablation ICP-MS. J Fish Biol 63:990–1007

Brophy D, Jeffries TE, Danilowicz BS (2004) Elevated manganese concentrations at the cores of clupeid otoliths: possible environmental, physiological, or structural origins. Mar Biol 144:779–786

Brothers EB, McFarland WN (1981) Correlations between otolith microstructure, growth, and life history transitions in newly recruited French grunts [*Haemulon flavolineatum* (Desmarest), Haemulidae]. Rapp P v Reun Cons Int Explor Mer 187:369–374

Buckel JA, Sharack BL, Zdanowicz VS (2004) Effects of diet on otolith composition in *Pomatomus saltatrix*, an estuarine piscivore. J Fish Biol 64:1469–1484

Campana SE (1999) Chemistry and composition of fish otoliths: pathways, mechanisms and applications. Mar Ecol Prog Ser 188:263–297

Campana SE (2005) Otolith science entering the 21st century. Mar Freshwater Res 56:485–495

Campana SE, Chouinard GA, Hanson JM, Frechet A (1999) Mixing and migration of overwintering Atlantic cod (*Gadus morhua*) stocks near the mouth of the Gulf of St. Lawrence. Can J Fish Aquat Sci 56:1873–1881

Campana SE, Chouinard GA, Hanson JM, Frechet A, Brattey J (2000) Otolith elemental fingerprints as biological tracers of fish stocks. Fish Res 46:343–357

Campana SE, Fowler AJ, Jones CM (1994) Otolith elemental fingerprinting for stock identification of Atlantic cod (*Gadus morhua*) using laser ablation ICPMS. Can J Fish Aquat Sci 51:1942–1950

Campana SE, Gagné JA (1995) Cod stock discrimination using ICPMS elemental assays of otoliths. In: Secor DH, Dean JM, Campana SE (Eds) Recent developments in fish otolith research. University of South Carolina Press, Columbia, South Carolina

Campana SE, Gagné JA, McLaren JW (1995) Elemental fingerprinting of fish otoliths using ID-ICPMS. Mar Ecol Prog Ser 122:115–120

Campana SE, Neilson JD (1985) Microstructure of fish otoliths. Can J Fish Aquat Sci 42:1014–1032

Campana SE, Oxenford HA, Smith JN (1993) Radiochemical determination of longevity in flyingfish *Hirundichthys affinis* using Th-228/Ra-228. Mar Ecol Prog Ser 100:211–219

Campana SE, Thorrold SR (2001) Otoliths, increments, and elements: keys to a comprehensive understanding of fish populations? Can J Fish Aquat Sci 58:30–38

Campana SE, Thorrold SR, Jones CM, Günther D, Tubrett M, ongerich HL, Jackson S, Halden NM, Kalish JM, Piccoli P, de Pontual H, Troadec H, Panfili J, Secor DH, Severin KP, Sie SH, Thresher R, Teesdale WJ, Campbell JL (1997) Comparison of accuracy, precision, and sensitivity in elemental assays of fish otoliths using the electron microprobe, proton-induced X-ray emission, and laser ablation inductively coupled plasma mass spectrometry. Can J Fish Aquat Sci 54:2068–2079

Campana SE, Zwanenburg KCT, Smith JN (1990) $^{210}Pb/^{226}Ra$ determination of longevity in redfish. Can J Fish Aquat Sci 47:163–165

Chesney EJ, McKee BM, Blanchard T, Chan LH (1998) Chemistry of otoliths from juvenile menhaden *Brevoortia patronus*: evaluating strontium, strontium/calcium and strontium isotope ratios as environmental indicators. Mar Ecol Prog Ser 171:261–273

Chittaro PM, Fryer BJ, Sale PF (2004) Discrimination of French grunts (*Haemulon flavolineatum* Desmarest, 1823) from mangrove and coral reef habitats using otolith microchemistry. J Exp Mar Biol Ecol 308:169–183

Chittaro PM, Usseglio P, Fryer BJ, Sale PF (2005) Using otolith microchemistry of *Haemulon flavolineatum* (French Grunt) to characterize mangroves and coral reefs throughout Turneffe Atoll, Belize: difficulties at small spatial scales. Estuaries 28:373–381

Choat JH, Robertson DR (2002) Age-based studies. In: Sale PF (Ed) Coral reef fishes: dynamics and diversity in a complex ecosystem. Academic Press, San Diego

Christensen JN, Halliday AN, Lee D-C, Hall CM (1995) In situ Sr isotopic analysis by laser ablation. Earth Planet Sci Lett 136:79–85

Clarke LM, Friedland KD (2004) Influence of growth and temperature on strontium deposition in the otoliths of Atlantic salmon. J Fish Biol 65:744–759

Cowen RK, Paris CB, Srinivasan A (2006) Scaling of connectivity in marine populations. Science 311:522–527

Denit K, Sponaugle S (2004) Growth variation, settlement, and spawning of gray snapper across a latitudinal gradient. Trans Am Fish Soc 133:1339–1355

de Pontual H, Lagardere F, Amara R, Bohn M, Ogor A (2003) Influence of ontogenetic and environmental changes in the otolith microchemistry of juvenile sole (*Solea solea*). J Sea Res 50:199–210

DeVries DA, Grimes CB, Prager MH (2002) Using otolith shape analysis to distinguish eastern Gulf of Mexico and Atlantic Ocean stocks of king mackerel. Fish Res 57:51–62

de Vries MC, Gillanders BM, Elsdon TS (2005) Facilitation of barium uptake into fish otoliths: influence of strontium concentration and salinity. Geochim Cosmochim Acta 69:4061–4072

Dorval E, Jones CM, Hannigan R, van Montfrans J (2005) Can otolith chemistry be used for identifying essential seagrass habitats for juvenile spotted seatrout, *Cynoscion nebulosus*, in Chesapeake Bay? Mar Freshwater Res 56:645–654

Dove SG, Gillanders BM, Kingsford MJ (1996) An investigation of chronological differences in the deposition of trace metals in the otoliths of two temperature reef fishes. J Exp Mar Biol Ecol 205:15–33

Durrant SF, Ward NI (2005) Recent biological and environmental applications of laser ablation inductively coupled plasma mass spectrometry (LA-ICP-MS). J Anal At Spectrom 20:821–829

Edmonds JS, Caputi N, Morita M (1991) Stock discrimination by trace-element analysis of otoliths of orange roughy (*Hoplostethus atlanticus*), a deep-water marine teleost. Aust J Mar Freshwater Res 42:383–390

Edmonds JS, Lenanton RCJ, Caputi N, Morita M (1992) Trace elements in the otoliths of yellow-eye mullet (*Aldrichetta forsteri*) as an aid to stock identification. Fish Res 13:39–52

Edmonds JS, Moran MJ, Caputi N, Morita M (1989) Trace element analysis of fish sagittae as an aid to stock identification: pink snapper (*Chrysophrys auratus*) in Western Australian waters. Can J Fish Aquat Sci 46:50–54

Edmonds JS, Steckis RA, Moran MJ, Caputi N, Morita M (1999) Stock delineation of pink snapper and tailor from Western Australia by analysis of stable isotope and strontium/calcium ratios in otolith carbonate. J Fish Biol 55:243–259

Eggins SM, Kingsley LPJ, Shelley JMG (1998) Deposition and elemental fractionation processes during atmospheric pressure laser ablation sampling for analysis by ICP-MS. Appl Surf Sci 129:278–286

Elsdon TS, Gillanders BM (2002) Interactive effects of temperature and salinity on otolith chemistry: challenges for determining environmental histories of fish. Can J Fish Aquat Sci 59:1796–1808

Elsdon TS, Gillanders BM (2003) Relationship between water and otolith elemental concentrations in juvenile black bream *Acanthopagrus butcheri*. Mar Ecol Prog Ser 260:263–272

Elsdon TS, Gillanders BM (2004) Fish otolith chemistry influenced by exposure to multiple environmental variables. J Exp Mar Biol Ecol 313:269–284

Elsdon TS, Gillanders BM (2005) Alternative life-history patterns of estuarine fish: barium in otoliths elucidates freshwater residency. Can J Fish Aquat Sci 62:1143–1152

Farrell J, Campana SE (1996) Regulation of calcium and strontium deposition on the otoliths of juvenile tilapia, *Oreochromis niloticus*. Comp Biochem Physiol A 115:103–109

Fassett JD, Paulson PJ (1989) Isotope dilution mass spectrometry for accurate elemental analysis. Anal Chem 61:643A–649A

Fisher R, Bellwood DR, Job SD (2000) Development of swimming abilities in reef fish larvae. Mar Ecol Prog Ser 202:163–173

FitzGerald JL, Thorrold SR, Bailey KM, Brown AL, Severin KP (2004) Elemental signatures in otoliths of larval walleye Pollock (*Theragra chalcogramma*) from the northeast Pacific Ocean. Fish Bull (Wash DC) 102:604–616

Flik G, Verbost PM (1993) Calcium transport in fish gills and intestine. J Exp Biol 184:17–29

Forrester GE, Swearer SE (2002) Trace elements in otoliths indicate the use of open-coast versus bay nursery habitats by juvenile California halibut. Mar Ecol Prog Ser 241:201–213

Fowler AJ (1995) Annulus formation in coral-reef fish. In: Secor DH, Dean JM, Campana SE (Eds) Recent developments in fish otolith research. University of South Carolina Press, Columbia, South Carolina

Fowler AJ, Campana SE, Jones CM, Thorrold SR (1995a) Experimental assessment of the effect of temperature and salinity on elemental composition of otoliths using solution-based ICPMS. Can J Fish Aquat Sci 52:1421–1430

Fowler AJ, Campana SE, Jones CM, Thorrold SR (1995b) Experimental assessment of the effect of temperature and salinity on elemental composition of otoliths using laser ablation ICPMS. Can J Fish Aquat Sci 52:1431–1441

Gaetani GA, Cohen AL (2006) Element partitioning during precipitation of aragonite from seawater: a framework for understanding paleoproxies. Geochim Cosmochim Acta 70:4617–4634

Gaffey SJ, Bronnimann CE (1993) Effects of bleaching on organic and mineral phases in biogenic carbonates. J Sediment Petrol 63:752–754

Gallahar NK, Kingsford MJ (1992) Patterns of increment width and strontium: calcium ratios in otoliths of juvenile rock blackfish, *Girella elevata* (M.). J Fish Biol 41:749–763

Gallahar NK, Kingsford MJ (1996) Factors influencing Sr/Ca ratios in otoliths of *Girella elevata*: An experimental investigation. J Fish Biol 48:174–186

Gauldie RW (1996) Effects of temperature and vaterite replacement on the chemistry of metal ions in the otoliths of *Oncorhynchus tshawytscha*. Can J Fish Aquat Sci 53:2015–2026

Gauldie RW, Coote GC, Mulligan KP, West IF (1992) A chemical probe of the microstructural organization of fish otoliths. Comp Biochem Physiol A Comp Physiol 102:533–545

Gauldie RW, Fournier DA, Dunlop DE, Coote G (1986) Atomic emission and proton microprobe studies of the ion content of otoliths of chinook salmon aimed at recovering the temperature life history of individuals. Comp Biochem Physiol A Comp Physiol 84:607–616

Geffen AJ, Pearce NJG, Perkins WT (1998) Metal concentrations in fish otoliths in relation to body composition after laboratory exposure to mercury and lead. Mar Ecol Prog Ser 165:235–245

Gillanders BM (2001) Trace metals in four structures of fish and their use for estimates of stock structure. Fish Bull (Wash DC) 99:410–419

Gillanders BM (2002a) Connectivity between juvenile and adult fish populations: do adults remain near their recruitment estuaries? Mar Ecol Prog Ser 240:215–223

Gillanders BM (2002b) Temporal and spatial variability in elemental composition of otoliths: implications for determining stock identity and connectivity of populations. Can J Fish Aquat Sci 59:669–679

Gillanders BM, Kingsford MJ (1996) Elements in otoliths may elucidate the contribution of estuarine recruitment to sustaining coastal reef populations of a temperate reef fish. Mar Ecol Prog Ser 141:13–20

Gillanders BM, Kingsford MJ (2000) Elemental fingerprints of otoliths of fish may distinguish estuarine 'nursery' habitats. Mar Ecol Prog Ser 201:273–286

Gillanders BM, Kingsford MJ (2003) Spatial variation in elemental composition of otoliths of three species of fish (family Sparidae). Est Coast Shelf Sci 57:1049–1064

Gillanders BM, Sanchez-Jerez P, Bayle-Sempere J, Ramos-Espla A (2001) Trace elements in otoliths of the two-banded bream from a coastal region in the south-west Mediterranean: are there differences among locations? J Fish Biol 59:350–363

Grady JR, Johnson AG, Sanders M (1989) Heavy metal content in otoliths of king mackerel (*Scomberomorus cavalla*) in relation to body length and age. Contrib Mar Sci 31:17–24

Gray AL (1985) Solid sample introduction by laser ablation for inductively coupled plasma source-mass spectrometry. The Analyst 110:551–556

Gray JS (2002) Biomagnification in marine systems: the perspective of an ecologist. Mar Pollut Bull 45:46–52

Guido P, Omori M, Katayama S, Kimura K (2004) Classification of juvenile rockfish, *Sebastes inermis*, to *Zostera* and *Sargassum* beds, using the macrostructure and chemistry of otoliths. Mar Biol 145:1243–1255

Guilderson TP, Fairbanks RG, Rubenstone JL (1994) Tropical temperature variations since 20,000 years ago: modulating inter-hemispheric climate change. Science 263:663–665

Gunn JS, Harrowfield IR, Proctor CH, Thresher RE (1992) Electron probe microanalysis of fish otoliths: evaluation of techniques for studying age and stock discrimination. J Exp Mar Biol Ecol 158:1–36

Günther D, Frischknecht R, Heinrich CA, Kahlert HJ (1997) Capabilities of an argon fluoride 193 nm excimer laser for laser ablation inductively coupled plasma mass spectrometry microanalysis of geological materials. J Anal At Spectrom 12:939–944

Günther D, Heinrich CA (1999) Enhanced sensitivity in laser ablation-ICP mass spectrometry using helium-argon mixtures as aerosol carrier. J Anal At Spectrom 14:1363–1368

Gussone N, Böhm F, Eisenhauer A, Dietzel M, Heuser A, Teichert BAA, Reitnner J, Wörheide G, Dullo WC (2005) Calcium isotope fractionation in calcite and aragonite. Geochim Cosmochim Acta 69:4485–4494

Gussone N, Einsenhauer A, Heuser A, Dietzel M, Bock B, Böhm F, Spero HJ, Lea DW, Bijma J, Nägler TF (2004) δ^{44}Ca, δ^{18}O and Mg/Ca reveal Caribbean Sea surface temperature and salinity fluctuations during the Pliocene closure of the Central-American gateway. Earth Planet Sci Lett 227:201–214

Halliday AN, Lee D-C, Christensen JN, Rehkamper M, Yi W, Luo X, Hall CM, Ballentine CJ, Pettkke T, Stirling C (1998) Applications of multi-collector ICPMS to cosmochemistry, geochemistry, and paleoceanography. Geochim Cosmochim Acta 62:919–940

Halpern BS (2004) Are mangroves a limiting resource for two coral reef fishes? Mar Ecol Prog Ser 272:93–98

Hamer PA, Jenkins GP, Gillanders BM (2003) Otolith chemistry of juvenile snapper *Pagrus auratus* in Victorian water: natural chemical tags and their temporal variation. Mar Ecol Prog Ser 263:261–273

Hanson PJ, Koenig CC, Zdanowicz VS (2004) Elemental composition of otoliths used to trace estuarine habitats of juvenile gag *Mycteroperca microlepis* along the west coast of Florida. Mar Ecol Prog Ser 267:253–265

Hanson PJ, Zdanowicz VS (1999) Elemental composition of otoliths from Atlantic croaker along an estuarine pollution gradient. J Fish Biol 54:656–668

Harlan JA, Swearer SE, Leben RR, Fox CA (2002) Surface circulation in a Caribbean island wake. Cont Shelf Res 22:417–434

Hedges KJ, Ludsin SA, Fryer BJ (2004) Effects of ethanol preservation on otolith microchemistry. J Fish Biol 64:923–937

Hobbs JA, Yin QZ, Burton J, Bennett WA (2005) Retrospective determination of natal habitats for an estuarine fish with otolith strontium isotope ratios. Mar Freshwater Res 56:655–660

Hoff GR, Fuiman LA (1993) Morphometry and composition of red drum otoliths: changes associated with temperature, somatic growth rate, and age. Comp Biochem Physiol A 106:209–219

Hoff GR, Fuiman LA (1995) Environmentally induced variation in elemental composition of red drum (*Sciaenops ocellatus*) otoliths. Bull Mar Sci 56:578–591

Horlick G, Montaser A (1998) Analytical characteristics of ICPMS. In: Montaser A (Ed) Inductively coupled plasma mass spectrometry. Wiley-VHC, New York

Humphreys RL, Campana SE, DeMartini E (2005) Otolith elemental fingerprints of juvenile Pacific swordfish *Xiphias gladius*. J Fish Biol 66:1660–1670

Jeffries TE, Jackson SE, Longerich HP (1998) Application of a frequency quintupled Nd:YAG source (λ = 213 nm) for laser ablation inductively coupled plasma mass spectrometric analysis of minerals. J Anal At Spectrom 13:935–940

Jones GP, Planes S, Thorrold SR (2005) Coral reef fish larvae settle close to home. Curr Biol 15:1314–1318

Kalish JM (1989) Otolith microchemistry: validation of the effects of physiology, age and environment on otolith composition. J Exp Mar Biol Ecol 132:151–178

Kalish JM (1990) Use of otolith microchemistry to distinguish the progeny of sympatric anadromous and non-anadromous salmonids. Fish Bull (Wash DC) 88:657–666

Kalish JM (1991) Determinants of otolith chemistry: seasonal variation in the composition of blood plasma, endolymph and otoliths of bearded rock cod *Pseudophycis barbatus*. Mar Ecol Prog Ser 74:137–159

Kalish JM (2001) Use of the bomb radiocarbon chronometer to validate fish age. Final Report FRDC Project 93/109, Fisheries Research and Development Corporation, Canberra, Australia

Kennedy BP, Blum JD, Folt CL, Nislow KH (2000) Using natural strontium isotopic signatures as fish markers: methodology and application. Can J Fish Aquat Sci 57:2280–2292

Kennedy BP, Folt CL, Blum JD, Chamberlain CP (1997) Natural isotope markers in salmon. Nature 387:766–767

Kraus RT, Secor DH (2004a) Incorporation of strontium into otoliths of an estuarine fish. J Exp Mar Biol Ecol 302:85–106

Kraus RT, Secor DH (2004b) Dynamics of white perch *Morone americana* population contingents in the Patuxent River estuary, Maryland USA. Mar Ecol Prog Ser 279:247–259

Lange N, Swearer S, Sturner WQ (1994) Human postmortem interval estimation from vitreous potassium: an analysis of original data from six different studies. Forensic Sci Int 66:159–174

Lea DW (2003) Elemental and isotopic proxies of marine temperatures. In: Elderfield H (Ed) The oceans and marine geochemistry. Elsevier-Pergamon, Oxford

Lea DW, Mashiotta TA, Spero HJ (1999) Controls on magnesium and strontium uptake in planktonic foraminifera determined by live culturing. Geochim Cosmochim Acta 63:2369–2379

Limburg KE (1995) Otolith strontium traces environmental history of sub-yearling American shad *Alosa sapidissima*. Mar Ecol Prog Ser 119:25–35

Limburg KE (2001) Through the gauntlet again: demographic restructuring of American shad by migration. Ecology 82:1584–1596

Long AM, Wang WX (2005) Metallothionein induction and bioaccumulation kinetics of Cd and Ag in a marine fish *Terapon jarbua* challenged with dietary and waterborne Ag and Cu. Mar Ecol Prog Ser 291:215–226

Love KM, Woronow A (1991) Chemical changes induced in aragonite using treatments for the destruction of organic material. Chem Geol 93:291–301

Marriot CS, Henderson GM, Belshaw NS, Tudhope AW (2004) Temperature dependence of δ^{7}Li, δ^{44}Ca and Li/Ca during growth of calcium carbonate. Earth Planet Sci Lett 222:615–624

Martin GB, Thorrold SR (2005) Temperature and salinity effects on magnesium, manganese, and barium incorporation in otoliths of larval and early juvenile spot *Leiostomus xanthurus*. Mar Ecol Prog Ser 293:223–232

Martin GB, Thorrold SR, Jones CM (2004) Temperature and salinity effects on strontium incorporation in otolith of larval spot (*Leiostomus xanthurus*). Can J Fish Aquat Sci 61:34–42

Martin PA, Lea DW (2002) A simple evaluation of cleaning procedures on fossil benthic foraminiferal Mg/Ca. Geochem Geophys Geosyst 3:doi:10.1029/2001GC000280

McCulloch M, Cappo M, Aumend J, Muller W (2005) Tracing the life history of individual barramundi using laser ablation MC-ICP-MS Sr-isotopic and Sr/Ba ratios in otoliths. Mar Freshwater Res 56:637–644

McCulloch MT, Gagan MK, Mortimer GE, Chivas AR, Isdale PJ (1994) A high resolution Sr/Ca and δ^{18}O coral record from the Great Barrier Reef, Australia and the 1982–83 El Niño. Geochim Cosmochim Acta 58:2747–2754

Millar RB (1990) Comparison of methods for estimating mixed stock fishery composition. Can J Fish Aquat Sci 47:2235–2241

Miller JA, Shanks AL (2004) Evidence for limited larval dispersal in black rockfish (*Sebastes melanops*): implications for population structure and marine-reserve design. Can J Fish Aquat Sci 61:1723–1735

Milton DA, Chenery SR (1998) The effect of otolith storage methods on the concentrations of elements detected by laser-ablation ICPMS. J Fish Biol 53:785–794

Milton DA, Chenery SR (2001a) Sources and uptake of trace metals in otoliths of juvenile barramundi (*Lates calcarifer*). J Exp Mar Biol Ecol 264:47–65

Milton DA, Chenery SR (2001b) Can otolith chemistry detect the population structure of the shad hilsa *Tenualosa ilisha*? Comparison with the results of genetic and morphological studies. Mar Ecol Prog Ser 222:239–251

Milton DA, Chenery SR (2003) Movement patterns of the tropical shad hilsa (*Tenualosa ilisha*) inferred from transects of ^{87}Sr/^{86}Sr isotope ratios in their otoliths. Can J Fish Aquat Sci 60:1376–1385

Milton DA, Chenery SR, Farmer MJ, Blaber SJM (1997) Identifying the spawning estuaries of the tropical shad, terubok *Tenualosa toli*, using otolith microchemistry. Mar Ecol Prog Ser 153:283–291

Milton DA, Short SA, O'Neill MF, Blaber SJM (1995) Aging of 3 species of tropical snapper (Lutjanidae) from the Gulf of Carpenteria, Australia, using radiometry and otolith ring counts. Fish Bull (Wash DC) 93:103–115

Milton DA, Tenakanai CD, Chenery SR (2000) Can the movements of barramundi in the Fly River region, Papua New Guinea be traced in their otoliths? Est Coast Shelf Sci 50:855–868

Montaser A, Minnich MG, McLean JA, Liu H (1998) Sample introduction in ICPMS. In: Montaser A (Ed) Inductively coupled plasma mass spectrometry. Wiley-VHC, New York

Morris JA, Rulifson RA, Toburen LH (2003) Life history strategies of striped bass, *Morone saxatilis*, populations inferred from otolith microchemistry. Fish Res 62:53–63

Mugiya Y, Hakomori T, Hatsutori K (1991) Trace metal incorporation into otoliths and scales in the goldfish, *Carassius auratus*. Comp Biochem Physiol C 99:327–332

Mugiya Y, Satoh C (1997) Strontium accumulation in slow-growing otoliths in the goldfish *Carassius auratus*. Fish Sci 63:361–364

Mugiya Y, Uchimura T (1989) Otolith resorption induced by anaerobic stress in goldfish, *Carassius auratus*. J Fish Biol 35:813–818

Mulligan TJ, Martin FD, Smucker RA, Wright DA (1987) A method of stock identification based on the elemental composition of striped bass *Morone saxatilis* (Walbaum) otoliths. J Exp Mar Biol Ecol 114:241–248

Munro AR, McMahon TE, Ruzycki JR (2005) Natural chemical markers identify source and date of introduction of an exotic species: lake trout (*Salvelinus namaycush*) in Yellowstone Lake. Can J Fish Aquat Sci 62:79–87

Nagelkerken I, van der Velde G (2004) Are Caribbean mangroves important feeding grounds for juvenile reef fish from adjacent seagrass beds? Mar Ecol Prog Ser 274:143–151

Nagelkerken I, van der Velde G, Gorissen MW, Meijer GJ, van't Hof T, den Hartog C (2000) Importance of mangroves, seagrass beds and the shallow coral reef as a nursery for important coral reef fishes, using a visual census technique. Est Coast Shelf Sci 51:31–44

Nägler TF, Eisenhauer A, Müller A, Hemleben C, Kramers J (2000) The δ^{44}Ca–temperature calibration on fossil and cultured *Globigerinoides sacculifer*: new tool for reconstruction of past sea surface temperatures. Geochem Geophys Geosyst 1:doi:10.1029/2000GC000091

Okumura M, Kitano Y (1986) Coprecipitation of alkali metal ions with calcium carbonate. Geochim Cosmochim Acta 50:49–58

Outridge PM, Chenery SR, Babaluk JA, Reist JD (2002) Analysis of geological Sr isotope markers in fish otoliths with subannual resolution using laser ablation-multicollector-ICP-mass spectrometry. Environ Geol 42:891–899

Pannella G (1971) Fish otoliths: daily growth layers and periodical patterns. Science 173:1124–1127

Patterson HM, Kingsford MJ, McCulloch MT (2004a) Elemental signatures of *Pomacentrus coelestis* otoliths at multiple spatial scales on the Great Barrier Reef, Australia. Mar Ecol Prog Ser 270:229–239

Patterson HM, Kingsford MJ, McCulloch MT (2004b) The influence of oceanic and lagoonal plume waters on otolith chemistry. Can J Fish Aquat Sci 61:898–904

Patterson HM, Kingsford MJ, McCulloch MT (2005) Resolution of the early life history of a reef fish using otolith chemistry. Coral Reefs 24:222–229

Patterson HM, McBride RS, Julien N (2004c) Population structure of red drum (*Sciaenops ocellatus*) as determined by otolith chemistry. Mar Biol 144:855–862

Patterson HM, Swearer SE (2007) Long-distance dispersal and local retention of larvae as mechanisms of recruitment in an island population of a coral reef fish. Austral Ecol 32:122–130

Patterson HM, Thorrold SR, Shenker JM (1999) Analysis of otolith chemistry in Nassau grouper (*Epinephelus striatus*) from the Bahamas and Belize using solution-based ICP-MS. Coral Reefs 18:171–178

Proctor CH, Thresher RE (1998) Effects of specimen handling and otolith preparation on concentration of elements in fish otoliths. Mar Biol 131:681–694

Proctor CH, Thresher RE, Gunn JS, Mills DJ, Harrowfield IR, Sie SH (1995) Stock structure of the southern bluefin tuna *Thunnus maccoyii*: an investigation based on probe microanalysis of otolith composition. Mar Biol 122:511–526

Quinn TJ (1993) A review of homing and straying of wild and hatchery-produced salmon. Fish Res 18:29–44

Radtke RL (1985) Recruitment parameters resolved from structural and chemical components of juvenile *Dascyllus albisella* otoliths. Proc 6th Int Coral Reef Congr 5:397–401

Radtke RL (1989) Strontium-calcium concentration ratios in fish otoliths as environmental indicators. Comp Biochem Physiol A 92:198–194

Radtke RL, Kinzie RA III (1996) Evidence of a marine larval stage in endemic Hawaiian stream gobies from isolated high-elevated locations. Trans Am Fish Soc 125:613–621

Radtke RL, Kinzie RA III, Folsom SD (1988) Age at recruitment of Hawaiian [USA] freshwater gobies. Environ Biol Fish 23:205–214

Radtke RL, Townsend DW, Folsom SD, Morrison MA (1990) Strontium: calcium concentration ratios in otoliths of herring larvae as indicators of environmental histories. Environ Biol Fish 27:51–62

Rooker JR, Secor DH, Zdanowicz VS, de Metrio G, Relini LO (2003) Identification of Atlantic bluefin tuna (*Thunnus thynnus*) stocks from putative nurseries using otolith chemistry. Fish Oceanogr 12:75–84

Rooker JR, Secor DH, Zdanowicz VS, Itoh T (2001a) Discrimination of northern bluefin tuna from nursery areas in the Pacific Ocean using otolith chemistry. Mar Ecol Prog Ser 218:275–282

Rooker JR, Zdanowicz VS, Secor DH (2001b) Chemistry of tuna otoliths: assessment of base composition and postmortem handling effects. Mar Biol 139:35–43

Rosenheim BE, Swart PK, Thorrold SR, Eisenhauer A, Willenz P (2005) Salinity change in the subtropical Atlantic: secular increase and teleconnections to the North Atlantic Oscillation. Geophys Res Lett 32:L02603, doi:10.1029/2004GL021499

Rosenthal Y, Field PM, Sherrell RM (1999) Precise determination of element/calcium ratios in calcareous samples using sector field inductively coupled plasma mass spectrometry. Anal Chem 71:3248–3253

Roughgarden J, Gaines S, Possingham H (1988) Recruitment dynamics in complex life cycles. Science 241:1460–1466

Russo RE, Mao X, Gonzales JJ, Mao SS (2002) Femtosecond laser ablation ICP-MS. J Anal At Spectrom 17:1072–1075

Ruttenberg BI, Hamilton SL, Hickford MJH, Paradis G, Sheehy M, Standish JD, Ben-Tzvi O, Warner RR (2005) Elevated levels of trace elements in cores of otoliths and their potential use as natural tags. Mar Ecol Prog Ser 297:273–281

Ruttenberg BI, Warner RR (2006) Spatial variation in the chemical composition of natal otoliths from a reef fish in the Galápagos Islands. Mar Ecol Prog Ser 328:225–236

Sandin SA, Regetz J, Hamilton SL (2005) Testing larval fish dispersal hypotheses using maximum likelihood analysis of otolith chemistry data. Mar Freshwater Res 56:725–734

Searcy SP, Sponaugle S (2001) Selective mortality during the larval-juvenile transition in two coral reef fishes. Ecology 82:2452–2470

Secor DH (1992) Application of otolith microchemistry analysis to investigate anadromy in Chesapeake Bay striped bass *Morone saxatilis*. Fish Bull (Wash DC) 90:798–806

Secor DH, Campana SE, Zdanowicz VS, Lam JWH, Yang L, Rooker JR (2002) Inter-laboratory comparison of Atlantic and Mediterranean bluefin tuna otolith microconstituents. ICES J Mar Sci 59:1294–1304

Secor DH, Henderson-Arzapalo A, Piccoli PM (1995) Can otolith microchemistry chart patterns of migration and habitat utilization in anadromous fishes? J Exp Mar Biol Ecol 192:15–33

Secor DH, Piccoli PM (1996) Age- and sex-dependent migrations of striped bass in the Hudson River as determined by chemical microanalysis of otoliths. Estuaries 19:778–793

Secor DH, Rooker JR, Zlokovitz E, Zdanowicz VS (2001) Identification of riverine, estuarine, and coastal contingents of Hudson River striped bass based upon otolith elemental fingerprints. Mar Ecol Prog Ser 211:245–253

Secor DH, Zdanowicz VS (1998) Otolith microconstituent analysis of juvenile bluefin tuna (*Thunnus thynnus*) from the Mediterranean Sea and Pacific Ocean. Fish Res 36:251–256

Shen GT, Cole JE, Lea DW, Linn LJ, McConnaughey TA, Fairbanks RG (1992) Surface ocean variability at Galapagos from 1936–1982: calibration of geochemical tracers in corals. Paleoceanography 5:563–588

Shen KN, Tzeng WN (2002) Formation of a metamorphosis check in otolith of the amphidromous goby *Sicyopterus japonicus*. Mar Ecol Prog Ser 228:205–211

Shiao JC, Tzeng CS, Leu CL, Chen FC (1999) Enhancing the contrast and visibility of daily growth increments in fish otoliths etched by proteinase K buffer. J Fish Biol 54:302–309

Smith DC, Fenton GE, Robertson SG, Short SA (1995) Age determination and growth of orange roughy (*Hoplostethus atlanticus*): a comparison of annulus counts with radiometric ageing. Can J Fish Aquat Sci 52:391–401

Smith SV, Buddemeier RW, Redalje RD, Houck JE (1979) Strontium-calcium thermometry in coral skeletons. Science 204:404–407

Spencer K, Schafer DJ, Gauldie RW, DeCarlo EH (2000) Stable lead isotope ratios from distinct anthropogenic sources in fish otoliths: a potential nursery stock marker. Comp Biochem Physiol A 127:273–284

Stoll HM, Encinar JR, Alonso JIG, Rosenthal Y, Probert I, Klaas C (2001) A first look at paleotemperature prospects from Mg in coccolith carbonate: cleaning techniques and culture measurements. Geochem Geophys Geosyst 2:doi:10.1029/2000GC000144

Stransky C, Garbe-Schonberg CD, Günther D (2005) Geographic variation and juvenile migration in Atlantic redfish inferred from otolith microchemistry. Mar Freshwater Res 56:677–691

Sturgeon RE, Willie SN, Yang L, Greenberg R, Spatz RO, Chen Z, Scriver C, Clancy V, Lam JW, Thorrold S (2005) Certification of a fish otolith reference material in support of quality assurance for trace element analysis. J Anal At Spectrom 20:1067–1071

Swart PK, Thorrold S, Rubenstone J, Rosenheim B, Harrison CGA, Grammer M, Latkoczy C (2002) Intra-annual variation in the stable oxygen and carbon and trace element composition of sclerosponges. Paleoceanography 17:1045, doi:10.1029/2000PA000622

Swearer SE, Caselle JE, Lea DW, Warner RR (1999) Larval retention and recruitment in an island population of a coral-reef fish. Nature 402:799–802

Swearer SE, Forrester GE, Steele MA, Brooks AJ, Lea DW (2003) Spatio-temporal and interspecific variation in otolith trace-elemental fingerprints in a temperate estuarine fish assemblage. Est Coast Shelf Sci 56:1111–1123

Swearer SE, Shima JS, Hellberg ME, Thorrold SR, Jones GP, Robertson DR, Morgan SG, Selkoe KA, Ruiz GM, Warner RR (2002) Evidence of self-recruitment in demersal marine populations. Bull Mar Sci 70:251–271

Thorrold SR, Hare JA (2002) Otolith applications in reef fish ecology. In: Sale PF (Ed) Coral reef fishes: dynamics and diversity in a complex ecosystem. Academic Press, San Diego

Thorrold SR, Jones CM, Campana SE (1997) Response of otolith microchemistry to environmental variations experienced by larval and juvenile Atlantic croaker *Micropogonias undulatus*. Limnol Oceanogr 42:102–111

Thorrold SR, Jones CM, Campana SE, McLaren JW, Lam JWH (1998a) Trace element signatures in otoliths record natal river of juvenile American shad *Alosa sapidissima*. Limnol Oceanogr 43:1826–1835

Thorrold SR, Jones CM, Swart PK, Targett TE (1998b) Accurate classification of juvenile weakfish *Cynoscion regalis* to estuarine nursery areas based on chemical signatures in otoliths. Mar Ecol Prog Ser 173:253–265

Thorrold SR, Jones GP, Planes S, Hare JA (2006) Transgenerational marking of embryonic otoliths in marine fishes using barium stable isotopes. Can J Fish Aquat Sci 63:1193–1197

Thorrold SR, Latkoczy C, Swart PK, Jones CM (2001) Natal homing in a marine fish metapopulation. Science 291:297–299

Thorrold SR, Shuttleworth S (2000) In situ analysis of trace elements and isotope ratios in fish otoliths using laser ablation sector field inductively coupled plasma mass spectrometry. Can J Fish Aquat Sci 57:1232–1242

Thresher RE, Proctor CH, Gunn JS, Harrowfield IR (1994) An evaluation of electron-probe microanalysis of otoliths for stock delineation and identification of nursery areas in a southern temperate groundfish, *Nemadactylus macropterus* (Cheilodactylidae). Fish Bull (Wash DC) 92:817–840

Townsend DW, Radtke RL, Corwin S, Libby DA (1992) Strontium calcium ratios in juvenile Atlantic herring *Clupea harengus* L. otoliths as a function of water temperature. J Exp Mar Biol Ecol 160:131–140

Townsend DW, Radtke RL, Malone DP, Wallinga JP (1995) Use of otolith strontium : calcium ratios for hindcasting larval cod *Gadus morhua* distributions relative to water masses on Georges Bank. Mar Ecol Prog Ser 119:37–44

Townsend DW, Radtke RL, Morrison MA, Folsom SD (1989) Recruitment implications of larval herring overwintering distributions in the Gulf of Maine, [USA] inferred using a new otolith technique. Mar Ecol Prog Ser 55:1–13

Tsukamoto K, Nakai I, Tesch WV (1998) Do all freshwater eels migrate? Nature 396:635–636

Vázquez Peláez M, Costa-Fernández JM, Sanz-Medel A (2002) Critical comparison between quadrupole and time-of-flight inductively coupled plasma mass spectrometers for isotope ratio measurements in elemental speciation. J Anal At Spectrom 17:950–957

Victor BC (1983) Recruitment and population dynamics of a coral reef fish. Science 219:419–420

Walther BD, Thorrold SR (2006) Water, not food, contributes the majority of strontium and barium deposited in the otoliths of a marine fish. Mar Ecol Prog Ser 311:125–130

Warner RR, Cowen RK (2002) Local retention of production in marine populations: evidence, mechanisms, and consequences. Bull Mar Sci 70:251–271

Warner RR, Swearer SE, Caselle JE, Sheehy M, Paradis G (2005) Natal trace-elemental signatures in the otoliths of an open-coast fish. Limnol Oceanogr 50:1529–1542

Weber PK, Hutcheon ID, McKeegan KD, Ingram BL (2002) Otolith sulfur isotope method to reconstruct salmon (*Onchorhynchus tshawytscha*) life history. Can J Fish Aquat Sci 59:587–591

Wells BK, Thorrold SR, Jones CM (2000) Geographic variation in trace element composition of juvenile weakfish (*Cynoscion regalis*) scales. Trans Am Fish Soc 129:889–900

Wieser ME, Buhl D, Bouman C, Schweiters J (2004) High precision calcium isotope ratio measurements using magnetic sector multiple collector inductively coupled plasma mass spectrometer. J Anal At Spectrom 19:844–851

Willie S, Metser Z, Sturgeon RE (2005) Isotope ratio precision with transient sample introduction using ICP orthogonal acceleration time-of-flight mass spectrometry. J Anal At Spectrom 20:1358–1364

Woodhead J, Swearer S, Hergt J, Maas R (2005) In situ Sr-isotope analysis of carbonates by LA-MC-ICP-MS: interference corrections, high spatial resolution and an example from otolith studies. J Anal At Spectrom 20:22–27

Yoshinaga J, Morita M, Edmonds JS (1999) Determination of copper, zinc, cadmium and lead in a fish otolith certified reference material by isotope dilution inductively coupled plasma mass spectrometry using off-line solvent extraction. J Anal At Spectrom 14:1589–1592

Yoshinaga J, Nakama A, Morita M, Edmonds JS (2000) Fish otolith reference material for quality assurance of chemical analyses. Mar Chem 69:91–97

Zacherl DC, Paradis G, Lea DW (2003) Barium and strontium uptake into larval protoconchs and statoliths of the marine neogastropod *Kelletii kelletii*. Geochim Cosmochim Acta 67:4091–4099

Zlokovitz E, Secor DH, Piccoli PM (2003) Patterns of migration in Hudson River striped bass as determine by otolith microchemistry. Fish Res 63:245–259

9. TROPICAL OTOLITHS – WHERE TO NEXT?

BRIDGET S. GREEN[1], BRUCE D. MAPSTONE[2], GARY CARLOS[1], AND GAVIN A. BEGG[3]

[1]*Tasmanian Aquaculture and Fisheries Institute, University of Tasmania, TAS 7053, Australia*

[2]*Antarctic Climate & Ecosystems Cooperative Research Centre, Hobart, TAS 7005, Australia*

[3]*Bureau of Rural Sciences, Department of Agriculture, Fisheries and Forestry, Canberra, ACT 2601, Australia*

The formation of alternating zones of different composition and optical density (translucent, protein rich zones and opaque, calcium rich zones) as otoliths grow throughout fishes lives is the underpinning basis of ageing from otoliths. A consistent theme throughout this book is that methods for enumerating otolith growth increments to estimate fish age are well-established but there is a general lack of understanding of the fundamental processes of increment formation. It remains surprisingly unclear what are the endogenous process responsible for this zone formation and how those endogenous processes interact with exogenous influences. This issue is not unique to tropical environments but perhaps takes on greater importance in the tropics because of the great diversity of species for which ages are sought and the consequent high cost of detailed validation of ageing methods if required for every species (Choat et al., Chapter 2). Routine empirical validation of increment periodicity might be obviated if we understood better the mechanisms driving the formation of growth increments with clear periodic structure.

Endogenous and exogenous causes of variation in increment width and density and causes of variation in the optical properties of zones within otolith increments are all poorly understood, as are the reasons for occasional formation of sub-increments and accessory primordia. For example, there has been considerable work focussed on the relationship between environmental temperature and the formation of opaque or translucent zones in otolith increments but it also has been demonstrated that key physiological events, especially reproduction, also influence the composition and optical density of increment zones in at least some species. Such endogenous influences are likely to change with life history stage, further complicating interactions between endogenous and exogenous drivers of zone formation. Better understanding these relationships and the physiological processes of continuous otolith formation are likely to increase the amount and certainty of information derived from each otolith, perhaps diminishing the sample sizes necessary in age-based studies. Considerable benefit will be gained by linking the fundamental biochemical research on otolith formation to the applied fields of otolith processing for ageing and inference of life history characteristics. Resolving the processes that determine the structures within otoliths, therefore, promises considerable reward for further research effort and, in the long run, improved efficiency of research applications to fisheries management.

B.S. Green et al. (eds.), *Tropical Fish Otoliths: Information for Assessment, Management and Ecology*, Reviews: Methods and Technologies in Fish Biology and Fisheries 11,
DOI 10.1007/978-1-4020-5775-5_9,

The influence of the environment, particularly temperature, on otolith formation could be explored initially by a wide range of systematic sampling from related or similar species distributed across a range of latitudes in both hemispheres. Such sampling would not provide definitive evidence of cause and effect relationships but might reveal patterns from which to hypothesise such relationships and so design targeted experiments for subsequent research. There also are a few unique opportunities for examining more rigorously links between local environment and otolith growth in collections of specimens from populations derived from documented translocations. One example is the tropical snapper (*Lutjanus kasmira*), which has been successfully introduced to Hawaii from two southern hemisphere populations ostensibly for the purpose of fishery development. These translocations essentially comprise a large-scale, longitudinal experiment that provides an opportunity to examine increment formation in relation to seasonality, temperature and timing of spawning between locations where the seasons are out of phase. Similar studies are possible also with some invertebrates where there have been similar translocations across the Pacific basin, including pearl shell, trochus and the Green snail, *Turbo marmoratus*.

Concern for conservation is an increasingly important driver of biological research but also represents a growing constraint on research that depends on the death of large numbers of individuals. Robust back-calculation applications provide the potential to infer age-based growth metrics from analysis of considerably fewer otoliths than alternative methods but back-calculation methods are still at a relatively early stage of development and testing. Vigliola and Meekan, in Chapter 5, noted the imbalance between the many published studies of conventional age validation (over 200 in the past 10 years) and the handful of published studies reporting on the validation and testing of appropriate back-calculation models. Only one back-calculation study in 10 years was based on longitudinal data collected at the individual level under conditions of variable growth. Much research remains to be done to establish whether back-calculation methods might fulfil the promise they intuitively hold to provide reliable retrospective insights to the growth of fishes throughout life, under varying environmental conditions and across multiple life history stages.

An area of otolith processing which receives close attention in only a few published manuscripts is the sources of error in the interpretation of bands, whether annuli or daily. The 'accepted practice' of reading an otolith multiple times and taking the average or modal count, disregarding the differences, has become general practice, with little regard to the source of the error causing different counts or whether the criterion for accepting or discarding counts is appropriate for such sources of error. A number of laboratories document reader-dependent interpretation errors through cross-checking individual readers regularly (e.g., weekly), testing accuracy of reader interpretation against a standard library of otoliths of all the species being read, at different ages, and retraining and retesting on a regular basis (e.g. Central Ageing Facility, Victoria, Australia). This approach is an important component of quality control over ageing procedures but does not, alone, elucidate the sources of interpretation error or how best to accommodate them in assigning ages from disparate counts. Further research on the origins of errors in interpreting otolith increments is necessary to improve comprehension of the sources of those errors and increase the robustness and efficiency of routine ageing practices. Other sources of error in otolith interpretation include transcription error either at the point of recording the original age estimation or when entering data into a database. These two sources of error, however, increasingly are being bypassed by using integrated data management systems linked with image analysis systems to allow automatic upload of counts into databases.

Some of the biggest advances in technology in human endeavours have arisen through revolution of design and concept rather than the evolution of existing technologies. In this vein, there is emerging research on alternative methods for ageing animals, including fish, that might eventually diminish our dependence on analyses of otoliths. For example, the quantification of lipofuscin and measuring telomere length have been posited as promising methods for estimating age. The advantages and disadvantages of quantifying lipofuscin were examined in Chapter 6. Telomeres are regions of repetitive DNA at the end of chromosomes which are believed to protect the end of the chromosome from destruction. The lengths of telomeres in humans and other vertebrates decrease with age in a linear trajectory. Analysis of telomeres is possible from blood or biopsy samples obtained non-lethally from the organism, it is a genetic technique requiring some expertise in the preservation and extraction of DNA from tissue or blood as well as specialised laboratory equipment. Early studies in a number of fish species suggest a general trend for telomeres to decline in length with fish age, but this trend is not consistent across species and was not present in a number of chondrichthyan species (C. Izzo pers. com.[1]). Clearly, further work is required to verify whether such methods will prove useful or efficient replacements for otolith analysis.

Ageing marine organisms apart from the bony fishes is a relatively newer field of research than ageing teleosts, though there is a solid history of research for squids and sharks. As with otoliths, a more comprehensive understanding of the processes of growth of structures used for ageing invertebrates or chondrichthyans may reduce the need for costly and repetitive validation programs for every ageing study of each species. The hard structures available for age estimation in invertebrates are limited, however, and ageing of many groups will depend on development of new methods for resolving presence of chronometers in soft tissues, such as those noted above.

This text has examined a range of aspects of processing otoliths in tropical fish, including ageing using daily and annual increments, validation, the periodicity of increments, the microstructure of otoliths and how otolith microchemistry can record the environmental history of the fish. There is also a large section devoted to ageing other marine organisms and ageing fish using structures apart from otoliths. By bringing together this expertise into one volume we aimed to provide fisheries managers, practitioners and biologists in the tropics with a single reference for a wide range of ageing techniques for tropical fisheries and fish biology.

---

[1] Mr Christopher Izzo School of Earth and Environmental Science, University of Adelaide, Adelaide, SA, Australia, email c.izzo@adelaide.edu.au

GLOSSARY

Some of the definitions below are contributions from previous glossaries (Kalish et al. 1995, Panfili et al. 2002).

Accessory growth centre. A growth centre formed beyond the otolith core that leads to a new plane of growth and from which a new series of growth increments appears to emanate. Formation of these structures is often associated with life-history transitions such as metamorphosis. Accessory growth centres are often referred to as accessory primordial but the term accessory growth centre is preferred because these features are different structurally from primordia (e.g., they do not contain primordial granules). The term "secondary growth centre" also has been used.

Annulus (pl annuli) (also annual increment). Growth increment validated to be formed once per year.

Asterisci (singular asteriscus). Smallest of the three pairs of otoliths found in teleost fish, not generally used for age determination.

Back-calculation. The derivation of growth history of individuals from measurements of increments within otoliths. Correct application requires a model of the relationship between otolith growth and fish somatic growth and assumptions about the constancy of increment deposition.

Band (also ring). A general expression describing a regular histomorphological structure laid down during the growth of hard parts. The term is used variously to describe only D- or L-zones or the pairs of those zones, which together generally define growth increments, and so should either be replaced with more consistently specific terms (e.g., zone or increment) or used only with precise definition. See also increment and zone.

Calcified structure. Any bony part which, for ageing purposes, contains regular discernable growth increments that might be formed on a predictable timescale. Otoliths are the calcified structures most commonly used for ageing in fish, although scales, teeth, vertebrae and other skeletal structures also have been used. Invertebrate ageing typically makes use of hard structures such as external shells, spines, and ossicles. Statoliths, cuttlebones, pens and beaks from some molluscs also have been used, although the latter two are chitinous rather than calcified structures.

Check. A point or line of clear change in the appearance of the macrostructure or microstructure of an otolith or other calcified structure. Checks have been used to describe the change between successive otolith zones, successive pairs of opaque and translucent zones, or microincrements. Microstructural checks (e.g., hatching checks) often appear as high-contrast microincrements with a deeply etched D-zone or an abrupt change in the microstructural growth pattern. The term should be used only with precise definition.

Core. The area surrounding the primordium and bounded by the first prominent D-zone. Some fish possess one or more accessory growth centres, each of which also may appear as a core.

D-zone. That portion of an otolith micro-increment that appears dark when viewed under transmitted light and as a depressed region when acid-etched and viewed with a scanning electron microscope. This component of a micro-increment contains a greater amount of organic matrix and a lesser amount of calcium carbonate than the L-zone. D-Zones have been referred to as the discontinuous or matrix-rich zones in earlier works on daily increments but "D-zone" is the preferred term.

Daily increment. An increment formed over a 24-h period. In its general form, a daily increment consists of a D-zone and an L-zone. The term is synonymous with "daily growth increment" and "daily ring" but the term "daily ring" is misleading and inaccurate and should not be used. The term "daily increment" is preferred.

Discontinuity. See "Check".

Endolymph. Fluid medium in which the otolith is bathed inside each otolith sac. Elements present in the endolymph are incorporated onto the growing otolith surface.

Growth curve. A descriptive term for the mathematical function describing individual change in size (either linear dimension or mass) over time. Although there are many possible growth trajectories, some classes of curves are arbitrarily referred to by their general shape, as follows: flat-topped (also called table-topped, square, asymptotic or plateaued) – where an initial period of relatively rapid growth is followed by a relatively long period of little or no growth; indeterminate – characterised by life-long, continuous increase in size and a strong positive correlation between size and age over the entire lifespan; intermediate – reflecting increase in size over most of an individual's life but reaching an approximate asymptote relatively late in life, so resembling a blend of both of the proceeding patterns.

Increment (also growth increment). A general term describing the growth of an otolith (or other calcified structure) during a specific interval (e.g., daily or annually). The dimensions, chemistry and period of formation may vary widely, so usually requiring an adjective to specify the type of increment being discussed (hence, "daily increment" or "annual increment"). A primary (usually daily) increment consists of a D-zone and an L-zone, whereas an annual increment typically comprises an opaque zone and a translucent zone, each comprised of many daily increments (although these daily increments become increasingly difficult to resolve under light microscopy as fish age). There may also be secondary structures such as sub-daily increments and false and double zones within annual or daily increments.

L-zone. That portion of a micro-increment that appears light when viewed under transmitted light, and as an elevated region when acid-etched and viewed with a scanning electron microscope. The component of a micro-increment that contains a lesser amount of organic matrix and a greater amount of calcium carbonate than the D-zone. L-zones were referred to as "incremental zone" in earlier works on daily increments but "L-zone" is the preferred term.

Lapilli (singular lapillus). One of the three pairs of otoliths found in teleost fish, usually intermediate in size between the sagittae and asterisci.

Mark. A general expression describing a histomorphological structure or zone laid down during the growth of hard parts. Mark is at times used with the same meaning as "check", "increment" or "zone" and so should be used with precise definition or discarded in favour of more specific terms.

Micro-increment. An increment that is less than 50 μm in width, typically from one to 20 μm. The prefix "micro" serves to indicate that the object denoted is of relatively small size and that it can only be observed with a compound or electron microscope. Often used to describe daily and sub-daily increments. See increment.

Nucleus. The centre region of an otolith. See primordium.

Ossicle (Invertebrates). Plate-like elements which form the tests of echinoderms. Observation of growth increments in these structures has been used for ageing.

Otolith. The general term for any of the three pairs (asterisci, lapilli and sagittae) of calcified balance organs encapsulated within the base of the skull of teleost fish. The formation of these structures through discernable growth increments associated with specific, regular periods (e.g., daily or annual) is the basis for most age determination methods in fish.

Primordium (pl. primordia). The initial complex structure of an otolith. It consists of granular or fibrillar material surrounding one or more optically dense nuclei from 0.5 μm to 1.0 μm in diameter. Some fish (e.g., salmonids) develop multiple primordial in the early stages of otolith growth, which generally fuse during formation of the otolith core.

Resorption. The loss or erosion of part of a calcified structure through a physiological process, often resulting in either non-continuous growth of the structure or a variable and unreliable relationship between size and age of the structure.

Ring. See "Band".

Sagittae (singular sagitta). Largest of the three pairs of otoliths found in teleost fish and the structure most often used for age determination.

Statoliths. In cephalopods, a pair of internal calcified structures analogous to fish otoliths.

Sub-daily increments. An increment formed over a period of less than 24 h. See increment.

Sulcus acusticus (usually shortened to sulcus). A groove along the medial surface of the sagitta. A thickened portion of the otolithic membrane lies within the sulcus acusticus. The sulcus acusticus is often referred to in otolith studies because of the clarity of the increments near the sulcus in transverse sections of sagittae.

Transition zone. A region of change between two specific regions in otolith structure. A transition zone is identified sometimes as a region lacking structure or increments between two regions that have such structure. In other cases, a transition zone might be identified as a region of abrupt change in the form (e.g., width or contrast) of increments. Transition zones in otoliths often are associated with metamorphosis from larval to juvenile stages or with significant habitat changes, such as the movement from a pelagic to a demersal habitat or a marine to freshwater habitat. The term should be used only with precise definition.

Ultrastructure. The very fine-scale structure of tissues, typically observable only at very high levels of magnification, such as with electron microscopy.

Zone. A region of similar structure or optical density. Zones can be described by their optical appearance to an observer as opaque or translucent but appearance is dependent on light source, which always should be reported. Pairs of adjacent zones often form the periodic increments used to age fish. See also D-zone and L-zone.

WIP. Water insoluble proteins present in the otolith, considered to play a major structural role.

WSP. Water soluble proteins present in the otolith. Because of their solubility characteristics, these compounds interact with precipitation processes within the endolymph and are therefore considered mediators in the following aspects of otolith formation.

Crystal nucleation and growth – while crystal nuclei can form spontaneously in supersaturated solutions, they will re-dissolve unless their size exceeds a critical value. Mediation of initial nuclei by some WSP such as glycoproteins (which can act as a calcification inhibitor) may regulate accretion both initially and during later crystal growth of the whole otolith.

Crystal poisoning: during otolith formation foreign compounds such as phosphate groups, having a chemical affinity to aragonite (the calcium carbonate polymorph comprising fish otoliths), may be incorporated into lattice vacancies and interfere with the growth of the crystal. WSP such as phosphatase enzyme hydrolyses such non-calcium components at the site of mineralization thus destroying the crystal poison effect.

References

Kalish JM, Beamish RJ, Brothers EB, Casselman JM, Francis RICC, Mosegaard H, Panfili J, Prince ED, Thresher RE, Wilson CA, Wright PJ (1995) Standard terms for otolith research. In: Secor DA, Dean JM, Campana SE (Eds) Recent developments in fish otolith research. University of South Carolina Press, Columbia, SC, pp 723–729

Panfili J, de Pontual H, Troadec H, Wright PJ (2002) Manual of fish sclerochronology. IFREMER-IRD, Brest, France

INDEX

The manufacturer's authorised representative in the EU is Springer Nature Customer Service Centre GmbH, Europaplatz 3, 69115 Heidelberg, Germany. If you have any concerns regarding our products, please contact ProductSafety@springernature.com

Printed and bound by CPI Group (UK) Ltd, Croydon, CR0 4YY

07/01/2026

02030951-0008